# Engineering Materials

This series provides topical information on innovative, structural and functional materials and composites with applications in optical, electrical, mechanical, civil, aeronautical, medical, bio- and nano-engineering. The individual volumes are complete, comprehensive monographs covering the structure, properties, manufacturing process and applications of these materials. This multidisciplinary series is devoted to professionals, students and all those interested in the latest developments in the Materials Science field, that look for a carefully selected collection of high quality review articles on their respective field of expertise.

**Indexed at Compendex (2021) and Scopus (2022)**

Mahdi Darab · Amirmahmoud Bakhshayesh ·
Maryam Tajabadi

# Perovskites for Next-Generation Wearable Technology

Springer

Mahdi Darab
School of Metallurgy and Materials
Engineering
Iran University of Science and Technology
Tehran, Iran

Amirmahmoud Bakhshayesh
School of Metallurgy and Materials
Engineering
Iran University of Science and Technology
Tehran, Iran

Maryam Tajabadi
School of Metallurgy and Materials
Engineering
Iran University of Science and Technology
Tehran, Iran

ISSN 1612-1317 ISSN 1868-1212 (electronic)
Engineering Materials
ISBN 978-3-032-03930-9 ISBN 978-3-032-03931-6 (eBook)
https://doi.org/10.1007/978-3-032-03931-6



This Springer imprint is published by the registered company Springer Nature Switzerland AG
The registered company address is: Gewerbestrasse 11, 6330 Cham, Switzerland

# به نام آنکه ستایش را سزد

## ای عزیز

چگونه می‌توان نعمتت را سپاس گفت آنگاه که ما را همقدم کاروان مشتاقانی قرار دادی که در میان انبوه سؤال‌ها و چراها به دنبال نیم‌نگاهی از کمال کبریایی تو می‌گردند و آتش اشتیاق به دانستن و سوز نیاز به شناختن تو را با اشک قلم فرو می‌نشانند.

## و تو

آنقدر بزرگی که بر هر بضاعت اندک که از وجود بنده‌ای ناچیز به سوی تو جهت می‌گیرد، نظر می‌کنی و ما حیرت‌زده عظمت تو و شرمسار کوچکی گامهایمان، قدم در راهی نهاده‌ایم که آرزو داریم در هر منزلش حجم حضورت را احساس کنیم. چرا که تو آن غایت بی‌نهایتی که از رگ گردن نزدیکتر است و هر زمان دست یافتنی‌تر می‌نماید.

## بی‌نهایت نزدیک

حرکت ما را بپذیر و آنچنان کن که نظر لطف آن عزیزی که گیتی، قائم به وجود اوست، بدرقه راهمان باشد و همراهی مرکبش، نقطه اوج سفرمان؛ که اگر آنچه می‌کاویم و می‌آموزیم با تأیید آسمانیش متبرک نگردد، پرسه‌زدنی است بی‌حاصل و ما به تو پناه می‌بریم از بی‌حاصلی و بیهودگی.

*To those who seek understanding in a world full of questions, and who search not only for answers, but for meaning.*

*To those who are grateful to take part in the shared human effort to explore, to question, and to learn—*
*and who strive to offer what they discover in service of others and in the hope of a better world for all.*

# Preface

The convergence of advanced materials science and wearable technology stands at the threshold of a revolutionary transformation. As we go forward, the demand for intelligent, responsive, and seamlessly integrated wearable devices accelerates, driven by applications spanning healthcare monitoring, fitness tracking, augmented reality, and smart textiles. At the heart of this technological evolution lies a class of materials that promises to redefine the very foundations of wearable electronics: perovskites.

Perovskite materials, with their unique crystal structure and exceptional tunability, have emerged as one of the most promising candidates for next-generation wearable applications. Their remarkable properties—including high carrier mobility, excellent optical absorption, mechanical flexibility, and solution-processable synthesis—position them as ideal building blocks for the multifunctional, lightweight, and cost-effective devices that modern wearable technology demands.

This book represents a comprehensive exploration of how perovskite materials can address the fundamental challenges facing wearable electronics today. From the basic principles of wearable device design to the cutting-edge applications in medical diagnostics and fashion industry, we present a thorough examination of the current state and future potential of perovskite-based wearables.

The journey through these pages begins with establishing the foundational knowledge of wearable electronics and the material requirements that make certain compounds suitable for intimate human-device interfaces. We then move deeper into the fascinating world of perovskite materials, exploring their structural diversity, electronic properties, and the numerous advantages they offer over conventional materials in wearable applications.

The book progresses through essential characterization techniques that enable researchers and engineers to understand and optimize perovskite materials for specific wearable applications. We examine how these materials can be transformed into highly sensitive sensors capable of monitoring everything from vital signs to environmental conditions, as well as energy harvesting and storage devices that can power wearable systems sustainably.

Integration challenges—perhaps the most critical aspect of translating laboratory discoveries into commercial products—receive detailed attention, addressing the engineering solutions needed to incorporate perovskites into flexible, durable, and biocompatible wearable platforms. The final chapters showcase real-world applications, from medical monitoring systems that could transform personalized healthcare to innovative smart textiles that blur the line between fashion and technology.

This work is intended for a diverse audience: materials scientists seeking to understand the application potential of perovskites, electronics engineers developing next-generation wearable devices, biomedical researchers exploring new diagnostic and therapeutic tools, and fashion technologists interested in creating responsive, intelligent garments. Graduate students and researchers entering this interdisciplinary field will find both fundamental principles and practical insights to guide their work.

The authors have drawn upon extensive research from leading laboratories worldwide, presenting both established knowledge and emerging trends and future directions in perovskite wearable technology. Each chapter includes case studies, design considerations, and practical examples that bridge the gap between theoretical understanding and real-world implementation.

As we stand on the brink of an era where our clothing and accessories will possess unprecedented sensing, computing, and communication capabilities, the materials that enable these advances become increasingly crucial. Perovskites, with their extraordinary versatility and performance characteristics, represent a key to unlocking the full potential of wearable technology.

We invite readers to join us in exploring this exciting frontier, where materials science meets human-centered design, and where the next generation of wearable technology will emerge to enhance human health, performance, and quality of life in ways we are only beginning to imagine.

We welcome feedback from the scientific community and encourage readers to share their insights, suggestions, and possible corrections, which can be sent to the authors via the publisher.

Tehran, Iran

Mahdi Darab
Amirmahmoud Bakhshayesh
Maryam Tajabadi

# Acknowledgments

The completion of this book would not have been possible without the invaluable contributions, support, and encouragement of numerous individuals who have shaped our understanding of crystalline structures and particularly perovskite materials.

We are grateful to our teachers and mentors who laid the foundation for our academic pursuits and continue to inspire our work in materials science. Our sincere appreciation goes to our colleagues and collaborators worldwide who have shared their expertise, research findings, and insights. The interdisciplinary nature of this field has required collaboration across diverse areas, and the stimulating discussions and feedback from our peers have enriched every chapter.

We extend our gratitude to the publishers whose materials we have reproduced under Creative Commons licenses, as well as those who granted us permission for reproduction. Clear references have been provided throughout the book for all materials used with permission, including those from Elsevier, The American Physical Society, and the Ceramic Society of Japan.

Special recognition goes to Ms. Zahra AliAkbari, our talented designer, and Dr. Seyed Ali Amiri for the facilitation and operational supports.

We extend our appreciation to the editorial and production teams at Springer Nature for their patience and professionalism throughout the publication process.

Finally, we express our profound gratitude to our families for their unconditional love, unwavering support, understanding, and encouragement during this journey.

Tehran, Iran
July 2025

Mahdi Darab
Amirmahmoud Bakhshayesh
Maryam Tajabadi

# Contents

# List of Figures

# List of Tables

# Chapter 1
# Introduction to Wearable Electronics

The evolution of wearable electronics is a story of humanity's enduring drive to augment the body with tools that extend perception, enhance capabilities, and seamlessly integrate technology into daily life. The origins of wearable technology stretch as far back as the thirteenth century, when the invention of eyeglasses [1] offered a glimpse into the potential of body-worn devices to improve human function. For centuries, these innovations remained purely mechanical, assisting with sight, timekeeping, or navigation. Yet, with the advent of electronics and computing in the twentieth century, a profound transformation began—one that would redefine how we interact with ourselves, with others, and with the surrounding world. From the earliest electronic experiments in the 1960s to the fancy smartwatches, fitness trackers, and biometric garments of today, wearable electronics have evolved into a cornerstone of the modern technological landscape. This chapter explores that journey, tracing the path from elementary beginnings to intelligent, networked systems while highlighting the breakthroughs, applications, market trends and future directions that continue to shape this rapidly advancing field.

## 1.1 Brief Overview and History of the Field

Wearable electronics represent the seamless integration of mobile, personal, and wireless sensing and communication technologies with innovative textile, fiber, and polymer-based components. These sophisticated devices are elegantly incorporated into clothing and accessories, enabling comfortable, stylish and continuous wear while transforming human interaction with technology in profound ways [2].

Though they appear cutting-edge today, wearable technology's roots extend back decades, emerging from humanity's enduring desire to augment capabilities through tools that could be carried—or better yet, worn—on the body.



M. Darab et al., *Perovskites for Next-Generation Wearable Technology*, Engineering Materials, https://doi.org/10.1007/978-3-032-03931-6_1

The evolution of wearable technology tells a compelling story driven by fundamental human needs and adapting to ever-changing lifestyles. This journey combines technological advancement with human creativity. What began with relatively simple innovations like digital watches and hearing aids has transformed into an ecosystem of sophisticated electronic garments that can monitor an astronaut's vital signs or enhance an Olympic athlete's performance. Wearable electronics have progressively transformed technological interactions—yesterday's dream becoming today's everyday reality, with tomorrow promising even more remarkable integration into daily life.

The rise of the Internet of Things (IoT), supercharged by artificial intelligence and machine learning algorithms, has catalyzed this transformation. Today's wearable devices do not solely exist on bodies; they create an intelligent, responsive layer that connects users to digital networks, enabling continuous connectivity and real-time monitoring capabilities that were unimaginable just a generation ago. These technologies now augment human senses, track health metrics, enhance productivity, and even express individual identities.

As society ventures further into what was once considered science fiction, a new relationship with technology emerges—one in which the boundaries between the digital and physical worlds, and between technology and human experience, grow increasingly blurred. More than mere gadgets, wearable electronics have become extensions of the self—signaling a fundamental shift in how technology supports, augments, and integrates with human life.

Tracing the developmental arc of wearable electronics reveals not only technological milestones but also shifting cultural attitudes toward technology's role in daily life. Historical progression can be understood through several distinct yet overlapping phases, each characterized by particular technological capabilities and use cases. Figure 1.1 presents a timeline highlighting the key phases in the development of wearable technology.

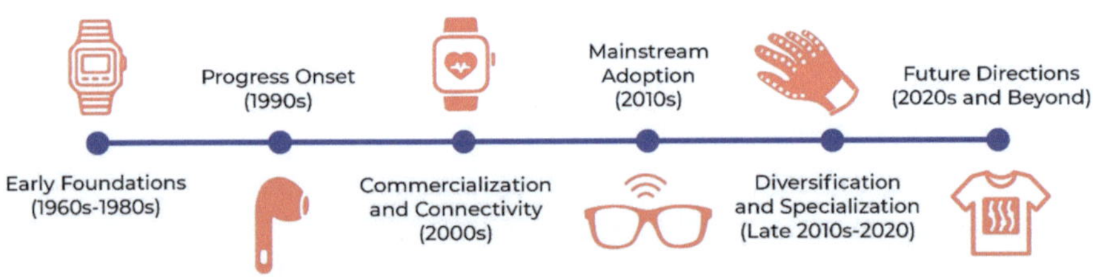


**Fig. 1.1** Schematic timeline for the evolution of wearable electronics

### 1.1.1 *Early Foundations (1960s–1980s)*

The conceptual origins of wearable electronics can be identified in the pioneering work of mathematicians Edward Thorp and Claude Shannon in 1961. Their collaborative development of a cigarette-pack sized computational device designed to predict roulette wheel outcomes represented one of the first documented instances of a wearable computer [3]. This work involved integrating critical features that are still fundamental to wearable computers today, such as sensing or information input, computing, and some form of actuation, feedback, or communication of retrieved information.

The 1970s witnessed the commercialization of early wearable electronic devices, with Pulsar's introduction of the calculator watch in 1975 standing as a significant milestone [4]. These devices, while primitive by contemporary standards, established important precedents in human–computer interaction and portable electronics miniaturization. During this period, wearable technology remained largely functional rather than fashionable, with minimal consideration given to aesthetic integration or user comfort [5].

The experimental foundations established during this era laid crucial groundwork for subsequent developments, particularly in terms of component miniaturization and power management—challenges that continue to influence wearable design paradigms in the present day.

### 1.1.2 *Progress Onset (1990s)*

The 1990s marked a transformative period for wearable electronics, with significant advances in both technical capability and conceptual development [6]. Steve Mann, often considered the "father of wearable computing," pioneered the WearComp series at MIT Media Lab [7], developing increasingly sophisticated head-mounted display systems that established foundational principles for continuous mobile computing. Also, Mann's concept of the "Digital Eye Glass" (also known as EyeTap) developed 35 years ago, is another notable example of his pioneering work in wearable computing [8].

This era also saw the formation of MIT's Wearable Computing Project under Alex Pentland, which explored context-aware computing and augmented reality applications [9].

Commercial applications began to emerge alongside academic research. In 1994, Seiko launched the Ruputer, a wrist-worn computer that featured a small LCD screen and basic applications [10, 11]. The head-mounted display and other body-mounted devices gained popularity among early wearable enthusiasts. By the late 1990s, companies like Xybernaut were marketing complete wearable computer systems, though these remained primarily focused on industrial and military applications rather than consumer markets [3, 12].

### 1.1.3 Commercialization and Connectivity (2000s)

The early 2000s witnessed increased emphasis on connectivity and practical application of wearable electronics. Bluetooth technology, standardized in 1999, became instrumental in enabling wireless connections between wearable devices and other electronics. This period saw the emergence of specialized fitness trackers, with companies like Polar and Garmin developing wrist-worn heart rate monitors and GPS devices targeting athletes and fitness enthusiasts.

The introduction of smartphones, particularly the iPhone in 2007, dramatically accelerated wearable technology development by providing powerful processing platforms and connectivity hubs. By 2009, companies like Fitbit were introducing dedicated activity trackers that synchronized with mobile applications, establishing the ecosystem model that would come to dominate consumer wearables [13].

### 1.1.4 Mainstream Adoption (2010s)

The 2010s represented the decade when wearable electronics transitioned from niche technology to mainstream consumer products. Several key developments have driven this transition.

Google's introduction of Google Glass in 2013 generated unprecedented public awareness of wearable technology. However, the device also sparked significant privacy issues. The ability to take photos and videos discreetly led to limitations and even bans in various public places like cafes, casinos, and movie theaters. These privacy concerns attracted the scrutiny of regulators [14, 15] and raised public debates about surveillance and data protection and contributed to the ultimate commercial challenges of the product that time [16, 17].

Concurrently, the smartwatch category gained significant traction with entries from major technology companies, notably the Apple Watch in 2015, which quickly became the market leader and established many design and functional standards for the category [18]. The Apple Watch has been noted for its technological innovations, particularly in health-related features such as heart rate monitoring, electrocardiograms, and physiological symptom monitoring [19]. Around the same time, other major tech players such as Samsung and later Huawei began expanding their mobile ecosystems to include wearable devices, setting the stage for broader integration of smartwatches and fitness bands into everyday life.

Fitness trackers evolved into more comprehensive health monitoring devices, incorporating advanced sensors to track sleep patterns, stress levels, and even blood oxygen saturation. Companies like Garmin, Fitbit, and Samsung expanded their offerings to address growing consumer interest in personal health metrics.

The integration of wearable technology into fashion gained momentum through collaborations between technology companies and fashion brands [20–22]. Partnerships like Intel with Opening Ceremony and Apple with Hermès helped bridge the gap between technology and fashion, addressing earlier criticisms regarding aesthetic limitations [23, 24].

### 1.1.5 Diversification and Specialization (Late 2010s–2020)

Recent years have witnessed both diversification of form factors and increased specialization of applications. Smart textiles incorporating conductive fibers have enabled clothing with embedded sensing capabilities. Recent advancements in smart and wearable textiles have significantly influenced daily life. These innovative materials have found applications across a broad spectrum, including wearable electronics, barrier membranes, phase change materials, shape memory fabrics, optical fibers, and other functional textiles. Designed to offer enhanced convenience, comfort, and a sense of technological integration, these textiles are steadily turning long-imagined visions of a smart lifestyle into everyday reality. These technologies offer promising applications across diverse domains, including healthcare, protective gear, security systems, communication technologies, and electronic textiles [25–27].

Medical applications have become increasingly significant, with FDA-approved wearables for conditions ranging from diabetes management to cardiac monitoring. Instances of approved diabetes related wearable technologies are implantable CGMs and integrated CGM-insulin pump systems, allowing for precise and automated diabetes management [28, 29]. Wearable devices like the Philips Actiwatch Spectrum Pro and the Preventice BodyGuardian have received FDA 510(k) clearance, underscoring their clinical credibility. These devices monitor heart rate (HR) and respiratory rate (RR) using single-lead electrocardiogram (ECG) recordings. The BodyGuardian, in particular, has demonstrated strong correlation with conventional HR measurement methods, confirming its reliability for continuous cardiac monitoring [30].

Academic and industrial research has expanded into technologies like flexible electronics and energy harvesting [31], allowing for more comfortable, longer-lasting wearable devices.

The emergence of hearables—intelligent in-ear devices—has established a new category of wearables that combine audio functionality with health monitoring and digital assistant capabilities [32]. While companies like Apple with AirPods and Bose with their hearing aids have demonstrated dedication in the field, challenges in wireless connectivity, battery life, and comfort surfaced in this period of wearables' history.

Virtual and augmented reality headsets underwent significant transformation and diversification during the 2010s, with Meta's Quest series and Microsoft's HoloLens representing key advances in materials technology for wearable computing [33]. The evolution from bulky, tethered devices to lightweight, standalone headsets was enabled by several key material innovations: advanced polymer composites [34] reduced weight while maintaining durability; thermal management systems incorporating phase-change materials prevented overheating [35]; and sophisticated optical materials, from aspherical Fresnel lenses to waveguide optics utilizing diffraction gratings [36], dramatically improved visual fidelity while reducing form factor.

These headsets also pioneered integration of flexible electronics using polyimide substrate circuit boards that could conform to curved surfaces, alongside development of novel interface materials addressing the critical human factors of comfort and hygiene. Display technology evolved from traditional LCDs to OLED panels on flexible substrates, offering superior contrast and reduced motion sickness, while power management systems incorporated lithium-polymer batteries with silicon-composite anodes to increase energy density by approximately 25%. Collectively, these material science advancements established the foundation for immersive wearable computing experiences that balanced the competing demands of performance, comfort, and battery life—setting the trajectory for subsequent innovations in wearable technology throughout the following decade [33].

### *1.1.6 Future Directions (2020s and Beyond)*

As wearable technology continues to evolve, several trends are shaping its route towards the future. Integration with artificial intelligence is enhancing the predictive and contextual capabilities of wearables, making them more proactive in providing relevant information and assistance. Advances in materials science and flexible electronics are enabling more comfortable, less obtrusive form factors that can be worn for extended periods without causing user fatigue or discomfort.

The convergence of wearables with broader technological ecosystems—including smart homes, autonomous vehicles, and ambient computing environments—suggests a future where wearable devices serve as personal interfaces to increasingly pervasive computing infrastructure [37]. These devices are becoming less standalone products and more integral components of comprehensive digital experiences that span multiple domains of daily life, especially for disabled and elderly citizens.

With ongoing miniaturization of sensors and improvements in power efficiency, wearable technology continues to move toward the original vision of seamless integration between computation and the human experience, realizing the conceptual foundations established by early pioneers six decades ago. Energy harvesting technologies—from kinetic motion to body heat conversion—are beginning to supplement traditional batteries, addressing one of the field's most persistent limitations.

Google's mid-2025 unveiling of its advanced augmented reality (AR) glasses demonstrated how rapidly these technologies are maturing. During a landmark demonstration, researcher Shahram Izadi showcased the system's real-time translation capabilities, speaking in his mother tongue, Persian, while his words were instantly converted to English for his colleague, who responded in Hindi. This interaction highlighted not only the sophistication of the embedded AI and natural language processing systems but also the critical materials science breakthroughs that enabled such functionality in a consumer-wearable form factor. The next smart glasses may incorporated ultra-lightweight graphene-composite display substrates, printed flexible circuits with sub-micron traces, and transparent conductive films with unprecedented conductivity-to-transparency ratios—all contributing to a device that maintained all-day battery life in a package weighing less than a few grams.

This convergence of computational power and materials innovation exemplifies how progress in wearable electronics fundamentally depends on advances across multiple scientific disciplines. As we transition into the next section examining applications and market trends, it becomes clear that the technical foundations established during the historical development of wearables have created a platform capable of supporting increasingly diverse and sophisticated use cases across healthcare, productivity, and entertainment domains.

Thanks to AI, the future of wearable technology is evolving beyond simple sensing and responding toward sophisticated predictive capabilities and autonomous intervention. Such concept in smart wearables may be similar to condition-based monitoring systems in industrial IoT that forecast mechanical failures before they occur. This transformation depends on three key technological pillars: advanced materials science delivering flexible, biocompatible sensors with superior sensitivity; edge-based artificial intelligence systems recognizing complex patterns while operating under strict power constraints; and novel actuating materials that can execute physical interventions based on predictive analyses. The convergence of these elements promises wearables that transcend passive monitoring to become anticipatory extensions of human capability, fundamentally altering our relationship with technology and our surroundings while remaining critically dependent on continued innovation in materials that can sense changes, process data intelligently, and respond physically to execute commands.

As these foundational technologies continue to be refined, their integration into practical systems is increasingly being observed across multiple domains. Applications are being realized in areas such as healthcare monitoring, athletic performance optimization, and occupational safety, where predictive capabilities and autonomous interventions offer significant advantages. In the following section, the current and projected future applications of wearable electronics (Fig. 1.2) will be examined, and the market trends influencing the adoption and advancement of these intelligent wearables will be discussed.

## 1.2 Core Applications of Wearable Electronics and Market Trends

Wearable electronics are being deployed across a wide range of applications, with their market demonstrating substantial growth over the past decade. In 2019, the global wearable technology market was valued at approximately $50 billion [38], with projections indicating an increase to $82 billion by 2023 [39]. Further projections suggest continued expansion, with the market anticipated to grow from $54.1

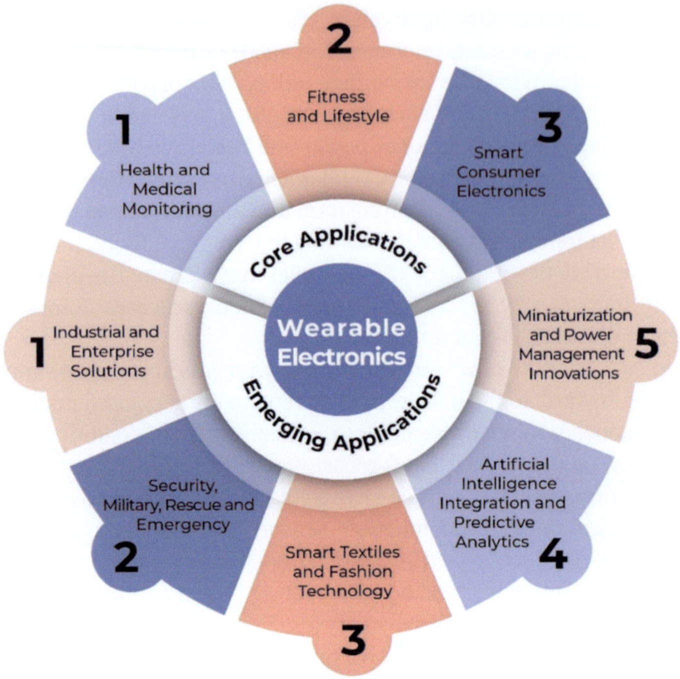


**Fig. 1.2** Current (core) and emerging (future) applications of wearable electronics

billion in 2024 to approximately $149.7 billion by 2032 [40]. These estimates underscore a robust growth trajectory and highlight the increasing integration of wearable technologies across consumer, medical, and industrial domains.

## *1.2.1 Health and Medical Monitoring*

Health and medical monitoring represent the most significant application area across industry analyses [41]. This sector comprises continuous vital signs monitoring including heart rate, blood oxygen saturation ($SpO_2$), electrocardiogram (ECG), blood pressure, and glucose tracking. An example of various wearable medical devices integrated into smart clothing is illustrated in Fig. 1.3.

The evolution toward clinical-grade wearables with FDA and CE regulatory clearances marks a critical transition from consumer wellness devices to legitimate medical instruments.

The integration of wearable devices with electronic health records (EHRs) for remote patient monitoring has demonstrated substantial clinical benefits, with studies showing a significant reduction in hospital readmissions and thus reduction of healthcare burdens [42]. Leading examples include the Apple Watch Series with

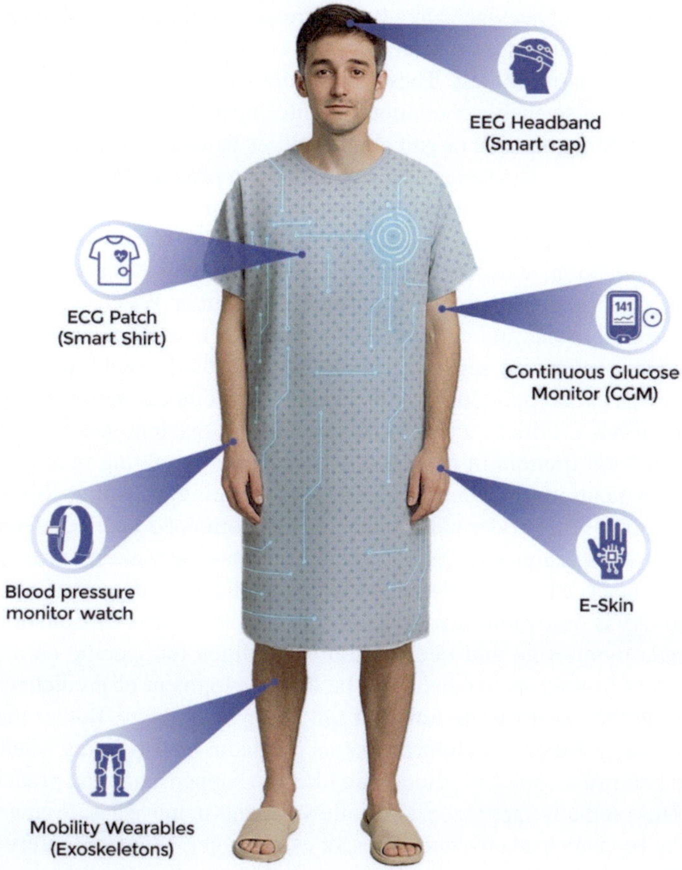


**Fig. 1.3** Conceptual illustration of various wearable medical technologies, including ECG patches (smart shirts), EEG headbands, continuous glucose monitors (CGMs), blood pressure monitor watches, mobility wearables (exoskeletons), and emerging e-skin technologies, all integrated within or connected to a smart medical garment

its comprehensive health tracking capabilities, Abbott's FreeStyle Libre for continuous glucose monitoring, and specialized devices like AliveCor for ECG monitoring. These devices represent a convergence of consumer accessibility with medical-grade accuracy.

In parallel, fall detection alarms have emerged as a critical component of remote patient monitoring, particularly for elderly populations and individuals at high risk of falls. Next-generation wearable fall alarms [43] are evolving beyond simple alert mechanisms, incorporating advanced motion sensors, machine learning algorithms, and real-time connectivity to automatically detect and report falls with increasing accuracy. Devices such as the Apple Watch and dedicated pendants from providers like Medical Guardian necklace already offer reliable fall detection, but newer solutions are being developed as discreet, multi-sensor systems integrated into belts, insoles, or even smart clothing. These innovations not only improve user comfort and adherence but also enable continuous monitoring and predictive analytics, which can flag deteriorating balance or gait abnormalities before a fall occurs. When integrated with EHRs, these systems allow clinicians to proactively intervene, thereby enhancing patient safety and reducing emergency hospitalizations.

Chronic disease management has become largely known as a particularly promising application, with specialized devices designed for conditions such as Parkinson's disease tremor tracking through the Emma Watch, epilepsy seizure detection via Embrace2, and comprehensive diabetes management systems. These devices leverage advanced sensor technologies to provide continuous monitoring capabilities that were previously only available in clinical settings. The Emma Watch, for instance, utilizes gyroscopic sensors and machine learning algorithms to counteract hand tremors in real-time, significantly improving quality of life for Parkinson's patients during daily activities like writing or eating. Similarly, the Embrace2 employs electrodermal activity sensors combined with accelerometers to detect convulsive seizures with high accuracy, automatically alerting caregivers and emergency contacts when episodes occur. These applications demonstrate how wearable technology is transitioning to specialized medical devices capable of providing clinical-grade monitoring and intervention capabilities for specific chronic conditions. The next frontier in this field will be the development of predictive capabilities, similar to the quest for the future of fall alarm technology. Rather than simply detecting events as they occur, future systems will focus on predicting when epileptic seizures or Parkinson's motor episodes are likely to happen, enabling proactive interventions. This predictive approach could allow patients to move to safe environments, caregivers to prepare protective measures, or even trigger automated safety protocols such as guiding patients to soft landing areas or alerting medical personnel before symptoms escalate. Such anticipatory healthcare represents a paradigm shift from reactive monitoring to preventive intervention, potentially reducing injury rates and improving overall patient outcomes.

An equally critical advancement lies in the development of non-invasive sensing technologies, which are paving the way for broader adoption of wearables in clinical contexts. Innovations such as optical glucose monitoring—designed to eliminate the need for finger-prick blood sampling—highlight this shift. Companies like Rockley Photonics are at the forefront, engineering wrist-worn spectrometers capable of detecting biomarkers including blood alcohol, hydration levels, and metabolic indicators through skin-based analysis. Efforts are being made to integrate optical sensors into wearable devices like contact lenses for continuous and non-invasive glucose monitoring as well [44]. By minimizing user discomfort and improving long-term compliance, these non-invasive sensing technologies are extending the reach of wearable devices into more continuous and passive forms of health monitoring, reinforcing their role in personalized and preventive medicine.

### 1.2.2 Fitness and Lifestyle

The fitness and lifestyle segment represents the foundation upon which the wearable electronics industry was built yet continues to evolve with increasingly sophisticated capabilities (Fig. 1.4). Modern fitness wearables extend far beyond simple step counting to include advanced metrics such as $VO_2$ max calculations, running dynamics analysis, sleep and circadian rhythm optimization, and stress recovery tracking through heart rate variability and skin temperature monitoring illustrates the comprehensive ecosystem of contemporary fitness wearables, showing how multiple specialized devices work synergistically to provide holistic performance monitoring. The integration of wrist-worn fitness trackers with specialized audio devices and smart textiles creates a seamless data collection network that captures physiological and biomechanical parameters across different body regions, enabling more accurate and comprehensive fitness analytics than any single device could provide alone.

WHOOP has emerged as a standout platform consistently mentioned across academic sources [45] for its ability to capture physiological metrics and provide personalized monitoring of training metrics, including strain, sleep, and recovery, while Garmin has established leadership in sport-specific analytics [46] and solar-powered device technology. The convergence of fitness and wellness applications with broader health platforms reflects the industry's movement toward comprehensive lifestyle monitoring rather than isolated activity tracking.

The consumer fitness segment demonstrates steady growth with increasing specialization in sport-specific applications. This specialization includes advanced motion capture systems for technique analysis and injury prevention, environmental adaptation features that optimize performance in extreme conditions through real-time physiological feedback, and the integration of smart apparel with embedded sensors for measuring muscle activity and form analysis.

### 1.2.3 Smart Consumer Electronics

As illustrated in Fig. 1.5, a typical user equipped with smart consumer electronics demonstrates the integration of consumer wearable electronics into daily life. Smart consumer electronics, including smartwatches, hearables, and AR/VR headsets, currently have the largest share of the total wearables market and the popularity and market size of smartwatches have been significant, with projections indicating growth from $28.5 billion in 2021 to $80.1 billion by 2028 [47].

This segment has expanded beyond basic notifications and connectivity to incorporate advanced features such as contactless payments, digital identity management, and sophisticated voice assistant integration.

Hearables, or smart earbuds, have emerged as a particularly dynamic category, offering real-time language translation, ambient sound filtering, and AI-powered

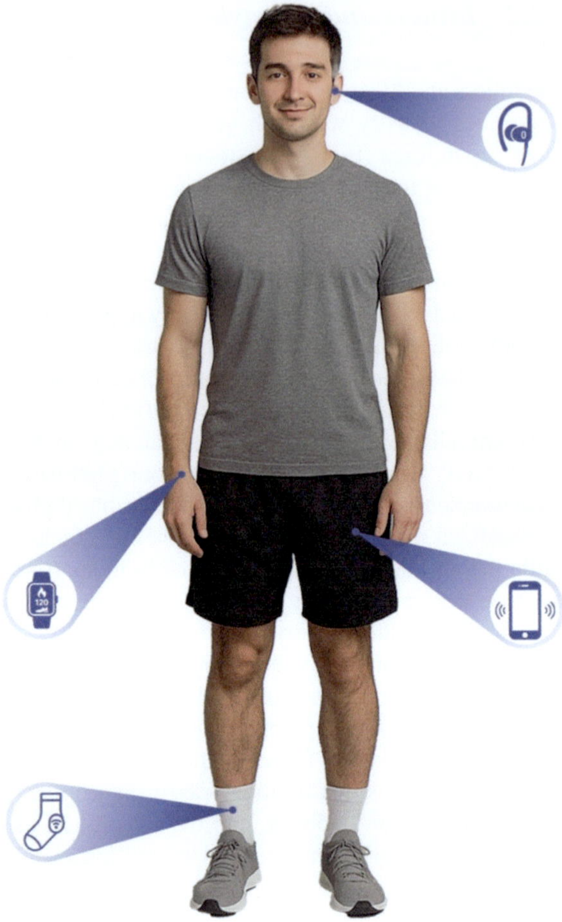


**Fig. 1.4** Everyday wearable technology for active lifestyle monitoring: workout earbud, wristband tracker, smart socks all connected via smartphone interface

personal assistance. The integration of these devices into broader technology ecosystems—connecting seamlessly with smartphones, smart homes, and autonomous vehicles—demonstrates the industry's evolution toward comprehensive digital integration rather than standalone device functionality in an analogous trend with the fitness and wellness applications.

Augmented and virtual reality applications represent the cutting edge of consumer wearables, with devices like Microsoft's HoloLens for enterprise applications, Meta's Quest series for gaming and training, Apple's Vision Pro for spatial computing and Google I/O 2025 for real-time interaction and translation and AI-assisted visual search. These platforms are driving the transition from traditional screen-based interaction towards immersive, contextual computing experiences.

**Fig. 1.5** Conceptual illustration of a user equipped with multiple consumer wearable electronics, including smart glasses, hearables, and a smartwatch, all connected through a central smartphone interface

## 1.3 Emerging Applications and Specialized Markets

### *1.3.1 Industrial and Enterprise Solutions*

Wearable technology is increasingly being adopted across industrial sectors such as manufacturing, logistics, and construction. Its integration into smart machinery, vehicular electronic sensors, and environmental monitoring systems—such as gas sensors—plays a critical role in enhancing operational efficiency, safety, and real-time data acquisition within industrial processes [48, 49].

Exoskeleton technology has progressed from experimental applications to practical implementation in logistics and manufacturing environments, with companies like Sarcos Robotics leading the development of productivity-enhancing wearable robotics [50, 51]. Augmented reality headsets have proven valuable for hands-free instruction delivery, remote expert guidance, and complex assembly processes in manufacturing environments.

Environmental monitoring capabilities integrated into industrial wearables provide real-time detection of hazardous gases, excessive noise levels, and dangerous temperature conditions [52–54]. Insurance companies are increasingly offering incentives for adopting wearable safety gear, recognizing the potential for significant reduction in workplace injuries and associated costs [55, 56].

Wearable radiation detection technologies play a critical role in environments where exposure to ionizing radiation poses significant risks. Lightweight and accurate wearable dosimeters are widely used for continuous personnel protection in defense, nuclear facilities, and medical settings [57]. Advanced devices, such as perovskite-based detectors, have demonstrated real-time monitoring capabilities for gamma radiation in industrial and medical applications [58]. In space exploration, crew-worn dosimeters like the Crew Active Dosimeter (CAD) on the International Space Station (ISS) are integral to radiation protection programs [59]. Additionally, in homeland security, compact and robust personal radiation detectors (PRDs) are employed by security personnel to detect photon-emitting radioactive materials, fulfilling critical requirements for counterterrorism operations [60, 61].

### 1.3.2 Security, Military, Rescue and Emergency Applications

All the applications with medical wearables and smart consumer electronics that can be used by civil people can also be used by security and military forces. There are, however, applications where it appears more militarily oriented than for civil use in wearable technologies. Smart helmets equipped with heads-up displays provide soldiers with real-time battlefield information, including GPS navigation, communications data, and threat detection alerts, while maintaining hands-free operation during combat [62]. Body-worn sensors monitor vital signs and physical condition of personnel, enabling effective communication [63] and the possibility to assess troop readiness and detect injuries or medical emergencies in real-time [64]. One recent groundbreaking application is contact lenses to see in the darkness using near-infrared (NIR) light [65]. A revolutionary study [65] has demonstrated the use of upconversion contact lenses (UCLs) that enable the human eye to perceive NIR light by converting it into visible wavelengths, allowing for real-time perception of NIR temporal and spatial signals without the need for bulky external devices such as night-vision goggles. The development of trichromatic UCLs (tUCLs) allows users to distinguish different NIR wavelengths as distinct visible colors, effectively enabling a form of NIR spatiotemporal color vision. In military and security contexts, this advancement could revolutionize covert operations, enhancing navigation, threat detection, and target identification in low-light or no-light environments while also opening the door to secure, invisible communication using encoded NIR signals only detectable by personnel equipped with such lenses. Advanced wearable systems also integrate thermal imaging and augmented reality capabilities directly into eyewear or helmet-mounted displays, giving military personnel enhanced visual capabilities in challenging environments. For security applications, smart uniforms embedded with RFID chips and biometric sensors help track personnel movement and verify identity, while wearable cameras provide continuous documentation for evidence collection and accountability. These technologies extend to specialized equipment like smart watches that can detect chemical or biological threats, and communication devices that automatically switch between secure networks based on location and threat level, transforming how military and security forces operate in the field.

### *1.3.3 Smart Textiles and Fashion Technology*

Smart textiles represented one of the most promising growth areas in wearable electronics in the beginning, and the market was expected to grow at a compound annual growth rate (CAGR) of 132% between 2016 and 2022 [66]. There is no solid report that could confirm that the anticipated CAGR materialized probably due to technological challenges, high production costs, poor market readiness and regulatory and standardization issues. However, the ambition to develop wearable electronics that deliver the same benefits as conventional devices at a more affordable price remains strong.

This segment involves temperature-regulating fabrics, LED-embedded clothing for dynamic visual displays, and the integration of conductive yarns and electronic fibers into traditional garments. Samsung's research into stretchable organic light-emitting diode (OLED) displays suggests new avenues for dynamic, conformable visual interfaces in wearable electronics.

The collaboration between technology companies and fashion houses, exemplified by Google's Jacquard partnership with Levi's, demonstrates the industry's recognition that successful wearable integration requires aesthetic appeal alongside functionality. Modular electronics that can be efficiently integrated into or removed from garments represent a sustainable approach to smart clothing that addresses both functionality and lifecycle management concerns. This is however still to be matured. However, this approach is still in its early stages and requires further advancement before widespread adoption.

### *1.3.4 Artificial Intelligence Integration and Predictive Analytics*

Artificial intelligence integration represents one of the most significant advancement trends across all wearable electronics applications. AI-powered predictive health analytics enable early disease detection, with examples including atrial fibrillation prediction through Apple Watch monitoring and diabetic retinopathy detection using Google's DeepMind technology. WHOOP's Strain Coach demonstrates how AI can provide personalized workout recommendations based on individual physiological data and recovery metrics.

The sophistication of AI applications in wearables extends beyond simple data analysis to include complex pattern recognition that can identify subtle changes in user behavior, physiological markers, and environmental conditions that may indicate health risks or performance inefficiencies. This capability transforms wearables from passive monitoring devices to active health and performance management tools.

### *1.3.5 Miniaturization and Power Management Innovations*

Continued miniaturization of wearable components, combined with improved power efficiency, enables the development of smaller, lighter devices with extended operational periods. This technological evolution represents a convergence of materials science, semiconductor engineering, and advanced manufacturing processes that are fundamentally reshaping the wearable electronics landscape.

Advances in battery technology are driving significant improvements in power density and charging capabilities. Graphene-based power storage systems offer exceptional energy density due to graphene's unique two-dimensional structure and high electrical conductivity, allowing for ultra-thin batteries that can be integrated into flexible substrates [67, 68]. These graphene-enhanced batteries demonstrate superior charge–discharge rates and extended cycle life compared to conventional lithium-ion technologies [69, 70].

Energy harvesting represents another critical advancement, with sophisticated systems capturing power from multiple environmental sources. Advanced photovoltaic cells integrated into wearable surfaces can efficiently convert ambient light into electrical energy, while piezoelectric generators harvest kinetic energy from body movement, heartbeat, and even respiratory motion. Thermoelectric generators capitalize on body heat differentials, creating continuous power generation from natural physiological processes. These combined harvesting approaches significantly reduce dependency on traditional charging methods, addressing one of the primary limitations of current wearable devices.

Perovskite materials have been recognized as transformative components in next-generation wearable displays and photovoltaic systems.

The tunable bandgaps (1.2–3.0 eV), high absorption coefficients ($>10^5$ $cm^{-1}$), and excellent charge transport characteristics with long diffusion lengths have made perovskites particularly attractive for wearable applications [71, 72], while their solution-processable nature enables seamless integration into textile substrates through low-temperature deposition techniques [73].

MicroLED display technology leverages these advanced materials to achieve unprecedented pixel density and energy efficiency. Individual LED elements measuring less than 100 μm enable displays with remarkable brightness levels while consuming minimal power. The fabrication of perovskite-enhanced microLEDs demands precise synthesis routes such as microware-assisted [74], sonochemical [75] or vapor deposition [76] synthesis as well as micro-molding process [77]. These techniques enable the control of crystal structure at the atomic level and optimize performance characteristics for specific wearable applications.

Another attractive application of wearable perovskites may be LiFi (Light Fidelity), especially for miniaturization and power-efficient communication in wearables. LiFi is a wireless technology that uses light instead of radio waves to transmit data quickly and securely.

Researchers have developed perovskite quantum dot-coated fibers that can both emit and detect light, enabling bidirectional LiFi communication [78]. These fibers are lightweight, flexible, and ideal for integration into smart wearable devices.

## 1.4 Future Outlook and Implications

The consolidated analysis of wearable electronics applications and market trends reveals an industry transitioning from novelty consumer devices to comprehensive lifestyle enhancement platforms spanning healthcare, entertainment, sports performance, augmented reality, and professional productivity. The strong consensus across industry sources regarding growth in diverse application domains, combined with advancing AI integration and ecosystem connectivity, suggests a future where wearable electronics become integral components of healthcare delivery, workplace safety, personal performance optimization, immersive entertainment experiences, and augmented reality interfaces.

The projected market expansion to approximately $150 billion by 2032 [40], driven by diversified demand including healthcare needs from aging populations, growing acceptance of technology-mediated health monitoring, expanding sports and fitness applications, emerging smart glasses and AR/VR markets, and increasing integration into consumer electronics ecosystems, indicates substantial opportunities for continued innovation across multiple sectors. Beyond traditional fitness tracking, the market encompasses smart glasses for augmented reality applications, advanced sports performance monitoring, and seamless integration with Internet of Things (IoT) ecosystems for smart home and workplace applications.

The evolution toward smart textiles, non-invasive sensing technologies, fall alarms for elderly population, and transparent computing interfaces promises to eliminate many current barriers to wearable adoption while expanding applications into previously inaccessible areas. Smart glasses represent a particularly significant growth vector, with applications ranging from navigation and translation to professional training and entertainment, while advanced sports wearables provide real-time biomechanical analysis and performance optimization for both professional athletes and recreational users.

### 1.4.1 Materials Innovation and Market Success

The advancement of wearable electronics fundamentally depends on continued materials science breakthroughs, particularly in flexible electronics [79], energy storage and self-powered [80], and display technologies [81, 82]. Next-generation materials including graphene-based conductors, advanced polymers for flexible substrates, and novel semiconductor compounds are enabling unprecedented form factors and performance characteristics. These materials innovations directly translate to improved user experiences through lighter, more durable, and more capable devices.

Perovskite materials development represents a potentially transformative force for the entire wearable technology market's future success. Their exceptional optoelectronic properties, solution-processable manufacturing compatibility, and remarkable versatility in display, photovoltaic, and sensing applications position perovskites as enabling materials for next-generation wearable devices. The successful commercialization of perovskite-based components could dramatically reduce manufacturing costs while simultaneously improving performance metrics across multiple device categories, from ultra-efficient microLED displays in smart glasses to high-performance photovoltaic cells for energy harvesting. As synthesis techniques for perovskite materials continue to mature and stability challenges are addressed, their widespread adoption could fundamentally reshape the economics and capabilities of wearable electronics, making sophisticated wearable technology accessible to broader consumer markets and enabling entirely new application categories that are currently limited by existing material constraints.

The intersection of perovskite materials and wearable technology entails promising frontiers in advanced materials science and electronics. As our world moves increasingly toward portable and personal electronic devices, the demand for flexible, efficient, and sustainable materials has never been greater. Perovskites, with their remarkable optoelectronic properties and versatile nature, are emerging as key candidates for next-generation wearable devices.

The unique crystal structure of perovskites allows for extensive ionic substitution and property tuning, making them highly adaptable for various wearable applications [73]. Their remarkable charge carrier mobility and long diffusion lengths, typically exceeding 100 nm in metal halide perovskites and hybrid organic–inorganic perovskites, enable efficient charge transport, which is essential for flexible electronic devices [83].

From the mechanical integrity viewpoint, the ability of organic–inorganic perovskites to maintain their performance under mechanical stress makes them particularly suitable for wearable technology [84]. Studies have demonstrated that certain perovskite compositions can withstand bending radii of less than 5 mm while retaining over 90% of their initial performance [85–87]. This mechanical flexibility, combined with their straight processability at relatively low temperatures (<150 °C), enables cost-effective fabrication methods compatible with flexible substrates commonly used in wearable devices [86]. The integration of perovskites into stretchable electronics has shown promising results, with devices maintaining functionality under strains of up to 20% and hundreds of cycles [88].

A significant advantage of perovskites in wearable applications lies in their exceptional power conversion efficiency in photovoltaic devices, which has rapidly increased from 3.8% in 2009 to over 25% in recent years [89, 90]. This high efficiency, coupled with their low weight and potential for semi-transparency, makes perovskites ideal for self-powered wearable devices that can harvest energy from ambient light. Moreover, their broad absorption spectrum and high absorption coefficients allow for the fabrication of ultra-thin devices, crucial for maintaining user comfort in wearable applications [91].

The environmental stability of perovskites has been significantly improved through recent developments in composition engineering and encapsulation techniques. Researchers have successfully demonstrated perovskite-based wearable sensors that maintain stable performance under various environmental conditions, including mechanical stability and long-term environmental stability for up to two months without encapsulation. Additionally, perovskite-based humidity sensors have shown reliable performance for over 35 days under high humidity conditions (90% RH) and temperature fluctuations [92, 93]. The incorporation of 2D/3D hybrid structures has shown particular promise in enhancing device stability while maintaining the superior optoelectronic properties that make perovskites attractive for wearable technology [94, 95]. These advancements suggest a bright future for perovskite-based wearable devices in applications ranging from health monitoring to augmented reality displays which readers may partially explore in this book.

## References


1. V. Ilardi, Renaissance vision from spectacles to telescopes, in *Optics InfoBase Conference Papers* (2004)
2. A.J. Martin, *Sensors and Computing Systems in Smart Clothing* (2022). https://doi.org/10.1016/B978-0-12-819526-0.00014-X
3. A. Reiss, O. Amft, *Design Challenges of Real Wearable Computers* (2015). https://doi.org/10.1201/b18703-27
4. K. Grifantini, Digital watches and pet rocks: the long-term value of an innovation often doesn't become apparent until it has gone through many product cycles-including buggy versions that annoy early adopters. Technol. Rev. **114**(2), 88 (2011)
5. M. Klasen-Memmer, H. Hirschmann, *Liquid Crystal Materials for Devices* (2016). https://doi.org/10.1007/978-3-319-14346-0_83
6. M. Malmivaara, *The Emergence of Wearable Computing* (2009). https://doi.org/10.1533/9781845695668.1.3
7. S. Mann, Phenomenal augmented reality: advancing technology for the future of humanity. IEEE Consum. Electron. Mag. **4**(4), 92–97 (2015). https://doi.org/10.1109/MCE.2015.2463312
8. S. Mann, M.A. Ali, R. Lo, H. Wu, FreeGlass for developers, 'haccessibility', and digital eye glass + Lifeglogging research in a (sur/sous) veillance society, in *International Conference on Information Society, i-Society 2013* (2013), pp. 48–53
9. Download, practice, done [The future we deserve], IEEE Spectr. **51**(6), 72 (2014). https://doi.org/10.1109/MSPEC.2014.6821626

10. V.G. Motti, K. Caine, Smart wearables or dumb wearables? Understanding how context impacts the UX in wrist worn interaction, in *SIGDOC 2016—34th ACM International Conference on the Design of Communication* (2016). https://doi.org/10.1145/2987592.2987606
11. S.-B. Yu, H. Yoon, S.-H. Park, K.-T. Lee, Motion UI: motion-based user interface for movable wrist-worn devices, in *IEEE International Conference on Consumer Electronics—Berlin, ICCE-Berlin* (2017), pp. 265–266. https://doi.org/10.1109/ICCE-Berlin.2017.8210647
12. O. Amft, P. Lukowicz, From backpacks to smartphones: past, present, and future of wearable computers. IEEE Pervasive Comput. **8**(3), 8–13 (2009). https://doi.org/10.1109/MPRV.2009.44
13. G. Shin et al., Wearable activity trackers, accuracy, adoption, acceptance and health impact: a systematic literature review. J. Biomed. Inform. **93**, 103153 (2019). https://doi.org/10.1016/J.JBI.2019.103153
14. J. Meese, *Google Glass and Australian Privacy Law: Regulating the Future of Locative Media* (2014). https://doi.org/10.4324/9781315887036-19
15. S.U. Noble, S.T. Roberts, *Through Google-Colored Glass(es): Design, Emotion, Class, and Wearables as Commodity and Control* (2016). https://doi.org/10.1016/B978-0-12-801872-9.00009-0
16. A. Kotsios, Privacy in an augmented reality. Int. J. Law Inf. Technol. **23**(2), 157–185 (2015). https://doi.org/10.1093/ijlit/eav003
17. T. Sobeih, N. Whittaker, L. Han, DIVINE: building a wearable device for intelligent control of environment using Google Glass, in *Proceedings—15th IEEE International Conference on Computer and Information Technology, CIT 2015, 14th IEEE International Conference on Ubiquitous Computing and Communications, IUCC 2015, 13th IEEE International Conference on Dependable, Autonomic and Secure Computing* (2015), pp. 1280–1285. https://doi.org/10.1109/CIT/IUCC/DASC/PICOM.2015.190
18. T. Caldwell, The quantified self: a threat to enterprise security? Comput. Fraud Secur. **2014**(11), 16–20 (2014). https://doi.org/10.1016/S1361-3723(14)70549-4
19. A.J.C. Trappey, A.K.C. Shen, M.C.J. Yu, *Explore Transdisciplinary Collaborations for Smart Device Development: Advancing Health-Related Smartwatch Designs for Social Impact*, vol. 60 (2024). https://doi.org/10.3233/ATDE240875
20. C. Carciani, L. Bagnaschino, *TechCouturism, an Alternative Showcase for New Fashion Designers* (2018). https://doi.org/10.4018/978-1-5225-5484-4.ch030
21. Y.-A. Lee, *Trends of Emerging Technologies in the Fashion Product Design and Development Process* (2022). https://doi.org/10.1007/978-3-030-91135-5_1
22. L. Donatiello, E. Morotti, G. Marfia, S. Di Vaio, Exploiting immersive virtual reality for fashion gamification, in *IEEE International Symposium on Personal, Indoor and Mobile Radio Communications, PIMRC* (2018), pp. 17–21. https://doi.org/10.1109/PIMRC.2018.8581036
23. A. Joy, Y. Zhu, C. Peña, M. Brouard, Digital future of luxury brands: metaverse, digital fashion, and non-fungible tokens. Strateg. Chang. **31**(3), 337–343 (2022). https://doi.org/10.1002/jsc.2502
24. A. Zoppelletto, E. Santini, C. Rossignoli, F. Ricciardi, Interfirm collaboration enhancing twin transition: evidence from the Italian fashion industry. IEEE Trans. Eng. Manag. **72**, 546–557 (2025). https://doi.org/10.1109/TEM.2025.3528260
25. L. Lou, W. Yu, S. Ramkumar, *Wearable and Smart Responsive Textiles* (2019). https://doi.org/10.1002/9781119325062.ch15
26. V. Koncar, *Smart Textiles and Their Applications* (2016). https://doi.org/10.1016/C2014-0-03772-0
27. A.A. Merati, *Advances in Research and Applications of Smart Textiles* (2017)
28. M. Condren, S. Sabet, L.J. Chalmers, T. Saley, J. Hopwood, Technology for augmenting type 1 diabetes mellitus management. J. Pediatr. Pharmacol. Ther. **24**(2), 99–106 (2019). https://doi.org/10.5863/1551-6776-24.2.99
29. J. Ambalavanan, D.I. CDCES, M.C. Lansang, Diabetes technology: a primer for clinicians. Cleve. Clin. J. Med. **91**(6), 353–360 (2024). https://doi.org/10.3949/ccjm.91a.23073
30. E.S. Izmailova et al., Continuous monitoring using a wearable device detects activity-induced heart rate changes after administration of amphetamine. Clin. Transl. Sci. **12**(6), 677–686 (2019). https://doi.org/10.1111/cts.12673

31. H. Chu, J. Xue, D. Luo, H. Zheng, Z. Li, *Advances in Wearable Multifunctional Devices Based on Human-Body Energy Harvesting* (2024). https://doi.org/10.1002/admt.202302068
32. C.K.H. Ne, J. Muzaffar, A. Amlani, M. Bance, *Hearables, In-Ear Sensing Devices for Bio-Signal Acquisition: A Narrative Review* (2021). https://doi.org/10.1080/17434440.2021.2014321
33. V. Bondarenko, M. Attawna, G.T. Nguyen, F.H.P. Fitzek, Motion sickness comparison of 2 XR approaches: camera + screen or hologram on transparent lens (Meta Quest 3 or Microsoft HoloLens 2), in *2024 International Electronics Symposium: Shaping the Future: Society 5.0 and Beyond, IES 2024—Proceeding* (2024), pp. 728–733. https://doi.org/10.1109/IES63037.2024.10665757
34. S. Drieschner, F. Kloiber, M. Hennemeyer, J.J. Klein, M.W. Thesen, High quality diffractive optical elements (DOEs) using SMILE imprint technique. Adv. Opt. Technol. **10**(1), 11–16 (2021). https://doi.org/10.1515/aot-2020-0053
35. M. Abrash, Creating the future: augmented reality, the next human-machine interface, in *Technical Digest—International Electron Devices Meeting, IEDM* (2021), pp. 1.2.1–1.2.11. https://doi.org/10.1109/IEDM19574.2021.9720526
36. J. Xiong, E.-L. Hsiang, Z. He, T. Zhan, S.-T. Wu, Augmented reality and virtual reality displays: emerging technologies and future perspectives. Light: Sci. Appl. **10**(1) (2021). https://doi.org/10.1038/s41377-021-00658-8
37. S.K. Behera, G. Saisudha, L. Vasundhara, J. Nasreen, SmartWear—an integrated watch for patients, in *3rd International Conference on Power, Energy, Control and Transmission Systems, ICPECTS 2022—Proceedings* (2022). https://doi.org/10.1109/ICPECTS56089.2022.10047278
38. E. Fritsch, Growth in wearable technologies. Screen Print. **110**(5), 22–26 (2019)
39. K. Eschen, J. Abel, Toward integrated wearable devices: shape memory alloy textiles shape memory alloy textiles: provide complex deformation and force profiles for novel wearable applications. Adv. Mater. Processes **178**(3), 63–66 (2020)
40. Z. Wang, L. Wei, Functional fibers redefine wearables. Opt. Photonics News **36**(3), 26–33 (2025). https://doi.org/10.1364/opn.36.3.000026
41. Y.S. Çelik, E. İlçe, B. Mesut, Y. Özsoy, An overview of wearable medical device applications. J. Res. Pharm. **28**(3), 722–732 (2024). https://doi.org/10.29228/jrp.734
42. M.S. Hiware, P. Jirvankar, Wearable technology and its effect on the provision of healthcare, in *2024 2nd DMIHER International Conference on Artificial Intelligence in Healthcare, Education and Industry, IDICAIEI 2024* (2024). https://doi.org/10.1109/IDICAIEI61867.2024.10842892
43. H. Su et al., One-step electrospinning synthesis of lead-free perovskite nanofiber-based triboelectric nanogenerator for wearable intelligent fall alarm. J. Colloid Interface Sci. **696**, 137836 (2025). https://doi.org/10.1016/J.JCIS.2025.137836
44. M. Elsherif, M.U. Hassan, A.K. Yetisen, H. Butt, Wearable contact lens biosensors for continuous glucose monitoring using smartphones. ACS Nano **12**(6), 5452–5462 (2018). https://doi.org/10.1021/acsnano.8b00829
45. V. Romano, M. Etim-Andy, An examination of WHOOP fitness band usage in college athletics. J. Coach. Sport. Sci. **2**(2), 46–52 (2023). https://doi.org/10.58524/jcss.v2i2.220
46. T. Szot, Evolution of sport wearable global navigation satellite systems' receivers: a look at the Garmin Forerunner series. Proc. Inst. Mech. Eng. P. J. Sport. Eng. Technol. (2024). https://doi.org/10.1177/17543371241237319
47. G.R. Wagner, K. Ibarra, A. Mathew, Thermal testing and analysis of on-the-market smartwatches, in *InterSociety Conference on Thermal and Thermomechanical Phenomena in Electronic Systems, ITHERM* (2024). https://doi.org/10.1109/ITherm55375.2024.10709507
48. C.-S. Lin, C.-H. Tai, R.-J. Chen, K.-M. Li, H.-T. Young, W.-T. Hsiao, Characterization analysis of 532 nm laser drilling of silicon-based glass heterogeneous integration composites. Int. J. Adv. Manuf. Technol. **123**(9–10), 3699–3710 (2022). https://doi.org/10.1007/s00170-022-10425-7

49. J. Shi, M. Wagih, S. Beeby, Highly conductive flexible printed PEDOT:PSS films for green humidity sensing applications, in *FLEPS 2022—IEEE International Conference on Flexible and Printable Sensors and Systems, Proceedings* (2022). https://doi.org/10.1109/FLEPS53764.2022.9781556
50. D. Ippolito, C. Constantinescu, C.A. Rusu, Enhancement of human-centered workplace design and optimization with Exoskeleton technology. Procedia CIRP 243–248 (2020). https://doi.org/10.1016/j.procir.2020.02.173
51. C. Constantinescu, R. Rus, C.-A. Rusu, D. Popescu, Digital twins of exoskeleton-centered workplaces: challenges and development methodology. Procedia Manuf. 58–65 (2019). https://doi.org/10.1016/j.promfg.2020.01.228
52. M. Haghi, K. Thurow, N. Stoll, A multi-layer multi-sensor wearable device for physical and chemical environmental parameters monitoring (CO and $NO_2$), in *Proceedings of the International Conference on Information and Digital Technologies, IDT 2017* (2017), pp. 137–141. https://doi.org/10.1109/DT.2017.8024285
53. J. Moon, B.-K. Ju, Wearable sensors for healthcare of industrial workers: a scoping review. Electronics (Switzerland) **13**(19) (2024). https://doi.org/10.3390/electronics13193849
54. M.N. Tahir, D. Falana, K.O. Daniel, D. Arevalo, J. Guanipatin, U. Hassan, A wearable multipurpose toxic gas-monitoring device for industrial applications, in *2023 IEEE 19th International Conference on Body Sensor Networks, BSN 2023—Proceedings* (2023). https://doi.org/10.1109/BSN58485.2023.10331298
55. Y. Chen, W. Qiu, R. Ou, C. Huang, A contract-based insurance incentive mechanism boosted by wearable technology. IEEE Internet Things J. **8**(7), 6089–6100 (2021). https://doi.org/10.1109/JIOT.2020.3033543
56. D. Soliño-Fernandez, A. Ding, E. Bayro-Kaiser, E.L. Ding, Willingness to adopt wearable devices with behavioral and economic incentives by health insurance wellness programs: results of a US cross-sectional survey with multiple consumer health vignettes. BMC Public Health **19**(1) (2019). https://doi.org/10.1186/s12889-019-7920-9
57. S. Dhanekar, K. Rangra, Wearable dosimeters for medical and defence applications: a state of the art review. Adv. Mater. Technol. **6**(5) (2021). https://doi.org/10.1002/admt.202000895
58. L. Basiricò et al., Wearable perovskite films for on-line monitoring of radiotracers in nuclear medicine. Adv. Mater. Technol. **10**(2) (2025). https://doi.org/10.1002/admt.202401111
59. R. Gaza et al., The importance of time-resolved personal dosimetry in space: the ISS crew active dosimeter. Life Sci. Space Res. (Amst.) **39**, 95–105 (2023). https://doi.org/10.1016/j.lssr.2023.08.004
60. D. Ginzburg et al., Personal radiation detector at a high technology readiness level that satisfies DARPA's SN-13-47 and SIGMA program requirements. Nucl. Instrum. Methods Phys. Res. A **784**, 438–447 (2015). https://doi.org/10.1016/j.nima.2014.12.117
61. A. Osovizky et al., SENTIRAD—an innovative personal radiation detector based on a scintillation detector and a silicon photomultiplier. Nucl. Instrum. Methods Phys. Res. A **652**(1), 41–44 (2011). https://doi.org/10.1016/j.nima.2011.01.027
62. S. Jalui, T. Hait, T. Hathi, S. Ghosh, Advanced military helmet aided with wireless live video transmission, sensor integration and augmented reality headset, in *Proceedings of the 4th International Conference on Communication and Electronics Systems, ICCES 2019* (2019), pp. 123–127. https://doi.org/10.1109/ICCES45898.2019.9002600
63. S. Raut, G. Jaisinghani, N. Doshi, G. Gaikwad, Smart warfare: designing intelligent vests for enhanced military operations, in *7th International Conference on Inventive Computation Technologies, ICICT 2024* (2024), pp. 1016–1022. https://doi.org/10.1109/ICICT60155.2024.10544719
64. D. Roberts et al., Testing and evaluation of a wearable augmented reality system for natural outdoor environments, in *Proceedings of SPIE—The International Society for Optical Engineering* (2013). https://doi.org/10.1117/12.2015621
65. Y. Ma et al., Near-infrared spatiotemporal color vision in humans enabled by upconversion contact lenses. Cell (2025). https://doi.org/10.1016/j.cell.2025.04.019

66. D. Zavec Pavlinić, The potential of wearables related in smart textiles | Potencijalno nosive elektronike u odnosu na pametni tekstil. Sigurnost **59**(3), 219–226 (2017). https://doi.org/10.31306/s.59.3.3
67. M. Ye, Z. Zhang, Y. Zhao, L. Qu, Graphene platforms for smart energy generation and storage. Joule **2**(2), 245–268 (2018). https://doi.org/10.1016/j.joule.2017.11.011
68. W. Liu et al., Mechanically flexible reduced graphene oxide/carbon composite films for high-performance quasi-solid-state lithium-ion capacitors. J. Energy Chem. **80**, 68–76 (2023). https://doi.org/10.1016/j.jechem.2023.01.031
69. F. Xiang et al., Scalable and rapid far infrared reduction of graphene oxide for high performance lithium ion batteries. Energy Storage Mater. **1**, 9–16 (2015). https://doi.org/10.1016/j.ensm.2015.06.001
70. S. Scaravonati et al., Combined capacitive and electrochemical charge storage mechanism in high-performance graphene-based lithium-ion batteries. Mater. Today Energy **24** (2022). https://doi.org/10.1016/j.mtener.2021.100928
71. C.C. Li, T. Yu Huang, Y.H. Lai, Y.C. Huang, C. Shan Tan, Lead-free perovskites for flexible optoelectronics. Mater. Today Electron. **8** (2024). https://doi.org/10.1016/j.mtelec.2024.100095
72. R. Zhao, Z. Gu, P. Li, Y. Zhang, Y. Song, Flexible and wearable optoelectronic devices based on perovskites. Adv. Mater. Technol. **7**(3) (2022). https://doi.org/10.1002/admt.202101124
73. M. Tabassum, Q. Zia, Y. Zhou, Y. Wang, M.J. Reece, L. Su, A review of recent developments in smart textiles based on perovskite materials. Textiles **2**(3), 447–463 (2022). https://doi.org/10.3390/textiles2030025
74. D.N. Feria, J.-W. Su, G.-H. Wu, Y.-T. Zeng, J.-T. Lian, T.-Y. Lin, Enhanced formation mechanism in microwave-assisted triple cation perovskites for stable ambient-fabricated perovskite transistors, in *Proceedings of SPIE—The International Society for Optical Engineering* (2024). https://doi.org/10.1117/12.3022415
75. C.-S. Jo et al., Solution-processed fabrication of light-emitting diodes using $CsPbBr_3$ perovskite nanocrystals. ACS Appl. Nano Mater. **3**(12), 11801–11810 (2020). https://doi.org/10.1021/acsanm.0c02334
76. K.K. Naik, *Synthesis and Characterization of Perovskite-Based QDs, 1D, 2D, and Hierarchical Nanomaterials* (2023). https://doi.org/10.1016/B978-0-323-99529-0.00003-5
77. H.J. An, M.S. Kim, J.-M. Myoung, Strategy for the fabrication of perovskite-based green micro LED for ultra high-resolution displays by micro-molding process and surface passivation. Chem. Eng. J. **453** (2023). https://doi.org/10.1016/j.cej.2022.139927
78. Q. Shan et al., Perovskite light-emitting/detecting bifunctional fibres for wearable LiFi communication. Light Sci. Appl. **9**(1), 163 (2020). https://doi.org/10.1038/s41377-020-00402-8
79. Y.J. Yang, S.G. Lee, T.-I. Kim, Wearable devices for biofluid monitoring in a body: from lab to commercialization. Korean J. Chem. Eng. (2024). https://doi.org/10.1007/s11814-024-00295-y
80. H. Afshar, F. Kamran, F. Shahi, Recent progress in energy harvesting technologies for self-powered wearable devices: the significance of polymers, Polym. Adv. Technol. **36**(4) (2025). https://doi.org/10.1002/pat.70187
81. D. Yin et al., Highly transparent and flexible fabric-based organic light emitting devices for unnoticeable wearable displays. Org. Electron. **76** (2020). https://doi.org/10.1016/j.orgel.2019.105494
82. J. Flemish et al., Microled device technology for low-power wearable displays, in *Digest of Technical Papers—SID International Symposium* (2022), pp. 478–480. https://doi.org/10.1002/sdtp.15526
83. H. Chen et al., Flexible optoelectronic devices based on metal halide perovskites. Nano Res. **13**(8), 1997–2018 (2020). https://doi.org/10.1007/s12274-020-2805-x
84. B. Farhadi, F. Zabihi, M. Tebyetekerwa, I. Lugoloobi, M. Omidi, A. Liu, Simulations of lead-free organic–inorganic perovskites under tensile/compressive loads and different force fields. Mol. Phys. **121**(1) (2023). https://doi.org/10.1080/00268976.2022.2150704

85. H. Liu et al., Selective chemical etching to remove 0D $Cs_4Pb(IBr)_6$ from mixed-dimensional perovskite surface achieving high efficiency flexible all-inorganic perovskite solar cells with superior mechanical durability. Adv. Energy Mater. **14**(46) (2024). https://doi.org/10.1002/aenm.202402142
86. M. Pandey et al., Efficient, hysteresis free, inverted planar flexible perovskite solar cells: via perovskite engineering and stability in cylindrical encapsulation. Sustain. Energy Fuels **3**(7), 1739–1748 (2019). https://doi.org/10.1039/c9se00153k
87. P.-C. Wang, V. Govindan, C.-H. Chiang, C.-G. Wu, Room-temperature-processed fullerene/$TiO_2$ nanocomposite electron transporting layer for high-efficiency rigid and flexible planar perovskite solar cells. Sol. RRL **4**(10) (2020). https://doi.org/10.1002/solr.202000247
88. H. Li et al., Kirigami-based highly stretchable thin film solar cells that are mechanically stable for more than 1000 cycles. ACS Nano **14**(2), 1560–1568 (2020). https://doi.org/10.1021/acsnano.9b06562
89. Y. Wang, Z. Cheng, J. Li, K. Lv, Z. Li, H. Zhao, Innovative approaches to large-area perovskite solar cell fabrication using slit coating. Molecules **29**(20) (2024). https://doi.org/10.3390/molecules29204976
90. K. Beepat, S. Kumar, A. Sharma, D.P. Sharma, D. Pathak, J.-M. Nunzi, Perovskite materials for photovoltaics: a review. EPJ Appl. Phys. **98** (2023). https://doi.org/10.1051/epjap/2023230023
91. L. Zhang et al., Advancing perspectives on large-area perovskite luminescent films. Energy Fuels **38**(18), 17343–17354 (2024). https://doi.org/10.1021/acs.energyfuels.4c03168
92. C. Pi et al., Highly stable humidity sensor based on lead-free $Cs_3Bi_2Br_9$ perovskite for breath monitoring. J. Mater. Chem. C Mater. **9**(34), 11299–11305 (2021). https://doi.org/10.1039/d1tc02339j
93. T. Zhang et al., Low-temperature processed inorganic perovskites for flexible detectors with a broadband photoresponse. Nanoscale **11**(6), 2871–2877 (2019). https://doi.org/10.1039/c8nr09900f
94. K.-T. Ho et al., Surface effect on 2D hybrid perovskite crystals: perovskites using an ethanolamine organic layer as an example, Adv. Mater. **30**(46) (2018). https://doi.org/10.1002/adma.201804372
95. S. Yang et al., Investigation of 3D and 2D/3D heterojunction perovskite-based phototransistors. Appl. Phys. Lett. **120**(26) (2022). https://doi.org/10.1063/5.0089840

# Chapter 2
# Materials for Wearable Electronics

This chapter presents a systematic review of material selection criteria, design principles, and emerging trends in the field of wearable electronics, with a particular emphasis on perovskite-based materials. The discussion is structured to first address fundamental material requirements and engineering considerations essential for the development of high-performance wearable devices. Key aspects include mechanical flexibility, biocompatibility, and environmental resilience under operational conditions. A detailed analysis is provided on the functional roles of conductors, semiconductors, and dielectrics in wearable electronic systems, establishing a foundation for the subsequent discussion on perovskites as a promising class of materials. The unique advantages of perovskites—such as their tunable optoelectronic properties, solution-processable fabrication, and compatibility with high-efficiency device architectures—are critically evaluated. However, significant challenges remain, including long-term stability under ambient conditions, scalability of manufacturing processes, and concerns regarding material toxicity. These limitations are examined in the context of current research efforts aimed at mitigating these issues. The chapter concludes with an outlook on future research directions, highlighting innovative strategies for the integration of perovskite materials into next-generation wearable technologies. Potential advancements in device performance, multifunctionality, and sustainable fabrication methods are discussed, offering insights into the transformative potential of perovskites in this rapidly evolving field.

## 2.1 Requirements and Design Considerations

The development of numerous products, processes, and devices that are integral to daily life relies significantly on the innovation of new materials or the enhancement of existing ones. Wearable electronic products are no exception, the design of which necessitates a careful balancing of design constraints and trade-offs that differ

 
M. Darab et al., *Perovskites for Next-Generation Wearable Technology*, Engineering Materials, https://doi.org/10.1007/978-3-032-03931-6_2

markedly from traditional design methodologies. The process of designing Wearable electronic products involves translating an innovative concept or market demand into comprehensive specifications that guide the manufacturing of the wearable device [1]. Each phase of this process requires critical decisions regarding the selection of materials and the techniques employed in their fabrication. Precisely defining the intended need is crucial at the outset. Initially, the design of wearable electronic products is adaptable, allowing for a wide range of material considerations. However, as the design process becomes more refined and begins to take shape, the criteria for selection become more stringent, thereby narrowing the pool of appropriate materials [2]. The design of wearable electronic products is governed by ten fundamental requirements and design considerations that interact dynamically [3–6], as illustrated in Fig. 2.1.

The first and foremost requirement is context which encompasses the diverse environments, circumstances, and conditions that impact the development, application, and user experience of wearable electronic devices. A comprehensive understanding of these contextual factors is essential to ensure that the product effectively meets user requirements and performs optimally in real-world applications. The following elements constitute key dimensions of context in the design of wearable electronic products [7–10]:

- Cultural Context: Distinct cultural values, norms, and practices can influence the reception and acceptance of a wearable electronic device. Designers must be cognizant of these cultural considerations to ensure that the product resonates with and is suitable for its intended audience.
- Physical Context: This dimension pertains to the physical environment in which the product will be utilized. Variables such as ambient lighting, noise levels, temperature, and spatial constraints can have a profound effect on both the design and functionality of the product.

**Fig. 2.1** Ten fundamental requirements for designing wearable electronic products

- Technological Context: The prevailing technological landscape, encompassing both hardware and software as well as connectivity options, delineates the feasible parameters for designing a wearable electronic product. Designers should remain attuned to current and emerging technologies that may enhance the user experience.
- Market Context: An understanding of the competitive landscape and prevailing market trends equips designers to identify potential opportunities, distinguish their product from competitors, and address unmet user needs.
- Temporal Context: This aspect takes into account timing as it relates to market trends, user demands, and the product's lifecycle. Designers must proactively consider how temporal changes may influence the product's relevance and adaptability over time.

The integration of material and process design is fundamentally linked to industrial design, which shapes the aesthetic and functional identity of wearable electronic products. The significance of industrial design lies in its ability to provide user satisfaction; thus, elements such as color, texture, and style must resonate with the preferences of the target demographic to ensure market acceptance. Moreover, selected materials must facilitate diverse design options and allow for user customization [11].

To create specific shapes, materials undergo various processes, which can be collectively categorized as manufacturing processes. This includes primary forming, material removal, finishing, and joining processes [12]. The choice of process is pivotal, as it influences the resulting shape, size, precision, and overall cost. The relationship between material specification and process is reciprocal: the defined shape constrains material and process selection, while the chosen manufacturing processes also imposes limitations on the materials and their potential shapes [13]. Cost-effectiveness is an indispensable consideration in the production of wearable electronic products, including two primary factors: material sourcing—encompassing the cost of materials and the feasibility of mass production without compromising quality—and the economic viability of the manufacturing methods employed.

Additionally, the properties of materials, or function, significantly influence the processes utilized. Function is central to the interaction between wearable electronic products and users, particularly in relation to their sensory, cognitive, and motor responses. Generally, wearable electronic products should exhibit the following essential properties [14–17]:

- Flexibility: Materials should be pliable to accommodate bodily movements and minimize discomfort.
- Lightweight: Materials should possess minimal weight to facilitate ease of movement for the wearer.
- Breathability: Materials should allow for moisture and heat dissipation to enhance comfort during extended use, particularly in garments.
- Durability: Materials must be robust enough to endure daily wear and the environmental stresses of moisture, dirt, and physical abrasion.

- Biocompatibility: Materials that come into direct contact with the skin should be non-irritating and appropriate for prolonged use.
- Temperature Resistance: Materials should retain their functional properties across a range of temperatures and humidity levels.
- Chemical Resistance: Materials must withstand exposure to sweat, oils, and other substances encountered in daily life.

Integration of electronics and power supply considerations are of pivotal importance in designing a wearable electronic device. It is essential for embedded electronics to be thin to ensure user comfort and maintain a flexible profile within wearable designs. Additionally, the concept of modularity—designing components that can be easily replaced or upgraded—is vital for extending the operational lifespan of these devices. Moreover, the materials utilized must support wireless communication technologies, such as Bluetooth and Near Field Communication (NFC), to facilitate seamless interaction with other devices [18, 19].

Wearable technologies can primarily be categorized into input and output devices, which necessitates customized system-on-chip (SoC) designs tailored to their distinct functionalities [20]. Input wearables, including fitness bands [21], heart rate monitors [22], and NFC rings [19], are characterized by their incorporation of smart sensors, minimal displays, and low power consumption. Their primary function revolves around data collection, which is subsequently filtered and transmitted to processing hubs, such as mobile devices, to generate actionable insights or alerts. The design of these devices emphasizes ultra-low power connectivity, typically employing high-performance 32-bit microcontroller units (MCUs) [23] in conjunction with baseband communication technologies [24]. Conversely, output wearables—such as smartwatches and smart glasses—offer instant information access through compact screens, thereby enhancing user interaction and facilitating task management [25]. The SoC design for these devices prioritizes efficient processor utilization while ensuring low energy consumption. This involves developing a versatile platform capable of performing tasks typically managed by smartphones or tablets. For instance, the chipset configuration for smart glasses may integrate low-power, high-performance central processing units (CPUs) alongside compact graphics processing units (GPUs), video transcoders, and robust connectivity processors. Collectively, both categories highlight the necessity of processing efficiency and power management in enhancing user experience and overall device performance [26, 27].

The consideration of power supply and battery integration constitutes a critical aspect in the development of wearable technology. It is essential to select materials that can accommodate thin, flexible batteries or incorporate energy harvesting technologies such as solar cells [28, 29]. Furthermore, designing for energy efficiency in both the materials used and the operation of the device is imperative to minimize battery depletion [30]. Contemporary wearable devices often exhibit suboptimal battery performance, typically lasting only two to three days, whereas an ideal duration would extend to weeks or even months. Designers must implement advanced power management strategies, including automatic clock gating and intelligent control of core power states. These approaches will enhance the energy efficiency of wearable electronic devices, facilitate lower latency feedback, and improve decision-making processes related to dynamic voltage and frequency scaling (DVFS) [31]. Additionally, to maintain prolonged "always on" functionality, wearables should conform to low power standards, with Bluetooth Low Energy (LE) emerging as a prominent option. Bluetooth LE presents a power-efficient solution for wireless connectivity, enabling devices to operate with minimal battery consumption over extended periods while ensuring compatibility with existing smartphones and tablets for uninterrupted communication [32, 33].

In addition, the effective design of wearable electronics necessitates an operation mode that prioritizes intuitiveness, minimizes user effort, and features an interface capable of clearly conveying the product's status and responses through visual, auditory, or tactile feedback mechanisms [7]. Devices should be designed for ease of use, incorporating touch sensitivity and feedback systems, while also accommodating adjustable fit and comfort features. Understanding the users—encompassing their demographics, behaviors, preferences, and challenges—is critical in this process. Designers must consider the contexts in which users will engage with the wearable electronic products and how interaction will occur. Additionally, a focus on optimizing both user and product interfaces is essential to ensure seamless functionality between wearable devices and other mobile technology [34].

Another fundamental aspect of wearable electronics design is the consideration of material impact on the environment. This environmental impact includes two key dimensions: sustainability, which involves the use of eco-friendly materials and processes to mitigate ecological harm, and recyclability, focusing on the ease of recycling materials at the end of the product lifecycle [35].

Furthermore, regulatory compliance is a crucial consideration; designers must remain cognizant of relevant laws, regulations, and ethical standards that govern product design, particularly regarding safety, privacy, and sustainability. This compliance involves adherence to safety standards—ensuring that materials meet regulatory requirements such as the Restriction of Hazardous Substances (i.e., RoHS) and the Registration, Evaluation, Authorization, and Restriction of Chemicals (i.e., REACH)—as well as rigorous testing and certification protocols addressing durability, electrical safety, and skin compatibility [36].

By integrating these considerations, designers can develop wearable electronics that are functional, comfortable, aesthetically pleasing, and environmentally sustainable. The relevance of each factor may vary based on the specific application and intended user demographic of the wearable electronic device.

## 2.2 Conductors, Semiconductors, and Dielectrics

Wearable electronic devices represent a burgeoning sector that is reshaping how we interact with technology and monitor our health. As we witness the proliferation of smartwatches, fitness trackers, and medical devices, understanding the materials that form the backbone of these innovations is crucial. The functionalities and performance of wearable electronics are fundamentally tied to the materials employed in their construction. Understanding the characteristics and applications of materials is essential for the continued innovation and evolution of this field. As we explore the blend of materials that can provide not just utility but also comfort and aesthetic appeal, we position ourselves at the forefront of the wearable electronics revolution. The future holds exciting possibilities as researchers and engineers push the boundaries of what wearable technology can achieve through advanced materials innovation. This section delves into the three primary categories of materials essential for wearable electronics—conductors, semiconductors, and dielectrics—exploring their properties, functions, and relevance to the wearable technology landscape.

A fundamental characteristic of all materials is their inherent resistance to the flow of electric current. Ohm's Law, a foundational principle in electrical engineering and physics, quantitatively describes the relationship among voltage ($V$), current ($I$), and resistance ($R$) in a conductive material, expressed as follows [37]:

$$I = \frac{V}{R} \tag{2.1}$$

To advance our understanding of a material's intrinsic resistance, it is imperative to introduce the concept of resistivity. Resistivity is defined mathematically by the relation below [38]:

$$\rho = R\left(\frac{A}{L}\right) \tag{2.2}$$

where $\rho$ represents resistivity, $L$ denotes the length of the sample, and $A$ signifies its cross-sectional area. This formula reveals that resistivity is an intrinsic property specific to each material, with unique values for different substances.

Resistivity is initially defined under the assumption of a homogeneous sample with uniform current flow, although this does not always apply in practice. To enhance our understanding, we can consider the presence of an electric field ($E$) within the sample when a potential difference is applied between two fixed points. This electric field enables current flow and establishes a relationship between resistivity and the current density ($J$) within the material. As a result, resistivity's reciprocal is defined as conductivity ($\sigma$), prompting a reformulation of Ohm's Law in terms of conductivity, as follows [39]:

$$J = \sigma E \tag{2.3}$$

Among solid materials, metals exhibit the highest conductivity, exceeding $10^5$ S/m. In contrast, the conductivity of semiconductors varies widely within the range of $10^{-6} < \sigma < 10^5$ (S/m), while dielectrics demonstrate very low conductivity, typically less than $10^{-6}$ (S/m) [40].

In the following sections, we will delve deeper into these distinct categories of materials, elucidating their unique properties and contributions to the advancing field of wearable electronics.

### 2.2.1 Conductors

Conductors are materials characterized by their capacity to facilitate the flow of electric current while exhibiting minimal resistive losses. In an ideal conductor, Ohm's Law accurately describes the linear relationship among voltage, current, and resistance, whereby an increase in voltage results in a proportional increase in current, and resistance remains constant across a specified range of voltages and currents [41]. At the quantum level, metals generally possess negligible bandgaps ($E_g \approx 0$), as the conduction band and valence band overlap, allowing for the free movement of electrons between these two energy bands [42]. This phenomenon is rooted in the principles of metallic bonding and quantum tunneling. Furthermore, the density of states (DOS) near the Fermi level of conductors is significantly high, which enhances the availability of charge carriers, demonstrated by Angle-Resolved Photoemission Spectroscopy (ARPES). As a result, these materials demonstrate excellent electrical conductivity and respond rapidly to external electric fields, facilitating the flow of electric current [43].

Historically, metals such as copper, silver, and aluminum have been the preferred materials for electrical connections due to their exceptional electrical conductivity and mechanical strength [44]. In the realm of wearable devices, conductors are not only responsible for transmitting electrical signals but also play a critical role in managing heat dissipation. A comprehensive understanding of thermal conductivity at the solid-state level is essential for designing systems that prevent overheating, thereby ensuring the longevity and reliability of wearable technology [45].

Recent advancements have facilitated the development of flexible and stretchable conductors, which are pivotal for the comfort and usability of wearable devices. These innovations enable devices to maintain optimal performance while withstanding mechanical stresses. Conductive polymers have emerged as a promising alternative to traditional metals, particularly because their inherent flexibility and advantageous properties make them well-suited for wearable applications. Notable polymers, including polyaniline [46], polypyrrole [47], and PEDOT:PSS [48], exhibit electrical conductivity comparable to that of conventional metals while also offering benefits such as reduced weight, enhanced flexibility, and simplified processing techniques. The ability to print conductive polymers onto a wide variety of substrates showcases their potential for integrating electrical functionalities into wearable textiles [49]. The lightweight nature of conductive polymers is particularly advantageous, aligning well with the design imperatives of comfort and flexibility in wearable technologies [50]. Moreover, the application of solid-state physics in materials science enables the development of conductive materials that are not only functional but also biocompatible, which is critical for devices intended for direct contact with skin or bodily fluids. For instance, smart clothing designed to monitor biometric data can effectively utilize conductive threads that emphasize user comfort while maintaining high performance standards [51].

Furthermore, advancements in nanotechnology have catalyzed the development of nanoconductors, which not only enhance electrical performance but also minimize weight and volume. These innovations are instrumental in the ongoing evolution of more efficient and comfortable wearable devices, thereby advancing the field of wearable electronics [52].

Conductors play a fundamental role in wearable electronics by serving as vital conduits for electrical signals and power distribution. Their significance extends beyond mere electrical connectivity; they are integral to effective power delivery and seamless communication between the various components of the device. In this context, conductors are especially critical in antennas and radio frequency (RF) circuits that facilitate wireless communication in wearable devices [53]. A thorough understanding of electromagnetic properties is essential for optimizing the performance of these components, as maintaining signal integrity within electronic circuits is paramount. This integrity directly impacts the reliability of the data collected from sensors. By employing principles of solid-state physics in conductor design, engineers can mitigate noise and interference, thereby enhancing the accuracy of signal transmission [54].

Moreover, the integration of conductors with semiconductors to form hybrid circuits is a common practice in wearable electronics [55]. The interfaces formed at the junctions of these materials are crucial determinants of the device's behavioral characteristics, influencing aspects such as charge transfer and energy efficiency. Therefore, engineers must carefully consider both the intrinsic properties of the materials and their interactions to innovate and improve the functionality of wearable devices.

### *2.2.2 Semiconductors*

Semiconductors form a distinct category of materials that exhibit electrical properties situated between those of conductors and insulators. Their electrical resistance can be significantly influenced by multiple variables, including temperature, impurity concentrations, and applied electric fields. Although Ohm's Law is applicable to describe the behavior of these materials, the current–voltage relationship can often be nonlinear, particularly under diverse operating conditions. For example, in semiconductors, the relationship may demonstrate a nonlinear response, with current increasing disproportionately in relation to voltage. Under high electric fields, resistance may decrease, leading to phenomena such as saturation current [56]. To accurately model the electrical characteristics of semiconductors, advanced equations such as the Shockley equation are frequently employed, notably within semiconductor devices like diodes and transistors [57]. Most semiconductors generally have an energy band gap below around 3 eV; however, certain materials, classified as wide bandgap semiconductors, exhibit band gap values exceeding this threshold, thus distinguishing them from insulators. Within DOS, a prominent gap exists where no electronic states are available at the band gap energy level. When external energy—such as thermal energy or incident photons—is supplied, electrons can be excited from the valence band to the conduction band, resulting in variable conductivity [58]. Figure 2.2 depicts the band gap values of perovskite materials considered promising for wearable electronic applications [59].

The advancement of organic semiconductors and printed electronics has enabled the design and manufacturing of flexible and stretchable devices. This capability is crucial for wearable technologies that must conform to the contours of the human body while ensuring comfort during movement. Insights from solid-state physics are instrumental in optimizing these materials, maintaining their functional properties even when subjected to mechanical strain. Organic semiconductors possess advantageous traits such as flexibility and lightweight compositions, rendering them highly suitable for wearable applications [60].

Recent explorations into two-dimensional materials, including transition metal dichalcogenides (TMDs) and graphene, have revealed unique band structures and tunable band gaps, leading to innovative electronic and optical properties. Molybdenum disulfide ($MoS_2$), a prominent TMD, exhibits exceptional electronic attributes that may drive future advancements in flexible electronics [61]. Similarly, graphene's remarkable physical and chemical properties position it as a groundbreaking material for developing advanced wearable devices, enhancing functionality, performance, and user experience across various applications [62].

Materials science plays a vital role in the miniaturization and integration of semiconductors into compact form factors, which is essential for wearable devices that prioritize user comfort and practicality. Advances in fabrication techniques, such as photolithography and chemical vapor deposition, have facilitated the production of highly integrated circuits that seamlessly integrate into wearable designs [63]. Semiconductors are fundamental for performing critical functions such as signal

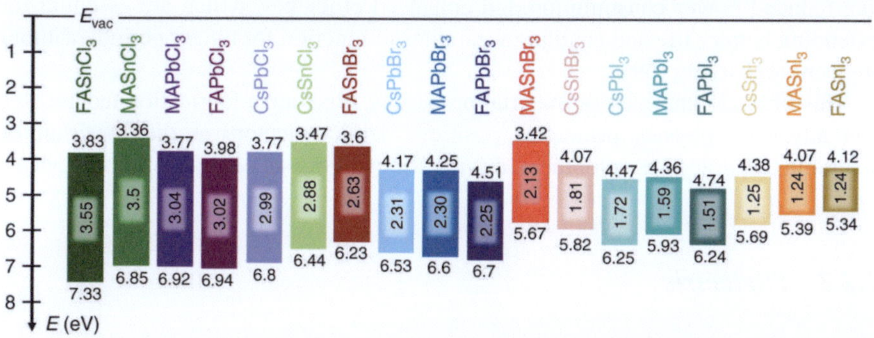


**Fig. 2.2** Schematic representation of the energy band structure for perovskite candidates for wearable device technologies. Reproduced from [59] published by Nature. Licensed under Creative Commons Attribution 4.0

processing, data storage, and battery management. As wearable technologies often require substantial data processing capabilities and local storage, the miniaturization of microprocessors and memory chips has become crucial for enabling real-time data analysis and storage, thereby enhancing the overall functionality of these devices [64]. The ability to modulate electrical conductivity through doping—by introducing specific impurities to create either p-type or n-type semiconductor materials—provides significant versatility within electronic circuits. Silicon remains the predominant semiconductor material in the electronics industry, including wearable technologies, due to its abundance, ease of processing, and favorable electrical characteristics [65]. Gallium arsenide (GaAs) is another noteworthy semiconductor, known for its superior electron mobility and efficiency in light-emitting applications. Devices constructed from GaAs find utility in environments that demand rapid signal processing and high-frequency operations, particularly in wireless communication systems designed for wearable devices [66].

Wireless communication technologies, such as Bluetooth and Near Field Communication (NFC), heavily rely on semiconductor-based transceivers. The principles of condensed matter physics underpin the operation of these communication interfaces, which must be designed for optimal efficiency in transmitting data over short distances while minimizing power consumption [67]. Moreover, many wearable devices incorporate a spectrum of sensors, including accelerometers, gyroscopes, and biosensors, all of which depend on semiconductor technology. The functionality of these sensors is governed by advanced physics principles, enabling them to detect physical changes and convert them into electronic signals. For instance, piezoelectric materials and semiconductor strain gauges are standard components in motion and health monitoring applications. These elements play a crucial role in tracking vital health metrics, such as heart rate and blood oxygen saturation, while facilitating data processing through integrated circuits [68].

Given that wearable electronics frequently operate on limited power sources, such as compact batteries or energy harvesting technologies, the selection of semiconductor materials is critical. Materials with low bandgap energies can facilitate reduced power consumption and enhanced efficiency, which are essential for extending battery life and enabling wearables to function for longer periods without frequent recharging [69].

This intricate interplay of material properties, innovative fabrication techniques, and advanced physics principles positions semiconductors at the forefront of wearable technology development, driving the evolution of next-generation devices.

### 2.2.3 *Dielectric*

Dielectric materials serve as insulators that exhibit negligible electrical conductivity under standard conditions. Characterized by exceptionally high resistivity, dielectrics prevent the flow of electric current when a voltage is applied, resulting in an almost imperceptible current in practical. The bandgaps of dielectric materials are significantly larger than those found in semiconductors. The DOS at the Fermi level is negligible, indicating a lack of free-moving charge carriers under normal conditions. This substantial bandgap restricts electrical conduction, as electrons require considerable energy to surmount the bandgap at room temperature. This behavior aligns with common observations that dielectrics are poor conductors of electricity. Dielectrics exhibit polarization in the presence of an electric field but do not allow significant current flow. It is crucial to note that, theoretically, even the most effective dielectric can conduct electricity if subjected to a sufficiently high potential difference; however, the resulting current is often negligible and of minimal practical relevance. As such, Ohm's Law may not apply in a conventional sense due to the insubstantial current produced in these materials. However, under the influence of elevated electric fields, dielectrics may undergo breakdown, resulting in electrical conduction that can lead to the failure of the dielectric materials [41, 42].

Dielectrics are essential for the storage of electrical energy within an electric field, which is characterized more by their permittivity than by conductivity. This property enables them to function effectively in capacitors and as insulators in electronic circuits, particularly in wearable technologies, where they facilitate necessary isolation and help minimize power losses. Dielectric materials serve as exceptional insulators and play a critical role in wearable electronics by isolating conductive components. This isolation prevents short circuits and promotes safe operation while enabling the utilization of electric fields for various functionalities, such as capacitive sensing. Common dielectric materials utilized in this context include ceramics, organic polymers, and hybrid organic–inorganic compounds [70].

Ceramics, such as barium titanate and lead zirconate titanate, are recognized for their high dielectric constants and thermal stability, making them well-suited for applications that require reliable performance, such as sensors and actuators in wearable devices. These dielectric materials can withstand a range of thermal conditions without degradation, ensuring longevity and dependability in diverse operating environments. Many wearable devices integrate sensors that rely on dielectric properties for operation, including piezoelectric sensors. A thorough understanding of the underlying principles of solid-state physics is essential for designing more efficient and sensitive sensors, which can enhance performance in applications such as motion tracking and biometric monitoring. The dielectric properties of materials significantly influence signal propagation speed and integrity in wearable devices. By optimizing these properties, engineers can minimize signal loss and crosstalk in electronic circuits, which is vital for the effective performance of applications like wireless communications and real-time data acquisition. In the context of wearable health monitoring devices, dielectrics are important not only for electrical functionality but also for ensuring biocompatibility. Research in solids has the potential to yield dielectric materials that pose minimal risk of adverse reactions when in contact with skin or bodily fluids [71].

Advanced dielectric materials, particularly organic and polymeric dielectrics, offer the necessary mechanical flexibility while retaining desirable electrical characteristics. This flexibility enhances comfort and wearability in devices. Polymeric materials, such as terephthalate [72] and polyimide [73], are preferred as dielectric substances due to their lightweight and flexible nature. Their versatility in processing facilitates innovative applications, including utilization in flexible printed circuit boards.

The performance of dielectric materials is generally evaluated based on their dielectric constant, or relative permittivity, which quantifies their ability to store electrical charge. As integral components in minimizing energy loss during electronic operations, dielectrics are especially critical in energy-harvesting systems within wearable devices. As the complexity and energy efficiency requirements of these devices escalate, the significance of dielectric materials becomes increasingly prominent. Ongoing research into nanodielectrics aims to boost energy storage capabilities and enhance the efficiency of electronic devices. Such advancements will markedly contribute to the evolution of autonomous wearable electronics that can achieve extended operational lifespans without the need for frequent recharging [74].

In addition, recent studies have delved into the properties of topological insulators, which exhibit conductive surface states protected by time-reversal symmetry. The band structure of these materials possesses intriguing characteristics that diverge from those of conventional semiconductors and conductors, opening new avenues for the development of advanced electronic components in wearable technologies [75].

## 2.3 Perovskites as Emerging Materials

Perovskites represent a highly promising category of materials currently under investigation for applications in wearable electronics. The term "perovskite" was first introduced in the early nineteenth century and is derived from a specific mineral, calcium titanium oxide, that was discovered in the Ural Mountains of Russia. These materials are characterized by a distinct crystal structure, which is recognized for its remarkable versatility and stability. The unique crystallographic configuration of perovskites allows for a vast array of chemical compositions, resulting in a diverse set of electronic, optical, and mechanical properties. This versatility positions perovskites favorably for a multitude of applications [76].

Initially recognized for their role in photovoltaic cells, perovskites exhibit exceptional electrical conductivity, mechanical flexibility, and tunable optical properties. These attributes render them particularly suitable for a wide range of applications in the field of engineering, particularly in the development of advanced wearable electronic devices [77].

### 2.3.1 Advantages of Perovskites in Wearable Electronics

Recent developments highlight the potential of perovskite materials in the realm of wearable electronics. Wearable devices, including fitness trackers, smartwatches, and advanced smart textiles, have fundamentally transformed the interaction between individuals and technology. These devices not only provide enhanced convenience and connectivity but also offer significant advancements in health monitoring, personalized user experiences, and data-driven insights. Perovskites, which can be engineered at the atomic level, allow for customization tailored to specific functional requirements. Notably, hybrid organic–inorganic perovskites, containing organic molecules within their structures, have demonstrated impressive optoelectronic properties. These attributes enable effective light absorption and emission capabilities, which are crucial for various applications.

One of the primary advantages of utilizing perovskites in wearable electronics is their potential for superior performance when compared to conventional materials. Their exceptional charge-carrier mobility facilitates rapid electronic signaling and processing, an essential feature for real-time health monitoring applications [78]. Additionally, the ability to process perovskites at low temperatures and employ straightforward solution-based fabrication techniques can substantially lower production costs. This cost-effectiveness is particularly beneficial for the widespread manufacturing of wearable devices, where accessibility and affordability are fundamental alongside functionality [79].

The lightweight and flexible characteristics of perovskite materials enhance their applicability in wearables, enabling devices to conform to diverse surfaces and interact seamlessly with the human body [80]. Furthermore, perovskites demonstrate a natural aptitude for light capture, positioning them as key candidates for integrating photovoltaic properties into wearables. This could lead to innovations such as smart textiles capable of monitoring physiological signals while simultaneously harnessing energy from ambient light or body movements, thereby reducing reliance on external power sources [81].

The convergence of these capabilities has the potential to propel the design of sustainable and high-performance wearable devices, marking a significant step forward in the field of wearable electronics.

### 2.3.2 Challenges Associated with Perovskites

The application of perovskite materials in the field of wearable electronics presents significant potential; however, several challenges hinder their commercial viability. A primary concern is the stability and durability of perovskite compounds. Many of these materials, especially organic–inorganic hybrids, exhibit vulnerability to environmental factors such as moisture, temperature fluctuations, and ultraviolet (UV) radiation. To ensure long-term reliability in practical wearable applications, extensive research is needed to enhance the stability of perovskite materials under diverse conditions [82].

Additionally, the environmental and health implications associated with certain lead-based perovskites pose another critical challenge. The toxicity of these compounds has prompted researchers to seek lead-free alternatives, although achieving comparable performance levels remains a notable obstacle. Current research is focused on innovating non-toxic materials, such as tin-based and bismuth-based perovskites, with the aim of balancing efficiency with safety considerations [83]. Continued efforts in material development and environmental assessment are essential for advancing the integration of perovskites into wearable technology.

### 2.3.3 Research Directions and Future Perspectives

The future of perovskites in wearable electronics seems promising, and ongoing research is uncovering new methods for improving their performance and applicability. As research in this area evolves, several key directions and future perspectives can be identified, as summarized in Table 2.1. By focusing on these research directions, the field of wearable electronics can leverage perovskite materials to create innovative, efficient, and flexible devices that enhance user experience and open new possibilities in health monitoring, energy harvesting, and smart textiles. Continued interdisciplinary collaboration between material scientists, engineers, and healthcare professionals will be essential for progressing in this promising area.

**Table 2.1** Key directions of perovskites research in wearable electronics applications

| No. | Research direction | Perspectives |
|---|---|---|
| 1 | Composition tuning | Exploring new compositions of perovskites to optimize their physical, electrical, and thermal properties for specific wearable applications [84] |
| 2 | Stability and durability | Developing stable compositions and protective coatings to enhance their longevity [85] |
| 3 | Integration with flexible substrates | Investigating ways to integrate perovskite materials with flexible and stretchable substrates to create conformable devices that can be worn comfortably on the skin or clothing without sacrificing performance [86] |
| 4 | Self-powered systems | Development of perovskite-based photovoltaic cells and thermoelectric generators that can harvest energy from the environment (like sunlight or body heat) for powering wearable devices, minimizing the need for external power sources [87] |

(continued)

**Table 2.1** (continued)

| No. | Research direction | Perspectives |
|---|---|---|
| 5 | Hybrid systems | Combining perovskites with other materials (e.g., conductive polymers) that enhance the performance of wearable electronics, such as increased flexibility, light-weight characteristics, and improved electronic properties [88] |
| 6 | Biosensing applications | Leveraging the unique electronic properties of perovskites for biosensing applications, including real-time health monitoring for biomarkers in sweat or sweat analysis for various medical conditions [89] |
| 7 | Wearable displays | Investigating the use of perovskites in display technologies for wearables, capitalizing on their high efficiency and color quality for applications in augmented reality glasses or smart clothing [90] |
| 8 | Advanced manufacturing techniques | Exploration of scalable and cost-effective manufacturing methods, such as inkjet printing, spray coating, and roll-to-roll processing, that enable the mass production of perovskite devices [91] |
| 9 | Connectivity and IoT integration | Developing perovskite-based devices that can interface seamlessly with the Internet of Things (IoT), facilitating the collection, processing, and transmission of health data for smart health monitoring systems [92] |
| 10 | Sustainability and eco-friendliness | Research into environmentally friendly manufacturing processes and the recyclability of perovskite materials to address sustainability concerns in the production of wearable devices [93] |
| 11 | Regulatory and standards considerations | Addressing regulatory challenges, safety standards, and user acceptability will be critical [94] |

## References


1. M. Ashby, *Materials Selection in Mechanical Design: Fourth Edition*, vol. 9780080952239 (Butterworth-Heinemann, 2010)
2. M. Ashby, K. Johnson, *Materials and Design: The Art and Science of Material Selection in Product Design* (2014). https://doi.org/10.1016/C2011-0-05518-7
3. T.S.S. Winston, O. Soboyejo, Advanced structural materials: properties, design optimization, and applications, in *Advanced Structural Materials: Properties, Design Optimization, and Applications* (2006)
4. M.M. Farag, *Materials and Process Selection for Engineering Design* (2020). https://doi.org/10.1201/9781003006091

5. P. Sanjarnia et al., *Bringing Innovative Wound Care Polymer Materials to the Market: Challenges, Developments, and New Trends* (2024). https://doi.org/10.1016/j.addr.2024.115217
6. M.R. Islam, S. Afroj, J. Yin, K.S. Novoselov, J. Chen, N. Karim, *Advances in Printed Electronic Textiles* (2024). https://doi.org/10.1002/advs.202304140
7. M. Shak Sadi, E. Kumpikaitė, *Advances in the Robustness of Wearable Electronic Textiles: Strategies, Stability, Washability and Perspective* (2022). https://doi.org/10.3390/nano12122039
8. G. Scandurra, A. Arena, C. Ciofi, *A Brief Review on Flexible Electronics for IoT: Solutions for Sustainability and New Perspectives for Designers* (2023). https://doi.org/10.3390/s23115264
9. S. Rao, P. Thankachan, B. Amrutur, M. Washington, P.K. Mony, Continuous, real-time monitoring of neonatal position and temperature during Kangaroo Mother Care using a wearable sensor: a techno-feasibility pilot study. Pilot Feasibility Stud. **4**(1) (2018). https://doi.org/10.1186/s40814-018-0293-5
10. A. Tandon et al., *Wearable Biosensors in Congenital Heart Disease: Needs to Advance the Field* (2023). https://doi.org/10.1016/j.jacadv.2023.100267
11. A.G.O. Mutambara, *Design and Analysis of Control Systems* (CRC Press, Boca Raton, 2024). https://doi.org/10.1201/9781032718880
12. G. Halevi, Selection of primary production processes, in *Process and Operation Planning* (Springer Netherlands, Dordrecht, 2003), pp. 87–104. https://doi.org/10.1007/978-94-017-0259-1_5
13. X. Hu, Y. Dou, J. Li, Z. Liu, *Buckled Structures: Fabrication and Applications in Wearable Electronics* (2019). https://doi.org/10.1002/smll.201804805
14. J. Wang et al., Sustainably powering wearable electronics solely by biomechanical energy. Nat. Commun. **7** (2016). https://doi.org/10.1038/ncomms12744
15. H. Park et al., Organic flexible electronics with closed-loop recycling for sustainable wearable technology. Nat. Electron. **7**(1) (2024). https://doi.org/10.1038/s41928-023-01078-9
16. Y. Jin et al., Estimation of joint torque in dynamic activities using wearable A-mode ultrasound. Nat. Commun. **15**(1), 5756 (2024). https://doi.org/10.1038/s41467-024-50038-0
17. B. Lee et al., High-performance compliant thermoelectric generators with magnetically self-assembled soft heat conductors for self-powered wearable electronics. Nat. Commun. **11**(1) (2020). https://doi.org/10.1038/s41467-020-19756-z
18. R. Lin et al., Digitally-embroidered liquid metal electronic textiles for wearable wireless systems. Nat. Commun. **13**(1) (2022). https://doi.org/10.1038/s41467-022-29859-4
19. M. Zulqarnain et al., A flexible ECG patch compatible with NFC RF communication. npj Flex. Electron. **4**(1) (2020). https://doi.org/10.1038/s41528-020-0077-x
20. L. Zhao, Y. Jia, Towards a self-powered ECG and PPG sensing wearable device, in *Proceedings of the Annual International Conference of the IEEE Engineering in Medicine and Biology Society, EMBS* (2021). https://doi.org/10.1109/EMBC46164.2021.9631062
21. G. Quer et al., Inter-individual variation in objective measure of reactogenicity following COVID-19 vaccination via smartwatches and fitness bands. NPJ Digit. Med. **5**(1) 2022. https://doi.org/10.1038/s41746-022-00591-z
22. N.S. Zheng et al., Sleep patterns and risk of chronic disease as measured by long-term monitoring with commercial wearable devices in the all of us research program. Nat. Med. **30**(9), 2648–2656 (2024). https://doi.org/10.1038/s41591-024-03155-8
23. J. Biggs et al., A natively flexible 32-bit Arm microprocessor. Nature **595**(7868) (2021). https://doi.org/10.1038/s41586-021-03625-w
24. K.K. Balasubramanian et al., Neural network-based bluetooth synchronization of multiple wearable devices. Nat. Commun. **14**(1) (2023). https://doi.org/10.1038/s41467-023-40114-2
25. H.C. Ates et al., *End-to-End Design of Wearable Sensors* (2022). https://doi.org/10.1038/s41578-022-00460-x
26. Y. Chen et al., All-analog photoelectronic chip for high-speed vision tasks. Nature **623**(7985) (2023). https://doi.org/10.1038/s41586-023-06558-8
27. E. Waisberg et al., Meta smart glasses—large language models and the future for assistive glasses for individuals with vision impairments. Eye **38**(6), 1036–1038 (2024). https://doi.org/10.1038/s41433-023-02842-z

28. D. Powell, A. Godfrey, *Considerations for Integrating Wearables into the Everyday Healthcare Practice* (2023). https://doi.org/10.1038/s41746-023-00820-z
29. D. Lv, Q. Jiang, Y. Shang, D. Liu, Highly efficient fiber-shaped organic solar cells toward wearable flexible electronics. npj Flex. Electron. **6**(1) (2022). https://doi.org/10.1038/s41528-022-00172-w
30. X. Huang et al., Garment embedded sweat-activated batteries in wearable electronics for continuous sweat monitoring. npj Flex. Electron. **6**(1) (2022). https://doi.org/10.1038/s41528-022-00144-0
31. S. Mohamed, H.A.A. Nomer, R. Yousri, A.W. Mohamed, A. Soltan, M.S. Darweesh, Energy management for wearable medical devices based on gaining–sharing knowledge algorithm. Complex Intell. Syst. **9**(6) (2023). https://doi.org/10.1007/s40747-023-01101-8
32. S. Yang et al., Flexible thermoelectric generator and energy management electronics powered by body heat. Microsyst. Nanoeng. **9**(1) (2023). https://doi.org/10.1038/s41378-023-00583-3
33. T. Zhang, J. Lu, F. Hu, Q. Hao, Bluetooth low energy for wearable sensor-based healthcare systems, in *2014 IEEE Healthcare Innovation Conference, HIC 2014* (2014). https://doi.org/10.1109/HIC.2014.7038922
34. K. Yang et al., Development of user-friendly wearable electronic textiles for healthcare applications. Sensors (Switzerland) **18**(8) (2018). https://doi.org/10.3390/s18082410
35. M. Dulal et al., *Toward Sustainable Wearable Electronic Textiles* (2022). https://doi.org/10.1021/acsnano.2c07723
36. Y. Wakabayashi, H. Matsui, K. Ikai, M. Hayashi, H. Wakabayashi, K. Yamamoto, Developing a practical method for validation of computerized systems integrated with smart and/or wearable devices for regulatory compliance of clinical trials. Ther. Innov. Regul. Sci. **51**(1) (2017). https://doi.org/10.1177/2168479016666585
37. E.T. Gilbert-Kawai, M.D. Wittenberg, *Essential Equations for Anaesthesia* (2014). https://doi.org/10.1017/cbo9781139565387
38. A. Binley, L. Slater, *Resistivity and Induced Polarization: Theory and Applications to the Near-Surface Earth* (2020). https://doi.org/10.1017/9781108685955
39. J. Herbert, Moulson AJ, *Electroceramics: Materials, Properties, Applications* (2003)
40. R.K. Pandey, *Fundamentals of Electroceramics: Materials, Devices, and Applications* (2018)
41. C. Kittel, *Introduction to Solid State Physics*, 8th edn. (Wiley & Sons, New York, NY, 2004)
42. R.E. Hummel, *Electronic Properties of Materials* (2011). https://doi.org/10.1007/978-1-4419-8164-6
43. T. Yu et al., The relevance of ARPES to high-Tc superconductivity in cuprates. NPJ Quantum Mater. **5**(1) (2020). https://doi.org/10.1038/s41535-020-0251-3
44. Z. Lin, T. Li, S. Yang, B. Ji, Z. Wang, Revolutionizing flexible electronics with liquid metal innovations. Device **2**(5), 100331 (2024). https://doi.org/10.1016/j.device.2024.100331
45. Y.S. Ju, *Thermal Management and Control of Wearable Devices* (2022). https://doi.org/10.1016/j.isci.2022.104587
46. B. Fang et al., Scalable production of ultrafine polyaniline fibres for tactile organic electrochemical transistors. Nat. Commun. **13**(1) (2022). https://doi.org/10.1038/s41467-022-29773-9
47. W. Wei, M. Wu, Y. Chen, A flexible and wearable strain sensor from polypyrrole-doped elastomers with dual functions in motion monitoring and thermotherapy. Chin. J. Chem. **41**(13) (2023). https://doi.org/10.1002/cjoc.202200823
48. Y.F. Wang et al., Fully printed PEDOT:PSS-based temperature sensor with high humidity stability for wireless healthcare monitoring. Sci. Rep. **10**(1) (2020). https://doi.org/10.1038/s41598-020-59432-2
49. P. Wang et al., *The Evolution of Flexible Electronics: From Nature, Beyond Nature, and to Nature* (2020). https://doi.org/10.1002/advs.202001116
50. S. Khan, S. Ali, A. Bermak, *Recent Developments in Printing Flexible and Wearable Sensing Electronics for Healthcare Applications* (2019). https://doi.org/10.3390/s19051230
51. L. Avellar, C. Stefano Filho, G. Delgado, A. Frizera, E. Rocon, A. Leal-Junior, AI-enabled photonic smart garment for movement analysis. Sci. Rep. **12**(1) (2022). https://doi.org/10.1038/s41598-022-08048-9

52. S. Gong et al., A wearable and highly sensitive pressure sensor with ultrathin gold nanowires. Nat. Commun. **5** (2014). https://doi.org/10.1038/ncomms4132
53. P. Sharma, X. Hui, J. Zhou, T.B. Conroy, E.C. Kan, Wearable radio-frequency sensing of respiratory rate, respiratory volume, and heart rate. NPJ Digit. Med. **3**(1) (2020). https://doi.org/10.1038/s41746-020-0307-6
54. J. Jung et al., Highly stretchable and transparent electromagnetic interference shielding film based on silver nanowire percolation network for wearable electronics applications. ACS Appl. Mater. Interfaces **9**(51) (2017). https://doi.org/10.1021/acsami.7b14626
55. K. Lim et al., Material and structural considerations for high-performance electrodes for wearable skin devices. Commun. Mater. **5**(1), 49 (2024). https://doi.org/10.1038/s43246-024-00490-8
56. R.P. Huebener, *Conductors, Semiconductors, Superconductors* (2015)
57. W. Shockley, The theory of p-n junctions in semiconductors and p-n junction transistors. Bell Syst. Tech. J. **28**(3) (1949). https://doi.org/10.1002/j.1538-7305.1949.tb03645.x
58. P. Wellmann, N. Ohtani, R. Rupp, *Wide Bandgap Semiconductors for Power Electronics: Materials, Devices, Applications* (2021). https://doi.org/10.1002/9783527824724
59. S. Tao et al., Absolute energy level positions in tin- and lead-based halide perovskites. Nat. Commun. **10**(1) (2019). https://doi.org/10.1038/s41467-019-10468-7
60. X. Xu, Y. Zhao, Y. Liu, *Wearable Electronics Based on Stretchable Organic Semiconductors* (2023). https://doi.org/10.1002/smll.202206309
61. J. Tang et al., Low power flexible monolayer $MoS_2$ integrated circuits. Nat. Commun. **14**(1) (2023). https://doi.org/10.1038/s41467-023-39390-9
62. M.R. Islam et al., Fully printed and multifunctional graphene-based wearable e-textiles for personalized healthcare applications. iScience **25**(3) (2022). https://doi.org/10.1016/j.isci.2022.103945
63. P. Wang et al., Well-defined in-textile photolithography towards permeable textile electronics. Nat. Commun. **15**(1) (2024). https://doi.org/10.1038/s41467-024-45287-y
64. J.V. Vaghasiya, C.C. Mayorga-Martinez, M. Pumera, *Wearable Sensors for Telehealth Based on Emerging Materials and Nanoarchitectonics* (2023). https://doi.org/10.1038/s41528-023-00261-4
65. J. Proud, B. Fan, R. Shook, J. Fernandez, *Wearable Device Made With Silicone Rubber and Electronic Components*, U.S. Patent No. 20160335632A1
66. Y. Yun, S. Moon, S. Kim, J. Lee, Flexible fabric-based GaAs thin-film solar cell for wearable energy harvesting applications. Sol. Energy Mater. Sol. Cells **246** (2022). https://doi.org/10.1016/j.solmat.2022.111930
67. S. Olenik, H.S. Lee, F. Güder, *The Future of Near-Field Communication-Based Wireless Sensing* (2021). https://doi.org/10.1038/s41578-021-00299-8
68. S.K. Gill et al., Consumer wearable devices for evaluation of heart rate control using digoxin versus beta-blockers: the RATE-AF randomized trial. Nat. Med. **30**(7), 2030–2036 (2024). https://doi.org/10.1038/s41591-024-03094-4
69. A. Ali, H. Shaukat, S. Bibi, W.A. Altabey, M. Noori, S.A. Kouritem, *Recent Progress in Energy Harvesting Systems for Wearable Technology* (2023). https://doi.org/10.1016/j.esr.2023.101124
70. Y. Guan et al., A plantar wearable pressure sensor based on hybrid lead zirconate-titanate/microfibrillated cellulose piezoelectric composite films for human health monitoring. Lab Chip **22**(12) (2022). https://doi.org/10.1039/d2lc00051b
71. Y.G. Kim, J.H. Song, S. Hong, S.H. Ahn, Piezoelectric strain sensor with high sensitivity and high stretchability based on kirigami design cutting. npj Flex. Electron. **6**(1) (2022). https://doi.org/10.1038/s41528-022-00186-4
72. G. Wang et al., Wearable supercapacitors on polyethylene terephthalate fabrics with good wash fastness and high flexibility. J. Power Sources **367** (2017). https://doi.org/10.1016/j.jpowsour.2017.09.047
73. H. Wang, Z. Zhao, P. Liu, X. Guo, A soft and stretchable electronics using laser-induced graphene on polyimide/PDMS composite substrate. npj Flex. Electron. **6**(1) (2022). https://doi.org/10.1038/s41528-022-00161-z

74. Z. Xu et al., High-performance dielectric elastomer nanogenerator for efficient energy harvesting and sensing via alternative current method. Adv. Sci. **9**(18) (2022). https://doi.org/10.1002/advs.202201098
75. M. Li et al., Filling the gap between topological insulator nanomaterials and triboelectric nanogenerators. Nat. Commun. **13**(1) (2022). https://doi.org/10.1038/s41467-022-28575-3
76. H.J. Snaith, *Perovskites: The Emergence of a New Era for Low-Cost, High-Efficiency Solar Cells* (2013). https://doi.org/10.1021/jz4020162
77. M.A. Green, A. Ho-Baillie, H.J. Snaith, The emergence of perovskite solar cells. Nat. Photonics **8**(7), 506–514 (2014). https://doi.org/10.1038/nphoton.2014.134
78. Y. Tang et al., Enabling low-drift flexible perovskite photodetectors by electrical modulation for wearable health monitoring and weak light imaging. Nat. Commun. **14**(1) (2023). https://doi.org/10.1038/s41467-023-40711-1
79. J. Fan, B. Jia, M. Gu, Perovskite-based low-cost and high-efficiency hybrid halide solar cells. Photonics Res. **2**(5) (2014). https://doi.org/10.1364/prj.2.000111
80. Y. Gao et al., *Flexible Perovskite Solar Cells: From Materials and Device Architectures to Applications* (2022). https://doi.org/10.1021/acsenergylett.1c02768
81. R. Sharif et al., *A Comprehensive Review of the Current Progresses and Material Advances in Perovskite Solar Cells* (2023). https://doi.org/10.1039/d3na00319a
82. N. Ahn, M. Choi, *Towards Long-Term Stable Perovskite Solar Cells: Degradation Mechanisms and Stabilization Techniques* (2024). https://doi.org/10.1002/advs.202306110
83. P. Su et al., Pb-Based perovskite solar cells and the underlying pollution behind clean energy: dynamic leaching of toxic substances from discarded perovskite solar cells. J. Phys. Chem. Lett. **11**(8) (2020). https://doi.org/10.1021/acs.jpclett.0c00503
84. N.K. Elangovan, R. Kannadasan, B.B. Beenarani, M.H. Alsharif, M.K. Kim, Z. Hasan Inamul, Recent developments in perovskite materials, fabrication techniques, band gap engineering, and the stability of perovskite solar cells. Energy Rep. **11** (2024). https://doi.org/10.1016/j.egyr.2023.12.068
85. T.A. Chowdhury, M.A. Bin Zafar, M. Sajjad-Ul Islam, M. Shahinuzzaman, M.A. Islam, M.U. Khandaker, *Stability of Perovskite Solar Cells: Issues and Prospects* (2023). https://doi.org/10.1039/d2ra05903g
86. O.S. Game et al., Direct integration of perovskite solar cells with carbon fiber substrates. Adv. Mater. **36**(20) (2024). https://doi.org/10.1002/adma.202209950
87. C. Perumal Veeramalai, S. Feng, X. Zhang, S.V.N. Pammi, V. Pecunia, and C. Li, Lead–halide perovskites for next-generation self-powered photodetectors: a comprehensive review. Photonics Res. **9**(6) (2021). https://doi.org/10.1364/prj.418450
88. K.H. Girish, K.A. Vishnumurthy, T.S. Roopa, *Role of Conducting Polymers in Enhancing the Stability and Performance of Perovskite Solar Cells: A Brief Review* (2022). https://doi.org/10.1016/j.mtsust.2021.100090
89. C. Bao et al., A multifunctional display based on photo-responsive perovskite light-emitting diodes. Nat. Electron. **7**(5), 375–382 (2024). https://doi.org/10.1038/s41928-024-01151-x
90. F. Chun et al., Multicolour stretchable perovskite electroluminescent devices for user-interactive displays. Nat. Photonics **18**(8), 856–863 (2024). https://doi.org/10.1038/s41566-024-01455-6
91. Perovskites take steps to industrialization. Nat. Energy **5**(1), 1–1 (2020). https://doi.org/10.1038/s41560-020-0552-6
92. M. Pirc, Ž Ajdič, D. Uršič, M. Jošt, M. Topič, Indoor energy harvesting with perovskite solar cells for IoT applications—a full year monitoring study. ACS Appl. Energy Mater. **7**(2), 565–575 (2024). https://doi.org/10.1021/acsaem.3c02498
93. H. Zhu et al., Environmentally friendly perovskite-based photovoltaics and luminescence: lead less or lead free. Sol. RRL **7**(24) (2023). https://doi.org/10.1002/solr.202300754
94. J. Li et al., Biological impact of lead from halide perovskites reveals the risk of introducing a safe threshold. Nat. Commun. **11**(1), 310 (2020). https://doi.org/10.1038/s41467-019-13910-y

# Chapter 3
# Perovskite Materials for Wearables

Perovskite materials exhibit a unique structural framework that enables systematic engineering of electronic, optical, and mechanical properties through precise compositional control and architectural modifications, making them particularly well-suited for addressing the complex materials requirements of wearable electronics. This chapter presents a systematic investigation of perovskite materials architecture and design principles specifically tailored for wearable device integration, building upon the fundamental advantages these materials offer for body-worn electronics. The discussion begins with a detailed classification of perovskite variants relevant to wearable applications, including halide, oxide, chalcogenide and double perovskites that extend compositional possibilities beyond traditional formulations. Attention is given to perovskite-polymer composites that bridge the gap between high-performance electronic materials and the mechanical requirements of textile integration.

The structural analysis explores how the fundamental crystal architecture of perovskites enables their remarkable property tunability, examining the relationship between atomic arrangements and the resulting electronic, optical, and mechanical characteristics that make these materials suitable for wearable platforms. The chapter then goes deep into the specific material properties most critical for wearable applications, with comprehensive coverage of band gap engineering strategies that enable precise control over optical absorption and emission characteristics. Mechanical property enhancement receives particular focus, detailing the strategies that can be systematically employed to achieve the flexibility and durability required for textile-integrated electronics while preserving the superior optoelectronic performance that distinguishes perovskites from conventional wearable materials.



M. Darab et al., *Perovskites for Next-Generation Wearable Technology*, Engineering Materials, https://doi.org/10.1007/978-3-032-03931-6_3

## 3.1 Introduction

Perovskite materials represent a transformative opportunity for advancing wearable technology, positioned to address the increasing demands for sophisticated, body-integrated electronic systems [1]. As consumer expectations evolve toward seamless integration of technology into daily life, the limitations of conventional semiconductor materials in wearable applications have become increasingly apparent, creating a compelling need for materials that can simultaneously deliver high performance, mechanical flexibility, and long-term stability.

The exceptional promise of perovskites for wearable applications stems from their unique combination of optoelectronic, mechanical [2], and processing properties that are rarely found together in a single material class. These crystalline structures exhibit remarkable tunability in their electronic and optical characteristics through compositional engineering, enabling bandgap adjustment across the visible spectrum and beyond. Simultaneously, perovskites can be processed into flexible, lightweight forms while maintaining their favorable optoelectronic properties [3]—a critical requirement for living body-worn devices that must conform to curved surfaces and withstand repeated mechanical deformation and movements.

The convergence of multiple advantageous characteristics positions perovskites as potentially revolutionary materials for next-generation wearable electronics. Their high efficiency in energy conversion processes, combined with inherently lightweight properties and self-powered operation capabilities through photovoltaic [4] and thermoelectric effects, addresses fundamental power management challenges in wearable systems. The flexibility [5] achievable through facile processing techniques enables integration with textile substrates, while their relatively low-cost manufacturing requirements and eco-friendly material compositions support sustainable production pathways [1, 6]. Additionally, certain perovskite formulations offer tunable transparency [7], enabling seamless and potentially invisible integration into clothing and accessories. This goes beyond the use of highly or semi-transparent perovskite-based materials, it also includes the ability to selectively adjust their level of transparency. For instance, translucent perovskite modules with transparency levels ranging from 7 to 37% have been recently developed using techniques like partial area ablation, offering a 'tunable transparency' for the wearable and general applications [8].

Unlike traditional semiconductors that require high-temperature processing and rigid substrates, perovskites can be integrated directly into textiles and flexible platforms using scalable manufacturing techniques. This multifaceted advantage profile establishes perovskites as prime candidates to overcome the fundamental challenges that have historically limited the widespread adoption of truly integrated wearable technologies. Figure 3.1 provides a comparative overview of the key suitability factors that make perovskite materials particularly attractive for wearable applications, illustrating the balanced excellence across multiple critical performance dimensions that distinguishes this material class from conventional alternatives.

This unique junction of processability, performance, and mechanical adaptability is rooted not only in the chemical versatility of perovskites but also in their inherently

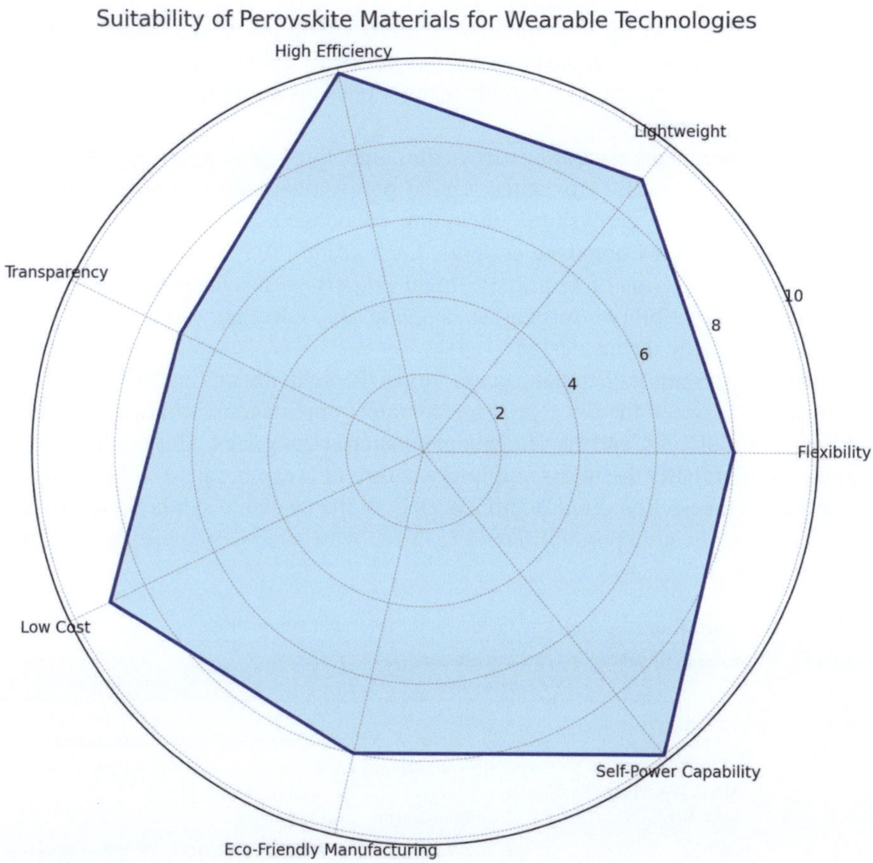


**Fig. 3.1** Comparative suitability factors of perovskite materials for wearable technology

flexible crystal structure. Understanding the underlying crystallography is essential to appreciate how compositional tuning leads to diverse functional properties—ranging from ferroelectricity to efficient light absorption—across different perovskite families.

Wearable electronic products demand a unique set of material properties to ensure comfort, functionality, and long-term reliability. Flexibility and stretchability are essential for accommodating continuous body motion, requiring materials that can withstand bending, folding, and conformal contact, particularly in the form of ultra-thin films or stretchable architectures. Environmental stability, including resistance to moisture, heat, and mechanical stress, is critical to device longevity. Additionally, biocompatibility is imperative for skin-contact applications, where non-toxic or encapsulated materials are preferred to minimize health risks.

These stringent demands are increasingly being met by the perovskites which offer a remarkable degree of tunable optoelectronic, mechanical, and chemical properties.

This versatility, combined with their compatibility with low-temperature, solution-based processing, has positioned perovskites as one of the most promising material platforms for next-generation wearable technologies, as highlighted in the previous literature review section.

Table 3.1 organizes wearable perovskites into three primary categories based on composition: Halide perovskites, Oxide perovskites, and a broader group of Emerging/Alternative perovskites, including Chalcogenides, Double perovskites, and perovskite-based Composite systems. Each group offers particular advantages and limitations in terms of optical/electrical properties, mechanical behavior, and application compatibility—serving as a guide for selecting suitable perovskite systems in wearable device design.

Figure 3.2 presents a systematic taxonomy of the field, illustrating the hierarchical relationships between the three primary domains: applications, synthesis and integration methodologies, and the fundamental material categories. This organizational framework highlights the interdisciplinary nature of wearable perovskite research, where material selection directly influences both the available synthesis routes and the ultimate application possibilities.

**Table 3.1** Categorization of perovskite materials for wearable applications

| Category | Example materials | Optical/electrical properties | Key traits | Typical wearable applications |
|---|---|---|---|---|
| Halide perovskites | $MAPbI_3$<br>$CsPbBr_3$<br>$FAPbCl_3$<br>$MA_{1-\alpha}FA_\alpha PbI_{3-\beta}X_\beta$ (X = Cl, Br) | • Direct bandgap (1.5–2.3 eV)<br>• High absorption<br>• Long carrier diffusion lengths | • Tunable optoelectronic properties<br>• Solution-processable<br>• Lightweight | • Solar cells<br>• Photodetectors<br>• Optical sensors |
| Oxide perovskites | $BaTiO_3$<br>$SrTiO_3$<br>$LaFeO_3$ | • Wide bandgap (~2–4 eV)<br>• High dielectric constant<br>• Piezoelectric and ferroelectric | • Chemically robust<br>• Biocompatible<br>• High-temperature stability | • Pressure/strain sensors<br>• Nanogenerators<br>• Memory devices |
| Chalcogenide perovskites | $BaZrS_3$<br>$SrHfSe_3$ | • Bandgap ~1.1–1.8 eV<br>• IR responsive<br>• Moderate mobility | • Lead-free<br>• Thermally stable<br>• Potential for flexibility | • IR sensors<br>• Thermoelectric generators<br>• PV devices |
| Double perovskites | $Cs_2AgBiBr_6$ $La_2MnNiO_6$ | • Bandgap ~1.8–2.4 eV (direct or indirect)<br>• Enhanced thermal/moisture stability | • Lead-free<br>• Tunable bandgap<br>• Better long-term stability | • Eco-friendly solar cells<br>• Stable photodetectors |
| Composites | $MAPbI_3$ in PDMS<br>$CsPbBr_3$@PU<br>$BaTiO_3$-PDMS blends | • Retains photoelectric functionality<br>• Flexible and stretchable electrical behavior | • Enhanced mechanical resilience<br>• Conformal integration<br>• Moisture barrier | • Stretchable sensors<br>• Wearable LEDs<br>• Biocompatible devices |

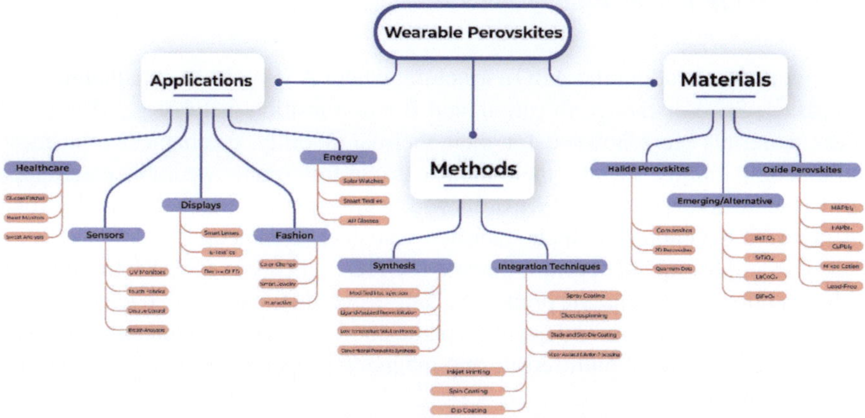


**Fig. 3.2** Holistic overview of wearable perovskite technologies: key applications, synthesis and integration methods, and material classifications

## 3.1.1 *Halide Perovskites*

Halide perovskites follow the general formula $ABX_3$, where A is a monovalent cation (organic, such as methylammonium ($CH_3NH_3$) [$MA^+$] or formamidinium ($CH(NH_2)_2$) [$FA^+$], or inorganic, like cesium [$Cs^+$]), B is a divalent metal cation (typically lead [$Pb^{2+}$] or tin [$Sn^{2+}$] or germanium [$Ge^{2+}$], and X is a halide anion (chloride, bromide, or iodide). These materials dominate wearable optoelectronics due to their outstanding light absorption, tunable bandgaps, and efficient charge transport. Their low-temperature solution processability enables cost-effective fabrication of flexible and printable devices. However, their susceptibility to moisture, oxygen, and thermal stress remains a key challenge for long-term wearable reliability.

The most widely studied halide perovskites include methylammonium lead iodide ($MAPbI_3$), the archetypal hybrid organic–inorganic perovskite with excellent photovoltaic performance, and cesium lead bromide ($CsPbBr_3$), an all-inorganic variant with enhanced stability and bright green emission. Mixed-cation and mixed-halide compositions, such as $MA_{1-\alpha}FA_{\alpha}PbI_{3-\beta}X_{\beta}$ (where $X = Cl$ or Br), allow fine-tuning of optoelectronic properties while maintaining structural integrity.

Halide perovskites are primarily employed in wearable solar cells, where their high-power conversion efficiency and lightweight nature enable seamless integration into textiles or accessories for energy harvesting. Their strong photo response also makes them ideal for wearable photodetectors in health monitoring and optical communication systems. Additionally, their tunable emission properties facilitate applications in flexible LEDs and displays, particularly when embedded in stretchable matrices.

### 3.1.2 Oxide Perovskites

Oxide perovskites adopt the $ABO_3$ structure, where A is typically an alkaline earth (e.g., $Ba^{2+}$, $Sr^{2+}$) or rare-earth cation, and B is a transition metal (e.g., $Ti^{4+}$, $Fe^{3+}$). These materials are renowned for their thermal stability, mechanical robustness, and biocompatibility, making them indispensable for durable wearable electronics. Unlike halide perovskites, oxides exhibit wide bandgaps, limiting their optoelectronic performance but enhancing their stability in harsh environments.

Key examples include barium titanate ($BaTiO_3$), a classic ferroelectric material with exceptional piezoelectric properties, and strontium titanate ($SrTiO_3$), known for its high dielectric constant and quantum paraelectric behavior. Lanthanum iron oxide ($LaFeO_3$) demonstrates multiferroic and magnetic properties, broadening its utility in sensing applications.

Oxide perovskites excel in wearable pressure and strain sensors, where their piezoelectric response converts mechanical motion into electrical signals for health and motion tracking. Their high dielectric constants also make them suitable for flexible capacitors and memory devices in wearable electronics. Additionally, their biocompatibility enables use in implantable or skin-contact sensors, while their thermoelectric properties support body heat harvesting in energy-autonomous wearables.

### 3.1.3 Chalcogenide Perovskites

Chalcogenide perovskites retain the $ABX_3$ structure but replace oxygen with heavier chalcogenides (sulfur [$S^{2-}$], selenium [$Se^{2-}$], or tellurium [$Te^{2-}$]). This substitution narrows their bandgap, enabling infrared (IR) sensitivity, while maintaining better thermal stability than halide perovskites. Their lead-free composition and reduced toxicity make them attractive for sustainable wearable technologies.

Prominent examples include barium zirconium sulfide ($BaZrS_3$), which exhibits strong IR absorption and good thermal stability, and strontium hafnium selenide ($SrHfSe_3$), a candidate for low-light photovoltaics.

Chalcogenide perovskites are particularly suited for IR sensors in health monitoring, where they can detect thermal signatures or blood oxygenation levels. Their narrow bandgaps also make them viable for indoor photovoltaic devices and thermoelectric generators, which convert body heat into usable energy for wearable electronics.

### 3.1.4 Double Perovskites

Double perovskites expand the conventional structure to $A_2B'B''X_6$, where two different B-site cations (e.g., $Ag^+$ and $Bi^{3+}$) provide enhanced compositional flexibility. This design mitigates lead toxicity while improving thermal and moisture stability, albeit often at the cost of reduced charge mobility due to indirect bandgaps.

Notable examples include cesium silver bismuth bromide ($Cs_2AgBiBr_6$), a lead-free alternative with reasonable optoelectronic performance, and lanthanum manganese nickel oxide ($La_2MnNiO_6$), which exhibits magnetic and catalytic properties.

Double perovskites are primarily explored for eco-friendly solar cells and photodetectors, where regulatory restrictions on lead necessitate safer alternatives. Their improved stability also supports long-term wearable sensors for continuous health monitoring.

### 3.1.5 Perovskite-Polymer Composites

Perovskite-polymer composites integrate perovskite crystals (e.g., $MAPbI_3$, $CsPbBr_3$, $BaTiO_3$) into stretchable polymers like polydimethylsiloxane (PDMS) or polyurethane. These hybrids retain the optoelectronic functionality of perovskites while gaining mechanical flexibility, stretchability, and environmental resistance—critical for wearable applications.

Common systems include $MAPbI_3$-PDMS for flexible solar cells and $CsPbBr_3$-polyurethane for stretchable LEDs. Oxide-polymer blends, such as $BaTiO_3$-PDMS, are also used in piezoelectric sensors.

Composites enable conformal and stretchable sensors that adhere to skin or textiles without cracking. They are also key to wearable displays and biocompatible devices, where mechanical resilience and moisture resistance are paramount.

### 3.1.6 Wearable Perovskites in Literature

Reviewing the body of research at the intersection of wearable technology and perovskite materials reveals several compelling trends in this rapidly evolving field. A search conducted in Scopus, a comprehensive repository of peer-reviewed scientific literature [9, 10], highlights a marked increase in research activity over the past decade. Specifically, the number of publications related to wearable perovskite technologies has grown significantly from 2015 to 2024, as illustrated in Fig. 3.3. Using the keywords "wearable + perovskites," the search identified 697 relevant documents published between 2015 and 2025, which serve as the foundation for the analyses presented in the following figures.

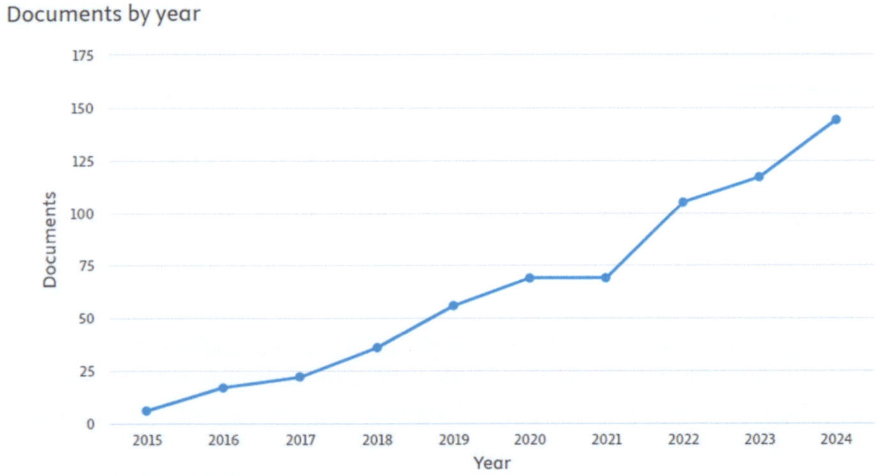


**Fig. 3.3** Number of scientific publications per year at the intersection of wearable technology and perovskite materials from 2015 to 2024

In 2015, the field was in its infancy, with fewer than 10 published papers. However, a steady upward trajectory followed, with significant development occurring in subsequent years. A dramatic rise occurred between 2021 and 2022, where the number of papers jumped from around 70 to over 100, reflecting renewed momentum likely fueled by advancements in material stability, device integration techniques, and emerging applications that now found a confident market footprint (also see the Diversification and Specialization section in Chap. 1). The trend continued robustly, peaking at approximately 145 publications in 2024, underscoring the field's rapid maturation and expanding interest from both academia and industry.

This trend reflects broader technological and societal shifts. Perovskite materials are increasingly being adapted for use in wearable sensors, energy harvesters, and health-monitoring devices. The integration of perovskites into flexible substrates and textiles aligns with the ongoing miniaturization and personalization of electronics, a key driver of wearable technology. Although the growth in publication numbers has continued into 2025, data from that year is not included in the graph, as the writing of this book was finalized during the first half of 2025, before a complete dataset could be compiled.

The sharp rise in recent publications underscores the emergence of wearable perovskite technologies as a highly active and rapidly evolving field of research. This momentum is driven by multidisciplinary collaboration across a range of scientific and engineering disciplines, including materials science, chemistry, physics, energy research, chemical engineering, biochemistry, and bioengineering.

Figure 3.4 illustrates the contribution of each field, presenting the percentage share of publications related to wearable perovskites across these disciplines.

The top ten most active countries in wearable perovskite research, home to leading universities and research institutions, are China, the United States, South

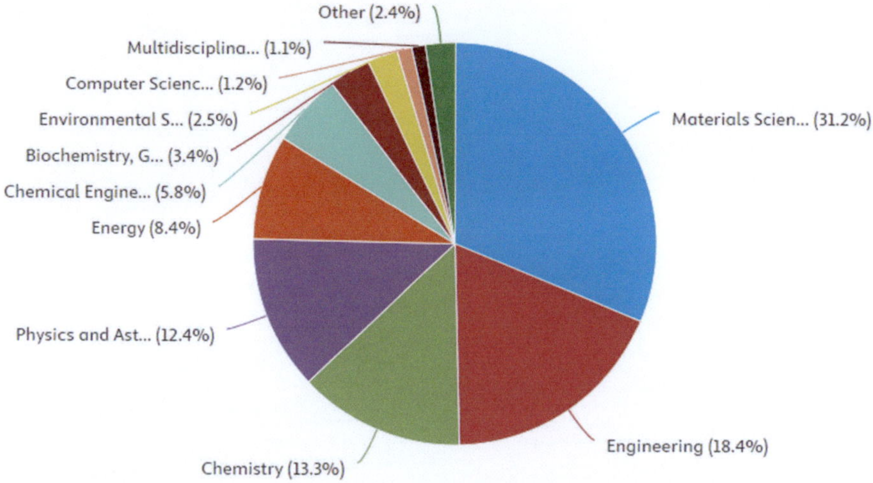


**Fig. 3.4** Subject-area breakdown of scientific publications on wearable perovskite technologies, based on Scopus data

Korea, India, Hong Kong, the United Kingdom, Singapore, Taiwan, Australia, and Japan (Fig. 3.5). China leads by a significant margin, clearly outpacing its closest competitor, the United States. However, the gap between the United States and the next group of countries, including South Korea and others in the top ten, is relatively small, with all of them producing a comparable volume of research publications in this field.

The keywords in the wearable perovskites' articles were also examined and mapped using VOSviewer [11] as shown in Fig. 3.6. Keyword analysis reveals a

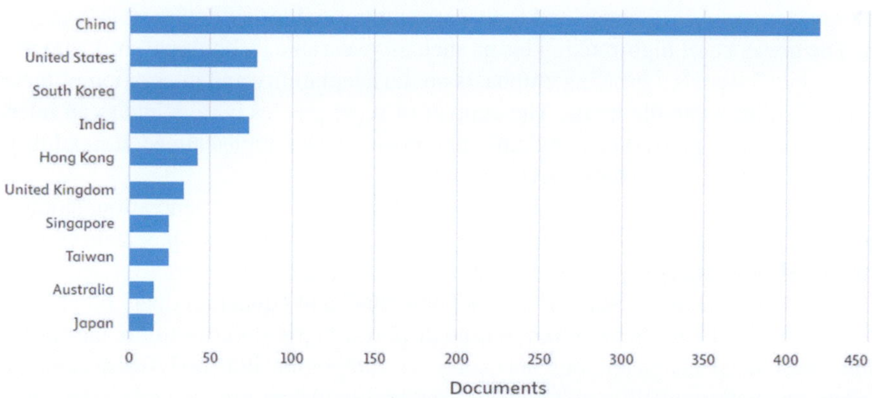


**Fig. 3.5** Top ten countries by number of scientific publications on wearable perovskite research (2015–2025)

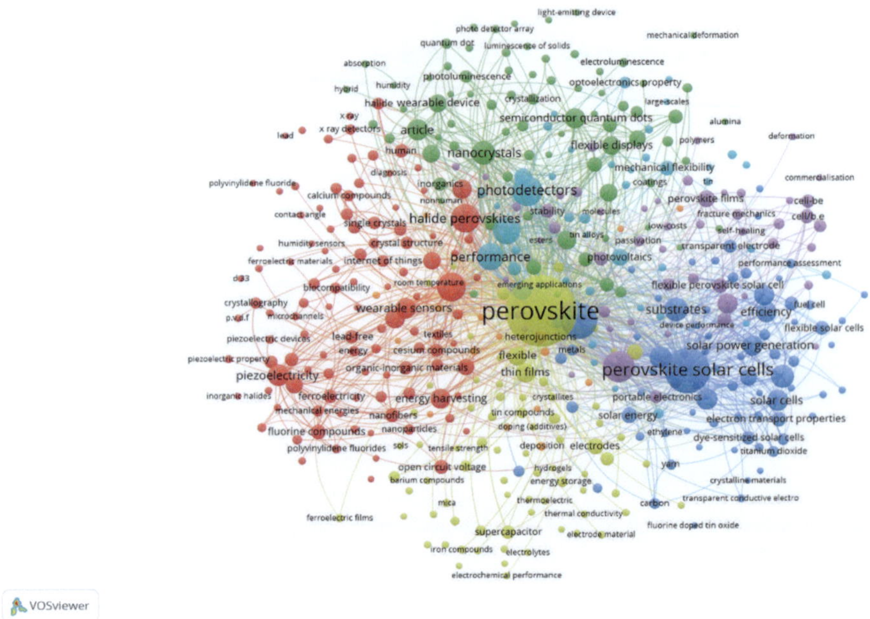


**Fig. 3.6** Network of the most repeated keywords for the field of wearable perovskites (2015–2025)

concentrated research effort centered on perovskite materials applications, particularly in solar cell technology. High-frequency terms such as "perovskite," "perovskite solar cells," "halide perovskites," and "photodetectors"—all with strong co-occurrence and link strength—underscore the sustained scientific interest in developing and optimizing perovskite-based optoelectronic devices. In parallel, keywords like "performance," "photons," and "nanocrystals" reflect ongoing efforts to enhance energy conversion efficiency and light-harvesting capabilities.

The presence of high-ranking terms such as "wearable technology" and "flexible electronics" signals a growing emphasis on the adaptability and integration of these materials into wearable items. The content of most articles here points to an intersection between perovskites and emerging device technologies aimed at flexibility, portability, and user-centered design.

The field is currently evolving toward a coupling of energy generation, storage, and intelligent wearable systems—each capitalizing on the unique optoelectronic and structural properties of perovskite compounds. In addition to energy-related applications, keyword trends such as "semiconductors" and "quantum dots" point to an active research focus on perovskite-based displays. These studies explore the optical and mechanical characteristics necessary for integration into body-worn devices, reflecting on the versatility of perovskite materials and their growing role in enabling multifunctional, compact, and user-adaptable electronics.

Looking forward, the research trajectory is expected to sharpen its focus on sustainability, particularly the development of lead-free, environmentally stable, and durable alternatives to conventional perovskites. As regulatory and environmental considerations gain prominence, innovation in green materials and the design of multifunctional, integrated systems will become increasingly critical.

Key research directions include enhancing the environmental robustness of perovskites, tailoring their mechanical properties for wearable formats, and developing transparent or semi-transparent modules for unobtrusive integration into textiles and accessories. This conjunction of high-performance materials with user-centric design is positioning perovskite-based wearable technologies at the forefront of next-generation electronics.

## 3.2 Crystal Structure of Perovskites

The fundamental appeal of perovskites lies in their highly adaptable crystal structure, described by the general formula $ABX_3$, where A and B are cations of different sizes and X is an anion. A unit cell of such simple cubic perovskite structure is presented in Fig. 3.7.

This deceptively simple structure provides extraordinary compositional flexibility, allowing researchers to fine-tune material properties by substituting different ions at each crystallographic site. The A-site can accommodate both single elements and molecular compounds, including alkali metals like cesium ($Cs^+$) and rubidium ($Rb^+$), alkaline earth metals such as barium ($Ba^{2+}$) and strontium ($Sr^{2+}$), as well as organic molecular cations like methylammonium ($MA^+$) and formamidinium ($FA^+$). Similarly, the B-site houses both elemental and compound cations, ranging from single metal ions like lead ($Pb^{2+}$), tin ($Sn^{2+}$), and titanium ($Ti^{4+}$) to more complex species, while the X-site contains anions including halides ($I^-$, $Br^-$, $Cl^-$) and oxides ($O^{2-}$).

For wearable applications, specific compositional choices become particularly relevant. Lead-based compositions like $CsPbI_3$ and $MAPbI_3$ offer excellent optoelectronic properties but raise biocompatibility concerns for skin-contact devices.

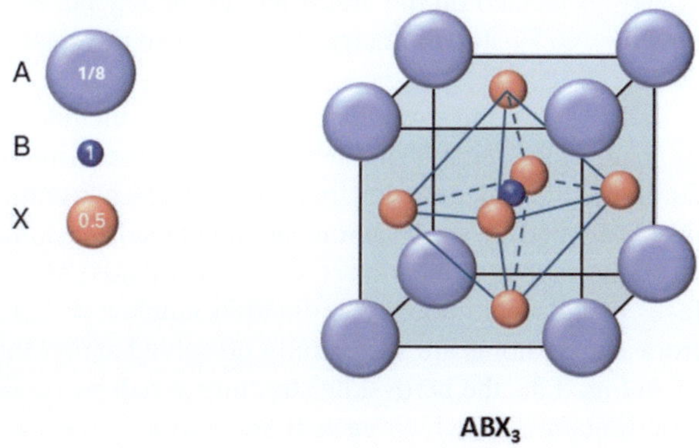


Fig. 3.7 Simple unit cell of an ideal cubic $ABX_3$ perovskite structure

Tin-based alternatives such as $MASnI_3$ provide lead-free solutions, though with reduced air stability. Mixed A-site compositions like $Cs_{0.2}FA_{0.8}PbI_3$ combine the thermal stability of cesium with the favorable electronic properties of formamidinium, creating materials optimized for the temperature variations encountered in wearable environments. Oxide perovskites like $BaTiO_3$ and $SrTiO_3$ excel in piezoelectric and dielectric applications for wearable sensors, while mixed halide systems such as $CsPbI_{3x}Br_x$ enable bandgap tuning for specific light-harvesting or emission requirements in wearable photonic devices.

In several textbooks, the evolution of how the perovskite structure is depicted is not explained in detail. As a result, readers with only basic knowledge of crystallography may find it difficult to fully grasp how atoms are arranged in a perovskite system.

To address this, a step-by-step construction of the perovskite lattice is outlined, as this understanding is essential for interpreting the structure–property relationships that make perovskites, such as cesium lead iodide ($CsPbI_3$), highly attractive for wearable electronic applications.

The visual sequence in Fig. 3.8a–e is designed to gradually build intuition for the atomic configuration and coordination environment in the ideal cubic perovskite ($ABX_3$) structure.

In Fig. 3.8a, the construction begins with a simple cube in which A-site cations ($Cs^+$ in $CsPbI_3$) are positioned at the eight corners. These monovalent alkali metal cations are relatively large and are shared among eight neighboring unit cells, resulting in a fractional contribution of 1/8 per corner, which eventually adds up to 1 A atom per unit cell.

In Fig. 3.8, the B-site cation ($Pb^{2+}$ in $CsPbI_3$) is placed at the center of the cube. This lead ion is fully enclosed within the unit cell and therefore contributes one B atom per unit cell.

In Fig. 3.8c, X-site anions ($I^-$ in $CsPbI_3$) are introduced at the centers of each cube face. These halide ions are shared between adjacent unit cells, and with six face-centered positions each contributing half an atom, they provide a total of 3 X atoms of real contribution per unit cell, completing the $ABX_3$ stoichiometry. At this stage, the cube edges are removed to enhance clarity and focus on the spatial arrangement of atoms. Readers are encouraged to examine the fractional atomic contributions (1/8, 1, and ½) labeled on the atoms and in the legend, as these visual cues help clarify how the $ABX_3$ stoichiometry emerges from the spatial arrangement within the unit cell.

In Fig. 3.8d, the structure is expanded to include eight connected unit cells, and the cube lines remain omitted for better visual interpretation. This broader view facilitates the understanding of coordination environments: each B-site cation ($Pb^{2+}$) is coordinated by six $I^-$ anions in an octahedral geometry (coordination number = 6), while each A-site cation ($Cs^+$) is coordinated by twelve $I^-$ anions located on the faces of adjacent cubes (coordination number = 12). The fractional contributions from shared atoms are structurally preserved across this extended array.

In Fig. 3.8e, the perovskite structure is redrawn using a more familiar octahedral representation, in which each B-site cation resides at the center of an octahedron formed by six X-site anions. The A-site cations are placed in the voids between

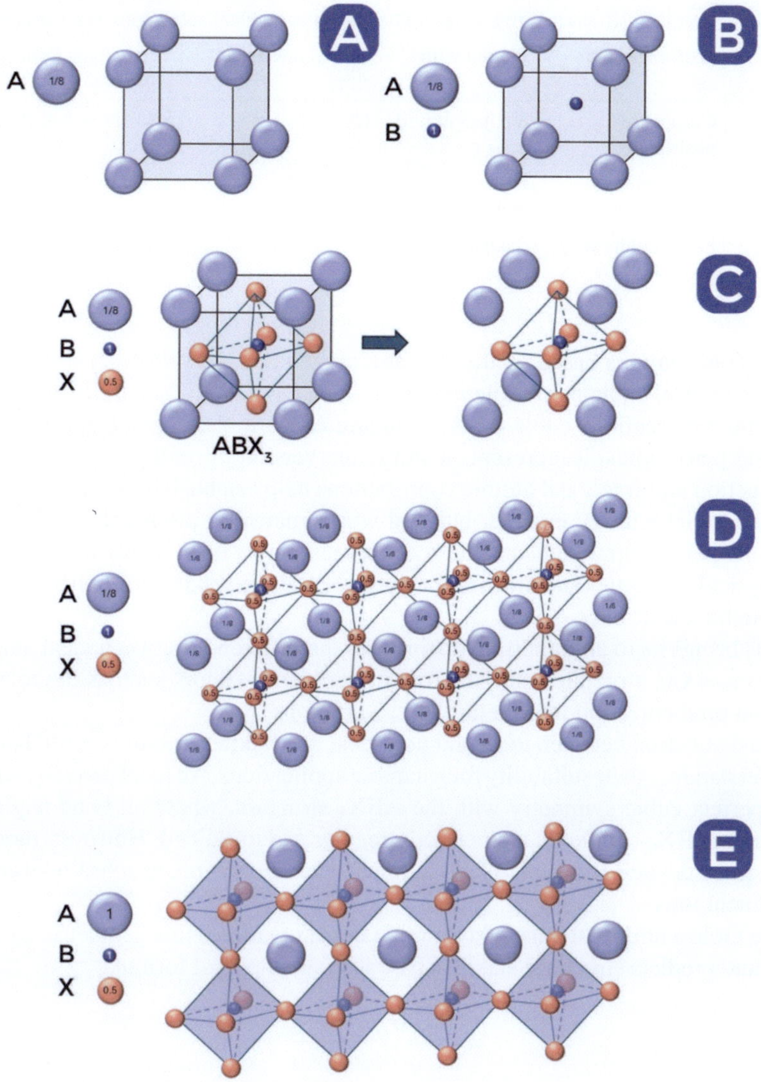


**Fig. 3.8** Step-by-step construction of the ideal cubic perovskite structure ($ABX_3$)

these octahedra. Although only one A atom per unit cell is symbolically represented, it corresponds to eight corner-sharing positions (each contributing 1/8) in the full lattice. This compact visualization is widely used in literature to emphasize the $BX_6$ connectivity, structural symmetry, and the three-dimensional network that defines perovskite materials.

To further clarify how the overall $ABX_3$ stoichiometry arises from shared atomic positions within the unit cell, Table 3.2 provides a concise overview of the fractional

**Table 3.2** Atomic positions and fractional contributions in an ideal cubic perovskite structure

| Atom | Position in unit cell | Shared with | Contribution per atom | Total atoms (×no. of positions) |
|---|---|---|---|---|
| A | Corners (8 positions) | 8 adjacent unit cells | 1/8 | $8 \times 1/8 = 1$ A atom |
| B | Body center (1 position) | None (fully enclosed) | 1 | $1 \times 1 = 1$ B atom |
| X | Face centers (6 positions) | 2 adjacent unit cells | ½ | $6 \times ½ = 3$ X atoms |

atomic contributions based on the spatial arrangement and symmetry of each atom type in the ideal cubic perovskite lattice.

While the idealized cubic $ABX_3$ structure offers a clear foundation for understanding perovskites, it represents a simplified version of reality. The step-by-step construction and fractional atomic contributions help establish how the basic geometry and stoichiometry emerge, but real-world materials often exhibit deviations from this perfect arrangement. These deviations play a crucial role in determining the physical properties and practical performance of perovskites in applications such as wearable electronics.

This brings us to an important distinction: the difference between ideal and non-ideal perovskite structures, and the role of geometric factors such as the tolerance factor in predicting and interpreting these variations.

The distinction between ideal and non-ideal perovskite structures is fundamental to understanding their suitability for wearable applications. An ideal perovskite maintains perfect cubic symmetry with the $ABX_3$ structure, where all bond angles are 90° and the $BX_6$ octahedra are perfectly regular and unstained. However, most real perovskites deviate from this ideal geometry due to size mismatches between the constituent ions.

The Goldschmidt tolerance factor ($t$) [12–14] quantifies this geometric compatibility and predicts structural stability based on an empirical formula:

$$t = \frac{r_A + r_X}{\sqrt{2}[r_B + r_X]} \tag{3.1}$$

where $r_A$, $r_B$, and $r_X$ are the ionic radii of A-site, B-site, and X-site ions, respectively.

Different ranges of the tolerance factor correspond to certain structural outcomes [12] in perovskite materials:

- $t = 1$: Represents an ideal cubic perovskite structure, though this perfect symmetry is rarely achieved in practice.
- $0.9 < t < 1$: Typically yields cubic or slightly distorted structures, often retaining high symmetry.
- $0.8 < t < 0.9$: Leads to octahedral tilting, resulting in tetragonal or orthorhombic distortions.

- $t < 0.8$: Associated with highly distorted structures or transition to non-perovskite phases. For $t < 0.71$ the structure is not considered perovskite [13].
- $t > 1$: Favors the formation of hexagonal, layered, or other non-perovskite configurations.

When the tolerance factor deviates from unity, the structure accommodates the size mismatch through octahedral tilting. The $BX_6$ octahedra rotate and *tilt* while maintaining corner-sharing connectivity, creating distorted but stable structures. Octahedral tilting, which directly impacts mechanical and electronic properties crucial for wearable applications, follows systematic patterns described by Glazer notation [15].

This common tilting system includes $a^0a^0a^0$ which basically deals with an ideal cubic structure with no tilting and includes $a^-a^-a^-$ (out-of-phase tilting about all three axes) and $a^-a^-c^+$ (mixed out-of-phase and in-phase tilting). The $a^-a^-a^-$ system creates maximum structural flexibility by allowing octahedra to rotate cooperatively under external stress, making it ideal for bendable wearable devices.

In contrast, $a^-a^-c^+$ tilting provides intermediate flexibility while maintaining better electronic connectivity along the c-axis, suitable for applications requiring both mechanical compliance and efficient charge transport.

The degree of tilting correlates directly with mechanical deformability. Structures with larger tilt angles can accommodate greater mechanical strain through further octahedral rotation without bond breaking. This relationship explains for example why $CsPbI_3$ with its $a^-a^-c^+$ tilting pattern (tilt angles ~15–20°) shows enhanced mechanical flexibility compared to cubic phases.

Tilting patterns also determine electronic band dispersion and charge transport pathways. Out-of-phase tilting reduces orbital overlap between adjacent octahedra, potentially increasing bandgaps and reducing carrier mobility. However, this trade-off becomes acceptable in wearable applications where mechanical durability often outweighs peak electronic performance.

Temperature-dependent tilting behavior is critical for wearable stability. Many perovskites undergo tilt transitions near body temperature (37 °C), potentially affecting device performance. Understanding these transitions allows optimization of compositions that maintain consistent tilting patterns across the operational temperature range of wearable devices.

For wearable technology applications, non-ideal, tilted perovskites are paradoxically more advantageous than their ideal counterparts. This counterintuitive notion stems from the unique mechanical and stability requirements of body-worn devices that differ fundamentally from conventional rigid electronics.

The pre-existing octahedral tilts in non-ideal perovskites create a structure that is inherently more accommodating to external mechanical stress. When subjected to bending or stretching forces typical in wearable applications, these tilted structures can redistribute stress through further octahedral rotation without breaking chemical bonds. When subjected to bending or stretching forces typical in wearable applications, these tilted structures can redistribute stress through further octahedral rotation without breaking chemical bonds as shown schematically in Fig. 3.9.

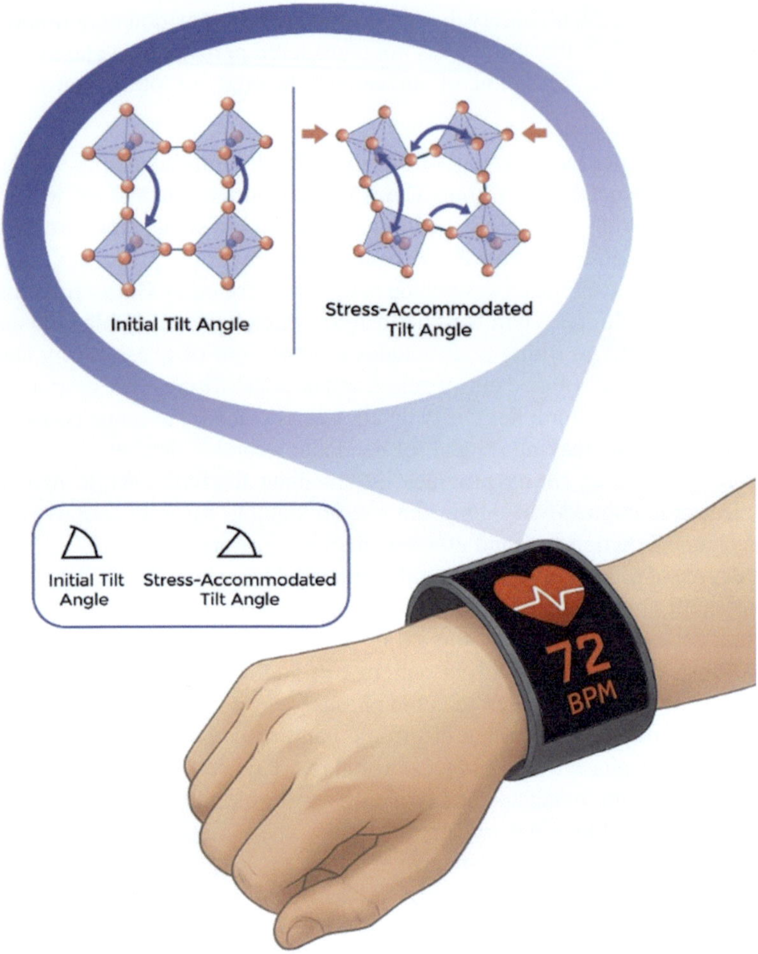


**Fig. 3.9** Schematic illustration of stress redistribution in tilted perovskite structures under mechanical deformation

The existing structural distortions act as built-in flexibility, allowing the material to deform reversibly under the mechanical stresses encountered in wearable devices. This mechanism is fundamentally different from conventional semiconductors, where mechanical deformation typically leads to irreversible damage through bond breaking or crack propagation.

Non-ideal perovskites often exhibit enhanced thermal and chemical stability compared to ideal cubic structures. The octahedral tilting can stabilize the structure against phase transitions and decomposition pathways, crucial for wearable devices that must function reliably across varying environmental conditions including body heat, humidity, and temperature fluctuations. This enhanced stability arises because the tilted configuration often represents a lower energy state that is less susceptible to thermally activated degradation processes.

The degree of octahedral tilting directly influences electronic and optical properties, enabling simultaneous optimization of multiple performance parameters. By engineering specific tolerance factors through compositional control, researchers can optimize bandgaps, charge transport properties, and mechanical responses simultaneously. This design flexibility is particularly valuable for wearable applications where multiple performance criteria must be balanced, such as maintaining adequate photovoltaic efficiency while ensuring mechanical durability under repeated flexing cycles.

$CsPbI_3$ (cesium lead iodide) perfectly illustrates the advantages of non-ideal perovskites for wearable applications. With a tolerance factor of approximately $t \approx 0.81$, $CsPbI_3$ adopts a tilted orthorhombic structure at room temperature with significant octahedral tilting. This non-ideal structure initially seemed problematic since the material has poor optoelectronic properties compared to the ideal cubic black phase stable, only above 320 °C. However, this apparent drawback becomes an asset for wearable technology. The inherent structural flexibility from octahedral tilting allows $CsPbI_3$ films to withstand mechanical deformation without catastrophic failure. When processed as quantum dots or stabilized through surface engineering, $CsPbI_3$ maintains its beneficial tilted structure while recovering excellent optoelectronic properties. These $CsPbI_3$ quantum dots, when incorporated into flexible polymer matrices, create mechanically robust films ideal for wearable displays and sensors.

For optoelectronic applications, the $ABX_3$ perovskite structure typically incorporates an A-site cation chosen from organic species such as methylammonium (MA) or formamidinium (FA), or from inorganic cations like $Rb^+$ or $Cs^+$. The B-site cation is usually $Pb^{2+}$ or $Sn^{2+}$, while the X-site anion is commonly $Cl^-$, $Br^-$, or $I^-$. Composed of these ions, the optoelectronic properties of wearable perovskites must be carefully engineered to meet the specific requirements of flexible, body-integrated devices. These properties can be broadly categorized into optical characteristics, electrical transport behavior, and stability considerations—each playing a critical role in determining device performance and longevity. The mechanical behavior of perovskites is intrinsically linked to their crystallographic structure, as the same ionic bonding and lattice arrangements that enable exceptional optoelectronic tunability also govern how these materials respond to mechanical deformation, thermal cycling, and environmental stresses encountered in wearable environments. Understanding and optimizing mechanical properties becomes crucial for ensuring that the carefully engineered optical and electrical characteristics remain stable under the repeated bending, stretching, and twisting motions inherent to body-worn applications.

The crystallographic structure plays an important role for specific characteristics favorable for optoelectronic wearable applications. A very representative example here is again $CsPbI_3$ as a leading all-inorganic perovskite material being widely investigated for next-generation wearable optoelectronic devices, including flexible solar cells, photodetectors and light emitting diodes (LED) [16]. Its functional performance is tightly governed by its crystallographic phase, which dictates both optical activity and environmental stability.

The α-$CsPbI_3$ phase is a black, cubic perovskite structure with a direct bandgap of ~1.74 eV, offering excellent light absorption and charge transport properties. It is the most desirable phase for applications like photovoltaics, LEDs, and photodetectors in wearable systems [17]. However, it is metastable at room temperature and generally forms at temperatures above 310 °C [18]. Stabilizing this phase under flexible, low-temperature processing conditions is a major research focus.

In contrast, the δ-$CsPbI_3$ phase is yellow, optically inactive, and thermodynamically stable at room temperature. Its wide indirect bandgap and poor optoelectronic properties render it undesirable for wearable devices. However, understanding this phase is critical for developing phase-stabilization strategies that prevent α-to-δ conversion.

The β-phase (tetragonal) and γ-phase (orthorhombic) are intermediate structures that may form at lower temperatures and under specific processing conditions. While less optically ideal than the α-phase, these can bridge stability and performance—particularly the γ-phase, which exhibits better optoelectronic properties than the δ-phase and can be used where full α-phase stabilization is challenging. Table 3.3 summarize the crystallographic phases of $CsPbI_3$, their colors, bandgap, stability at room, temperature and their optoelectronic properties relevant to wearable applications.

The table above highlights why so much current research focuses on stabilizing the α-phase through compositional engineering, surface treatments, or encapsulation.

**Table 3.3** Crystallographic phases of $CsPbI_3$ and their relevance to wearables

| Phase | Crystal structure | Color | Bandgap (eV) | Stability at room temp. | Optoelectronic properties | Wearable application |
|---|---|---|---|---|---|---|
| α-$CsPbI_3$ | Cubic perovskite | Black | ~1.74 | Metastable | Excellent; direct bandgap | Highly desirable for photodetectors and solar cells |
| β-$CsPbI_3$ | Tetragonal | Blackish | ~1.74 | Less stable | Intermediate | Useful in phase-transition control and partial stabilization |
| γ-$CsPbI_3$ | Orthorhombic | Dark | (~1.9–2.0) | Stabilizable at low T | Better than δ-phase | Promising for flexible devices where α-phase is hard to retain |
| δ-$CsPbI_3$ | Orthorhombic, non-perovskite | Yellow | Indirect (~2.8) | Stable | Poor; optically inactive | Undesirable |

The dramatic differences in optical properties between $CsPbI_3$ phases stem from fundamental changes in crystal structure that directly alter the electronic band structure and quantum confinement effects. The α-phase adopts a cubic perovskite structure where $PbI_6$ octahedra are corner-connected in three dimensions, creating a continuous inorganic framework. This connectivity allows for strong orbital overlaps between adjacent lead and iodine atoms, forming extended electronic states that can delocalize across the entire crystal. In contrast, the δ-phase crystallizes in an orthorhombic structure where $PbI_6$ octahedra are arranged in isolated chains or clusters with significantly reduced connectivity between neighboring octahedra.

The electronic band structure arises from the overlap of atomic orbitals, primarily the Pb 6s and 6p orbitals forming the conduction band and I 5p orbitals contributing to the valence band. In the α-phase, the three-dimensional connectivity creates strong Pb–I–Pb orbital overlap, leading to highly dispersive energy bands with significant bandwidth. The conduction band minimum and valence band maximum are well-separated in energy, resulting in the characteristic ~1.73 eV bandgap.

The δ-phase's reduced dimensionality dramatically alters this picture. The isolated or weakly connected $PbI_6$ units create much less orbital overlap between neighboring octahedra, resulting in narrower, less dispersive bands. More critically, the different coordination environment and longer Pb–I bond distances in the orthorhombic structure shift the relative positions of the Pb and I energy levels, widening the bandgap to ~2.8 eV.

Particle size critically influences the optoelectronic properties of $CsPbI_3$ through quantum confinement effects and phase stabilization mechanisms. The δ-phase inherently exhibits pseudo-quantum confinement due to its reduced structural dimensionality, where electronic states localize within individual $PbI_6$ octahedra rather than delocalizing across the crystal lattice. This localization increases the effective bandgap and transitions the excitonic behavior from extended Wannier excitons (characteristic of the α-phase) to more localized Frenkel-type excitons.

Quantum dot (QD) formation exploits these size-dependent effects to enhance optical properties. δ-phase QDs demonstrate superior optical absorption compared to α-phase counterparts under identical laser excitation conditions [19], indicating that structural dimensionality reduction can compensate for the typically unfavorable optical characteristics of the δ-phase. Conversely, α-phase $CsPbI_3$ nanocrystals exhibit enhanced thermodynamic stability when confined below 5.6 nm [20], where quantum confinement effects simultaneously blue-shift the bandgap and suppress the phase transition to the δ-phase.

This size-dependent phase stabilization demonstrates that quantum confinement effects can fundamentally alter the relative thermodynamic stability of different polymorphs, offering a pathway to access the desirable optoelectronic properties of the α-phase while maintaining structural integrity at ambient conditions. For wearable applications, phase selection depends on the specific application requirements, while the size-dependent tunability of optoelectronic properties provides a powerful design tool that allows engineers to customize $CsPbI_3$ performance for diverse use cases rather than being constrained by the bulk material properties of each phase.

This structure–property relationship explains why the black α-phase is desirable for solar cells (strong absorption, efficient charge separation) while the yellow δ-phase is unsuitable (poor absorption in visible range, strong exciton binding preventing charge separation). However, the δ-phase's stability and wide bandgap make it potentially useful for UV detection or blue-light applications where its unique properties become advantageous rather than detrimental.

The phase-dependent optical properties of $CsPbI_3$ thus represent a textbook example of how crystal structure determines electronic structure, which in turn dictates optical behavior—demonstrating why materials scientists spend considerable effort on phase stabilization strategies to access desired properties while maintaining structural stability. This chapter will further examine the fundamental properties of perovskites—optical, electrical, and mechanical—within the context of their crystallographic structure and the synthesis methodologies that govern their formation.

## 3.3 Fundamental Properties of Perovskites for Wearable Applications

Having established the foundational crystal structures and surveyed the current state of perovskite research in wearable applications and the categorization of high potential wearable perovskite materials, it becomes essential to examine the specific material properties that enable these remarkable capabilities. The performance of any wearable perovskite device depends fundamentally on the interplay between three critical property domains: optical characteristics that determine light interaction and energy conversion efficiency, electrical transport properties that govern charge generation, collection and transportation, and mechanical stability that brings about integration with flexible substrates and resilience under human motion.

These three property domains are inherently interdependent, as illustrated in Fig. 3.10, where their convergence defines the application-specific performance envelope for wearable devices. This analysis assumes that the material's chemical stability and biocompatibility are already established, as they are fundamental prerequisites for body-worn applications.

Modifications to enhance one characteristic inevitably influence the others, requiring careful optimization to achieve the delicate balance necessary for successful wearable implementation. While chemical stability and biocompatibility are essential prerequisites for any body-worn application, ensuring baseline safety and long-term reliability, the primary design challenges center on simultaneously optimizing optical, electrical, and mechanical performance within the constraints imposed by wearable form factors and operational environments.

The following sections cover each of these fundamental property domains and their specific requirements for wearable perovskite applications.

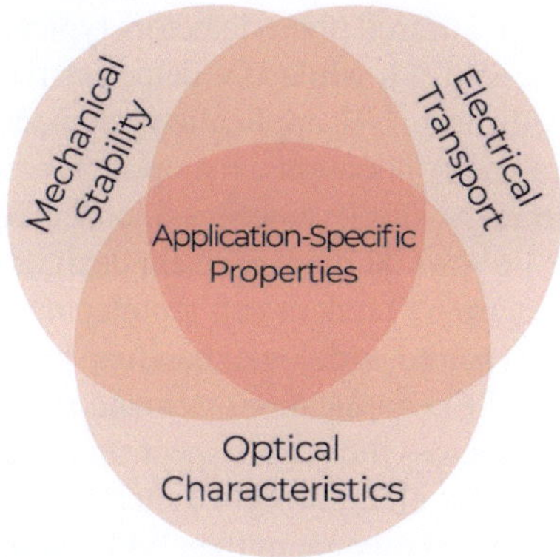


**Fig. 3.10** Interdependent property requirements for wearable perovskite devices

## *3.3.1 Optical and Electrical Properties*

The optical and electrical properties of perovskites are fundamentally interconnected, arising from the same underlying electronic band structure and defect states within the crystal lattice. Rather than treating these characteristics as independent material attributes, their coupled nature necessitates a unified discussion that recognizes how modifications to enhance optical performance inevitably influence electrical transport behavior, and vice versa. This interdependence becomes particularly critical in wearable applications, where devices must simultaneously achieve efficient light interaction and reliable charge transport while maintaining these properties under mechanical deformation.

The electronic band structure, which determines a material's optical absorption and emission properties, also controls the availability and mobility of charge carriers critical for electrical performance. While specific aspects of the band structure were discussed in Chap. 2 and are not repeated here, our focus remains on bandgap engineering for perovskite applications in wearable technology. Strategies that enhance photon absorption inherently influence carrier generation and transport, while defect states that drive non-radiative recombination also act as charge traps, affecting overall device efficiency. This dual function of electronic states underscores the need for careful material optimization to balance both optical and electrical performance in wearable perovskite devices.

Understanding this property coupling provides insight into why perovskites offer such remarkable versatility for wearable devices—the same compositional and structural engineering approaches can simultaneously optimize multiple performance characteristics. This integrated perspective guides the design of application-specific perovskite formulations that balance competing requirements across both optical and electrical domains.

Different optoelectronic applications require materials with specific bandgaps and optical properties to achieve optimal performance across diverse spectral ranges.

Near-infrared photodetectors typically utilize narrow bandgap materials around 1.0–1.2 eV [21], while UV detectors require wide bandgap materials exceeding 3.0 eV, with deep UV applications demanding bandgaps of 3.2–3.4 eV [22]. For photovoltaic applications, the optimal bandgap depends on the solar spectrum utilization strategy: single-junction silicon solar cells operate effectively with bandgaps around 1.1 eV, while more efficient designs often target 1.3–1.4 eV to better match the solar spectrum. Tandem cell architectures employ top cells with bandgaps of 1.6–1.8 eV [23] paired with lower bandgap bottom cells to maximize spectral coverage. Display and LED applications demand precise bandgap control to achieve pure color emission across the visible spectrum, requiring photoluminescence quantum yields above 10% for efficient light generation, noting that hybrid perovskite films are reported to have shown more than 90% photoluminescence quantum efficiency [24]. Regardless of application, the absorption coefficient must exceed 104 $cm^{-1}$ for efficient light harvesting in the thin films necessary for flexible device architectures, ensuring adequate optical absorption despite reduced material thickness.

The refractive index of perovskites represents a critical optical parameter that influences device architecture and performance in wearable applications. Methylammonium lead iodide ($MAPbI_3$), for instance, exhibits a refractive index ranging from approximately 2.6 at 300 nm to 2.0 at 800 nm [25], values considerably higher than conventional semiconductor materials used in flexible electronics. This elevated refractive index creates both opportunities and challenges for wearable device design: while it enhances light-matter interaction efficiency within thin active layers, it simultaneously introduces optical losses through total internal reflection (TIR) at interfaces with lower-index substrates and encapsulation materials commonly used in flexible devices [26–28]. The optical mode distribution in perovskite-based devices is directly governed by these refractive index contrasts, with simulations demonstrating that outcoupling efficiencies can exceed 20% despite the high refractive index when proper optical management strategies are employed [28–30].

For wearable photonic applications such as optical sensors and display technologies, the wavelength-dependent refractive index dispersion enables chromatic control and spectral filtering capabilities directly within the active material, reducing the need for additional optical components that would compromise mechanical flexibility. The refractive index also determines the critical angle for waveguiding in fiber-based wearable devices, where perovskite-integrated optical fibers enable full-duplex light fidelity (LiFi) communication systems that exploit the material's dual light-emitting and photo detecting capabilities within the same optical pathway [31].

Electrical transport properties determine how effectively charges can be generated, transported, and collected in wearable devices. Carrier mobility requirements vary significantly with application, from modest values above 1 $cm^2/V{\cdot}s$ for solar cells to higher mobilities exceeding 10 $cm^2/V{\cdot}s$ for fast photodetectors [32–34]. The balance between electron and hole mobilities becomes particularly important in LED applications, where efficient radiative recombination requires both carrier types to reach the active region. Defect density must remain below $101^6$ $cm^{-3}$ to minimize non-radiative recombination losses, while ionic conductivity should be suppressed below $10^{-8}$ S/cm to prevent electrochemical degradation during operation. Contact resistance with flexible electrodes represents a unique challenge, requiring values below 100 Ω $cm^2$ to maintain device efficiency despite mechanical deformation.

The stability requirements impose the most stringent constraints on material selection for wearable applications. Thermal stability must withstand body temperature variations and potential heating during operation, typically requiring less than 10% performance degradation after 100 h at 85 °C. Humidity resistance becomes critical in wearable environments, with targets of less than 20% degradation under 85% relative humidity conditions. Mechanical stability represents a unique challenge for wearables, demanding less than 15% performance loss after 1000 bend cycles at radii as small as 5 mm. Photostability under continuous illumination and UV exposure must be maintained to ensure consistent performance throughout the device lifetime.

The interdependence of these properties creates complex design challenges that often require trade-offs between competing requirements. Phase selection in materials like halide perovskites exemplifies this complexity, where the high-performance phases often offer superior optoelectronic properties but require stabilization strategies to maintain structural integrity. $CsPbI_3$ is a representative example of this dilemma to be discussed in the Mechanical Properties section. Quantum confinement effects in nanostructured perovskites provide additional tunability, allowing bandgap adjustment while potentially enhancing phase stability. For wearable applications, phase selection depends on the specific application requirements, while the size-dependent tunability of optoelectronic properties provides a powerful design tool that allows engineers to customize perovskites performance for diverse use cases rather than being constrained by the bulk material properties of each phase.

The precise control over optical and electronic properties demonstrated in these diverse applications underscores the importance of bandgap engineering as a fundamental tool for optimizing perovskite materials in wearable electronics. Unlike conventional rigid substrates, wearable devices impose additional constraints including mechanical flexibility, lightweight design, and operational stability under deformation conditions that demand sophisticated materials engineering approaches. Band gap engineering provides the necessary tunability to tailor perovskite properties for specific wearable applications while maintaining the structural integrity required for flexible device architectures [35]. The successful integration of perovskites into wearable devices ultimately requires a holistic approach that balances these optical, electrical, and stability requirements while considering the unique constraints of flexible, body-worn applications. This multifaceted optimization challenge drives ongoing research into compositional engineering, dimensional control, and encapsulation strategies that can unlock the full potential of perovskite materials for next-generation wearable technologies.

#### 3.3.1.1 Band Gap Engineering for Wearable Perovskites

Unlike conventional semiconductors with fixed electronic properties, perovskites offer a remarkable degree of compositional flexibility that enables precise control over their optoelectronic characteristics while maintaining the mechanical properties essential for wearable applications.

The fundamental advantage of perovskites lies in their ability to accommodate extensive compositional modifications without losing their crystal structure integrity. This compositional versatility translates directly into band gap tunability, allowing device engineers to optimize optical absorption, emission characteristics, and charge transport properties for specific wearable applications.

In the context of flexible optoelectronics, this tunability becomes particularly valuable. Flexible thin-film perovskite and organic semiconductor heterojunction [36] devices demonstrate the practical benefits of band gap engineering, achieving adjustable band-selective photodetection capabilities that can be optimized for specific sensing applications or environmental conditions. The ability to tune the spectral response of these devices opens new possibilities for wearable sensors that can adapt to different lighting conditions or target specific wavelength ranges for biomonitoring applications.

The development of flexible perovskite solar cells represents another compelling demonstration of how band gap engineering facilitates wearable device implementation. Through systematic compositional optimization, researchers have achieved flexible PSCs with high power conversion efficiency while maintaining the low manufacturing costs [37] that are essential for widespread adoption in consumer wearables. The combination of tunable optical properties, solution processability, and mechanical flexibility positions engineered perovskites as ideal candidates for next-generation wearable photovoltaic systems.

Beyond individual device performance, the band gap engineering of perovskites enables the design of integrated wearable systems where multiple optoelectronic functions can be achieved using variants of the same base material platform. This approach simplifies manufacturing processes, reduces material compatibility issues, and provides a pathway toward truly integrated wearable electronic systems that can harvest energy, detect environmental conditions, and communicate information using a unified material foundation.

This section examines different strategies employed engineer and tune band gap for perovskite materials that are most relevant to wearable applications. Figure 3.11 shows six primary methodologies have surfaced as particularly effective for achieving precise bandgap control in wearable perovskite systems: compositional engineering through halide and cation substitution, metal doping strategies that modify electronic band structure, strain engineering that exploits mechanical deformation effects, dimensional engineering utilizing quantum confinement principles, interface engineering for optimized charge transport, and additive engineering [38] for defect passivation and property enhancement.

Each approach offers distinct advantages and can be synergistically combined to achieve the multifunctional requirements of next-generation wearable optoelectronic devices. The selection and implementation of these engineering strategies depends critically on the target application requirements, substrate constraints, and desired operational parameters, making bandgap engineering an indispensable tool in the development of high-performance wearable perovskite technologies.

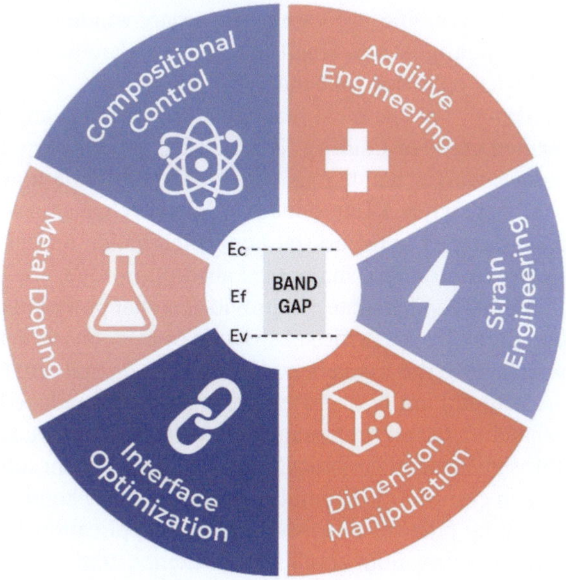


**Fig. 3.11** Strategies to perform band gap engineering for wearable perovskites

*Compositional Engineering* represents the most extensively utilized approach for precise band gap control in perovskite materials destined for wearable applications. This methodology involves systematic substitution of halide anions or organic/inorganic cations within the perovskite crystal structure to achieve desired optical and electronic properties. Halide mixing strategies, particularly through iodide-bromide substitution, enable continuous band gap tuning across the visible spectrum, making this approach particularly valuable for color-tunable displays and multi-spectral photodetectors in wearable devices. The incorporation of different organic cations, such as formamidinium (FA) or cesium (Cs) in place of methylammonium (MA), not only modifies the band gap but also enhances thermal stability and reduces ion migration, critical factors for wearable device longevity. Recent advances have demonstrated that mixed-cation perovskites exhibit superior mechanical flexibility and reduced hysteresis [39–41] compared to single-cation systems, making compositional engineering particularly suitable for flexible substrate applications.

Bismuth doping in tin iodide perovskites is another well-established method for effectively tuning the band gap, successfully reducing it from 1.3 to 0.8 eV [42]. This significant shift enhances properties such as light absorption, making the tunability of the material's properties promising for wearable applications.

Another example of doping for band gap engineering is homovalent substitution, where one ion is replaced by another with the same charge. For instance, substituting $Sr^{2+}$ for $Pb^{2+}$ in $CsPbBr_3$ increases the band gap from 2.29 to 2.64 eV [43], making the material absorb less in the visible range. This can be useful in applications where reduced light absorption is needed to prevent overheating or to control optical transparency. Such tuning allows designers to match the material's properties to specific wearable use cases, like smart windows or optical sensors. All example above highlights how small changes at the atomic level can lead to meaningful shifts in device behavior via band gap engineering.

Tuning via chemical reactions also includes anion exchange such as the documented cases of $CsPbX_3$ (X = Cl, Br, I) perovskite nanocrystals. Introducing KCl replaces larger halide ions with smaller chloride ions, leading to a blue shift in the band gap, while KI introduces iodide ions, causing a red shift [44]. This shift arises from changes in the crystal lattice and halide orbital overlap, directly affecting the electronic structure and optical absorption edge.

*Metal Doping* strategies provide another powerful avenue for band gap manipulation through systematic substitution of lead ions with alternative metal species while maintaining the fundamental perovskite crystal structure. Research has shown that bismuth doping in tin-based perovskites can reduce band gaps from 1.3 to 0.8 eV, expanding the absorption range into the near-infrared region crucial for certain wearable sensing applications. Conversely, homovalent substitutions such as strontium for lead in cesium lead bromide systems increase band gaps from 2.29 to 2.64 eV [43], providing pathways for wide-bandgap applications. Lead-free alternatives utilizing gallium-germanium combinations demonstrate tunable band gaps while addressing toxicity concerns associated with lead-based systems Band gap engineering in lead free halide cubic perovskites $GaGeX_3$ (X = Cl, Br, and I) based on first-principles calculation [45]. The incorporation of transition metals through controlled doping processes not only enables band gap engineering but also introduces magnetic properties and enhanced charge carrier mobility, particularly beneficial for multi-functional wearable devices.

*Strain Engineering* exploits the mechanical deformation inherent to wearable applications as a controllable parameter for band gap modulation. Unlike rigid device architectures, wearable perovskite systems experience continuous mechanical stress through bending, stretching, and conformational changes during operation. Research demonstrates that cesium tin iodide exhibits band gap reduction under tensile strain in tetragonal phases while showing increased band gaps under compressive conditions. Advanced flexible fabrication technologies enable precise control over strain distribution in perovskite films, allowing for spatially resolved band gap engineering across device areas [46]. The strain-induced modifications in electronic band structure are reversible and reproducible, making this approach particularly attractive for pressure-sensitive wearable sensors and mechanically responsive optoelectronic devices. Additionally, compressive strain in double perovskites such as cesium silver indium bromide enhances carrier mobilities through increased band dispersion, demonstrating the synergistic benefits of strain engineering for both optical and transport properties.

Temperature-induced polymorphism represents an additional dimension of band gap control in perovskite materials, particularly for halide perovskites. These compounds exhibit well-defined phase transitions that occur without alteration of chemical composition or bonding, driven by thermal energy changes. The characteristic phase sequence follows a predictable pattern: the high-temperature cubic ($\alpha$) phase transforms to the intermediate tetragonal ($\beta$) phase upon cooling, which subsequently transitions to the low-temperature orthorhombic (γ) phase with further temperature reduction [47–49]. Figure 3.12 shows crystal structure of the three polymorphs for $CsSnI_3$ perovskites in the same subsequent transition upon decreasing temperature, the bulk and surface Brillouin zones along with the symmetry k points of their crystal structures and corresponding band structures of each polymorph [49].

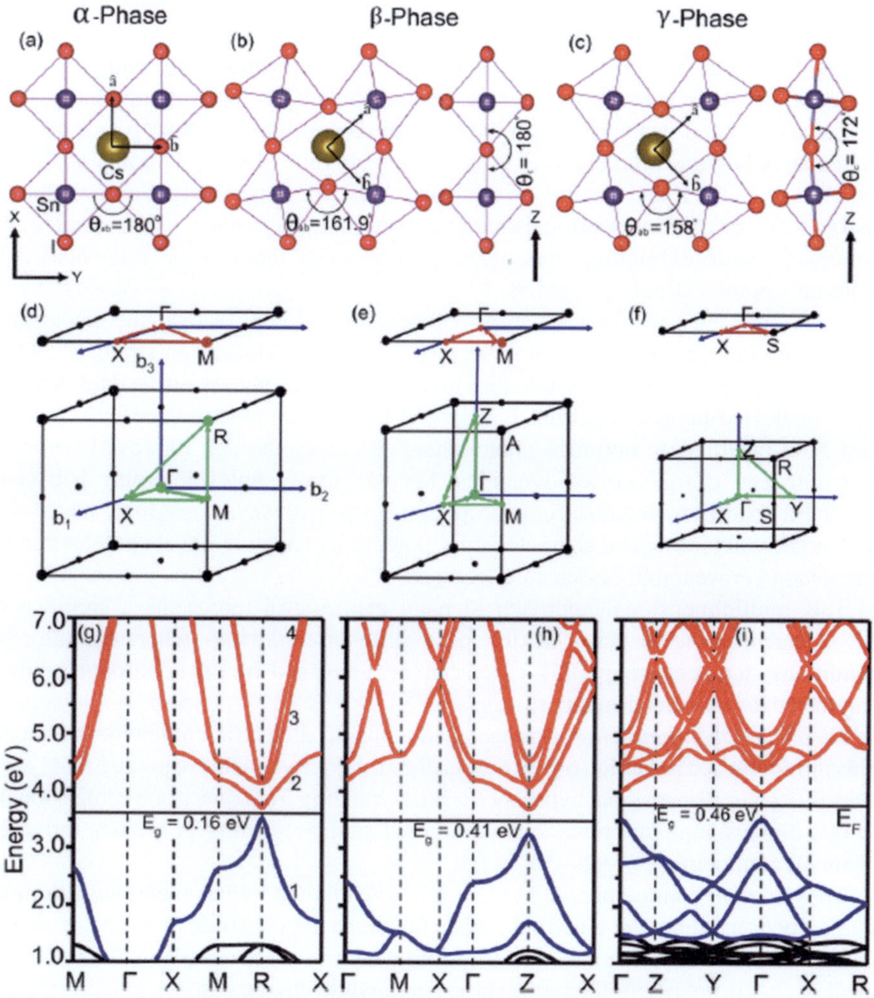


**Fig. 3.12** Crystal structures and electronic band structures of $CsSnI_3$ polymorphs: **a–c** cubic (α), tetragonal (β), and orthorhombic (γ) phases; **d–f** corresponding Brillouin zones with high-symmetry k-points; **g–i** computed ground-state band structures. Reproduced from [49] with permission from The American Physical Society

The structural complexity of perovskite polymorphs directly correlates with the degrees of freedom available for band gap engineering. $CsSnI_3$ exemplifies this relationship through its three distinct temperature-dependent phases: the high-symmetry cubic (α) phase, the tetragonal (β) phase characterized by octahedral rotation (angle $\theta_{ab}$) in the xy plane, and the orthorhombic (γ) phase featuring octahedral tilting in both the xy plane and along the z-axis. In the cubic phase, epitaxial strain effects are characterized solely by the c/a lattice parameter ratio. However, the transition to lower-symmetry phases introduces additional structural parameters, including both c/a and b/a ratios, as well as octahedral rotation and tilting angles $\theta_{ab}$ and $\theta_c$.

These structural transformations are accompanied by corresponding changes in the electronic band structure, where the Brillouin zone high-symmetry points evolve systematically—the time-reversal invariant momentum (TRIM) R point of the α phase maps to the Z and Γ points of the β and γ phases, respectively. Computational analysis reveals distinct electronic characteristics for each phase, with Sn-*s* and Sn-*p* dominated bands exhibiting phase-dependent behavior that directly influences the material's optoelectronic properties.

This expanded parameter space provides enhanced control over the electronic band structure, enabling more sophisticated band gap optimization strategies. The increased structural flexibility inherent to low-symmetry phases offers dual advantages for device engineering. First, it allows for precise tuning of valence and conduction band positions to optimize photovoltaic efficiency through improved spectral matching and charge carrier dynamics. Second, the complex interplay between structural distortions and electronic properties opens pathways to engineer materials with nontrivial topological characteristics, potentially enabling novel optoelectronic phenomena in wearable device architectures.

This multidimensional approach to band gap control represents a significant advancement in the design of flexible perovskite-based devices, where operational conditions, temperature in this case, can be leveraged as an additional tuning parameter for optoelectronic performance.

*Dimensional Engineering* leverages quantum and dielectric confinement effects through controlled reduction of perovskite dimensionality to achieve precise band gap control. Lower-dimensional hybrid perovskites, ranging from quasi-one-dimensional to two-dimensional structures, enable tailored semiconducting properties through quantum confinement effects [50, 51].

The quantum confinement effect in perovskite quantum wires manifests through size reduction, where the electronic wave functions are constrained in one or more spatial dimensions. In the case of perovskite quantum wires (PeQWs), the confinement occurs primarily in the radial dimension, while charge carriers remain free to move along the wire axis. This quasi-one-dimensional confinement leads to discrete energy levels and enhanced oscillator strength, resulting in improved radiative recombination efficiency. The energy gap tuning can be expressed through the effective mass approximation [52]:

$$E_{\text{gap}} = E_{\text{bulk}} + \frac{\hbar^2\pi^2}{2\mu L^2} \tag{3.2}$$

where $E_{bulk}$ represents the bulk material band gap, $\mu$ is the reduced mass of the electron–hole pair, and $L$ is the characteristic confinement dimension.

The $\hbar^2\pi^2/2\mu L^2$ term specifically originates from solving the time-independent Schrödinger equation for a particle in a one-dimensional infinite potential well [52], where the ground state wave function yields the lowest allowed energy contribution to the total band gap. Here, $\mu$ represents the reduced mass calculated as:

$$\mu = \frac{m_e m_h}{m_e + m_h} \tag{3.3}$$

For electron–hole pairs, $L$ is the confinement dimension, and the characteristic $1/L^2$ dependence serves as the fundamental signature of quantum confinement effects. This well-established quantum mechanical result has been extensively validated experimentally in semiconductor nanostructures since the 1980s and provides the theoretical foundation for precise band gap engineering in perovskite systems.

The practical implications of this relationship become evident when examining specific perovskite compositions under varying degrees of confinement. For $CsPbBr_3$ quantum wires with a bulk band gap of 2.36 eV [53] and reduced mass $\mu \approx 0.15m_0$, dimensional scaling from 7 to 4 nm [54] wire diameters increases the confinement energy, shifting emission from 540 nm (green) to 461 nm (blue) [55]. It is shown that the particles with diameter larger than 7 nm exhibits characteristics of bulk $CsPbBr_3$ [54]. This quantitative relationship enables systematic color tuning across the entire electromagnetic spectrum through precise control of nanoscale dimensions.

The incorporation of bulky organic spacer molecules creates quantum wells with tunable barrier heights, allowing for systematic band gap engineering from ultraviolet to near-infrared regions. Recent developments in perovskite quantum wires have demonstrated full-color emission capabilities essential for advanced wearable display applications [56].

The fabrication of high-quality PeQW arrays has been revolutionized through the development of porous alumina membrane template techniques, addressing long-standing challenges in uniform coating and crystallization control. This innovative approach employs ultrasmall pore sizes (~5 nm) fabricated on aluminum fiber substrates through roll-to-roll solution-coating processes, enabling the controlled growth of PeQW arrays while minimizing unwanted thin-film formation on template surfaces. The confined crystallization environment within PAM channels promotes spatially uniform solvent vaporization and metal halide perovskite crystallization, resulting in quantum wires with exceptional structural uniformity and enhanced photoluminescence efficiency through combined quantum confinement and surface passivation effects.

The resulting fiber light-emitting diodes (Fi-LEDs) demonstrate remarkable full-color emission capabilities with peaks at 625 nm (red), 512 nm (green), and 490 nm (sky-blue), exhibiting excellent mechanical flexibility and compatibility with textile fabrication processes. The quantum-confined PeQWs achieve outstanding photoluminescence and electroluminescence efficacy due to the synergistic effects of dimensional confinement and the three-dimensional porous alumina membrane structure, which provides effective surface passivation and reduces non-radiative recombination pathways. This breakthrough enables the creation of complex display architectures ranging from simple two-dimensional patterns to sophisticated three-dimensional lighting structures with excellent fluorescence uniformity, paving the way for advanced wearable display technologies that can be seamlessly integrated into textile applications while maintaining great optical performance under mechanical deformation.

Another size manipulation technique is to prepare two dimensional structures out of the perovskite bulk materials. Two-dimensional perovskite systems exhibit enhanced moisture stability and reduced ion migration compared to three-dimensional counterparts, addressing key reliability concerns for wearable electronics. The ability to control layer thickness at the molecular level provides high precision in band gap tuning while maintaining the beneficial properties of hybrid organic–inorganic systems [57]. The layered structure of 2D perovskites incorporates large organic spacer cations that create natural barriers between inorganic perovskite layers, effectively isolating the photoactive components from environmental degradation while simultaneously acting as dielectric confinement layers that enhance quantum confinement effects. This structural arrangement enables precise control over the quantum well thickness ($n$-value) through synthetic manipulation of the perovskite layer number, where single-layer ($n = 1$) structures exhibit the strongest confinement and highest band gaps, while increasing n-values gradually reduces the band gap toward bulk 3D perovskite values. The organic spacer layers, typically consisting of long-chain alkylammonium or aromatic ammonium cations, provide additional functionality by introducing hydrophobic barriers that significantly improve device stability under humid conditions and mechanical stress—critical requirements for textile-integrated applications [58]. Furthermore, the reduced dimensionality suppresses ion migration pathways that plague 3D perovskites, as the organic barriers limit ionic conductivity perpendicular to the perovskite planes, resulting in enhanced operational stability and reduced hysteresis in device performance. The synthetic versatility of 2D perovskites allows for the incorporation of various organic spacers with tailored properties, enabling simultaneous optimization of electronic, optical, and mechanical characteristics while maintaining processability from solution-based fabrication techniques compatible with large-area textile manufacturing processes.

*Interface Engineering* focuses on optimizing the electronic band alignment and charge transport characteristics at heterojunctions between perovskite active layers and adjacent transport materials. Interfacial modifications through low-dimensional perovskite capping layers or surface dipole engineering significantly improve both efficiency and stability of flexible perovskite devices [59, 60].

The introduction of interfacial buffer layers enables fine-tuning of energy level alignment, reducing recombination losses and improving charge extraction efficiency. Bilayer interface engineering through two-dimensional/three-dimensional perovskite heterostructures provides scalable approaches for large-area wearable applications [61]

Surface passivation strategies using organic molecules or inorganic compounds reduce trap state densities while simultaneously modifying surface band bending, effectively tuning the apparent band gap for device optimization. The development of flexible-compatible interface materials that maintain their properties under mechanical deformation represents a critical advancement for wearable perovskite technologies.

The optimization of interfacial energy band alignment between perovskite absorbers and charge transport layers has emerged as a critical determinant of device performance, with proper band offset engineering enabling efficient charge extraction while minimizing interfacial recombination losses [62]. Recent advances in interface modification through dipole molecules, such as trioctylphosphine oxide (TOPO), demonstrate the ability to create controlled surface band bending that improves energy level matching and enhances charge transfer efficiency in flexible perovskite devices [63].

Two-dimensional materials, including graphene derivatives, transition metal dichalcogenides, and MXenes, have shown exceptional promise as interfacial layers due to their dangling bond-free surfaces and tunable electronic properties that provide both excellent charge transport and environmental protection [64, 65].

The incorporation of 2D perovskite capping layers on 3D perovskite films creates heterostructures that simultaneously achieve defect passivation and improved charge extraction, with studies showing enhanced open-circuit voltages exceeding 1.2 V through double-sided surface passivation approaches [66, 67]. These interfacial engineering strategies are particularly crucial for flexible and wearable applications, where mechanical stress can disrupt charge transport pathways and create additional defect states that degrade device performance.

The development of mechanically robust interface materials signifies a fundamental challenge for wearable perovskite technologies, as traditional interfacial layers often exhibit poor adhesion and mechanical properties under repeated deformation cycles. Interfacial engineering in this realm addresses the needs for both perovskite flexible solar cells [60] and perovskite LEDs [68].

Self-healing interfacial systems utilizing host–guest interactions between organic spacer molecules and perovskite layers have demonstrated remarkable resilience, maintaining device functionality even after 4000 bending cycles while preserving the beneficial electronic properties of 2D/3D heterostructures [69].

Advanced interface engineering through embedding 2D perovskite layers within 3D frameworks creates mechanically flexible architectures that achieve power conversion efficiencies exceeding 20% in flexible devices while maintaining long-term stability under environmental stress [70].

The strategic selection of functional spacer cations in mixed-dimensional perovskite systems enables simultaneous optimization of charge transport, defect passivation, and mechanical flexibility, providing a pathway toward commercially viable wearable photovoltaic and optoelectronic devices [71]. The mentioned interfacial techniques complement dimensional engineering approaches, offering additional pathways for optimizing optoelectronic performance while maintaining the mechanical robustness required for textile integration.

*Additive Engineering* includes small molecular additives or processing aids to modify perovskite crystallization kinetics, defect densities, and resulting electronic properties without fundamentally altering the crystal structure [72]. Recent advances in additive engineering for wide-bandgap perovskites demonstrate significant reductions in open-circuit voltage losses, crucial for high-efficiency wearable solar cells [38].

Lewis base additives such as thiourea or pyridine coordinate with undercoordinated lead ions, passivating defect states and reducing non-radiative recombination pathways [73–75]. The incorporation of ionic additives modifies grain boundary properties and ion migration characteristics, enhancing operational stability under the variable environmental conditions encountered in wearable applications. Polymer additives provide dual functionality by improving mechanical flexibility while simultaneously influencing perovskite optoelectronic properties through intermolecular interactions. The selection and optimization of additives must consider compatibility with flexible substrates and processing temperatures, making this approach particularly valuable for roll-to-roll [76] manufacturing of wearable perovskite devices [77].

Advanced defect passivation strategies have evolved to target multiple defect sites simultaneously through carefully designed multifunctional additives [78, 79].

The implementation of double-site defect passivation approaches using fullerene derivatives demonstrates the potential for addressing both surface and grain boundary defects concurrently, resulting in enhanced charge carrier mobility and reduced trap-assisted recombination. Surface passivation with Lewis bases and acids has proven particularly effective for reducing the defect density of perovskite films, with post-treatment strategies including photo-curing processes that further minimize recombination centers [80]. These multifaceted approaches are essential for wearable applications where devices must maintain performance under mechanical stress and varying environmental conditions.

The strategic selection of organic ammonium salts for grain boundary passivation has also shown a critical advancement in additive engineering for flexible perovskite devices. Perfluorinated alkylammonium iodides have demonstrated exceptional effectiveness in simultaneous passivation of both surface and grain boundary defects, significantly improving long-term stability and efficiency [81]. The formation of low-dimensional perovskite structures at grain boundaries through these additives creates effective barriers against moisture ingress and ion migration, two primary degradation mechanisms in wearable applications. The compatibility of these additives with low-temperature processing conditions makes them particularly suitable for flexible substrates that cannot withstand high-temperature annealing procedures typically required for rigid glass-based devices.

Engineering of additives, solvents, and surface passivation techniques has become increasingly sophisticated, addressing the fundamental challenge of achieving matched power conversion efficiency and stability in organic–inorganic perovskite solar cells [82]. The integration of natural molecule-based additives presents an environmentally sustainable approach to defect passivation while maintaining high device performance. For wearable perovskite photovoltaics, the development of biocompatible additives that can function effectively under skin-contact conditions opens new possibilities for truly integrated wearable energy harvesting systems. The continued advancement in additive engineering strategies, particularly those addressing the unique requirements of flexible substrates and ambient processing conditions, will be crucial for the commercial viability of wearable perovskite technologies.

### *3.3.2 Mechanical Properties*

While the optical and electrical properties of perovskites provide the foundation for their optoelectronic functionality, the mechanical characteristics determine whether these materials can successfully transition from rigid laboratory demonstrations to practical wearable devices that must withstand the dynamic stresses of real-world applications.

Perovskites demonstrate mechanical properties that are characterized by moderate elastic moduli and inherent brittleness. The elastic modulus of perovskite films typically ranges from 31.8 GPa to as low as 2.57 GPa depending on processing conditions and additives [83, 84] positioning them between rigid inorganic semiconductors and flexible organic materials. However, perovskites offer negligible resistance to fracture, failing cohesively below 1.5 $J/m^2$ [85], which represents a fundamental limitation for mechanical robustness.

The fracture energy of perovskite materials is particularly concerning for wearable applications. Standard perovskite films exhibit minimal fracture indicating extremely low resistance to crack propagation under mechanical stress. This brittleness stems from the ionic nature of the crystal lattice and the weak van der Waals interactions between organic and inorganic components in hybrid perovskites.

Despite their limitations, perovskites possess several mechanical characteristics that can be advantageous for wearable technology. The moderate elastic modulus allows for some degree of flexibility compared to traditional inorganic semiconductors. Through strategic modification with additives such as 5-ammonium valeric acid, the elastic modulus can be reduced from 31.8 to 2.57 GPa while simultaneously increasing fracture energy from 0.33 to 1.19 $J/cm^2$ [84], demonstrating the tunability of mechanical properties through compositional engineering.

The relatively low processing temperatures required for perovskite film formation enable deposition on flexible plastic substrates without thermal damage, facilitating the fabrication of bendable devices [86]. Additionally, the solution processability of perovskites allows for the incorporation of mechanical strengthening additives and the formation of composite structures that can enhance overall device flexibility.

The primary mechanical challenges of perovskites for wearable applications stem from their inherent brittleness and low fracture toughness. Perovskites suffer from poor thermomechanical stability that must be overcome before the mass commercialization of wearable perovskite is viable, with key sources of instabilities tied to high residual stresses accumulated during processing [87]. These residual stresses can lead to premature failure under the cyclic loading conditions typical of body-worn devices.

The low fracture energy presents significant challenges for device durability. Wearable devices must withstand repeated bending, twisting, and impact loads from daily use, yet the cohesive failure of perovskites at low stress levels makes them susceptible to catastrophic failure [84, 85]. The propagation of microscopic cracks initiated during processing or use can rapidly compromise device performance and reliability.

Furthermore, the sensitivity of perovskites to environmental factors such as moisture and temperature variations can exacerbate mechanical degradation [88–90]. The coupling between chemical instability and mechanical stress creates additional failure modes that are particularly problematic in wearable environments where exposure to varying humidity and temperature is inevitable.

Table 3.4 summarizes the main mechanical properties of perovskites and the implications of them for wearable application.

The following section of this chapter highlights the most promising and rapidly advancing research directions expected to drive future improvements in the mechanical properties of wearable perovskite materials. Although less impactful areas are not covered, the selected topics represent core strategies aimed at overcoming the critical challenges of flexibility, durability, and mechanical reliability in next-generation wearable electronics.

**Table 3.4** Summary of perovskite mechanical properties and their implication for wearable applications

| Mechanical property | Implication for wearables | Impact | Research trends |
|---|---|---|---|
| Low fracture toughness | Required to enhance durability under mechanical stress | Negative | Improved via organic sodium salts [91] or polymeric additives [92] alkyl group additives, grain boundary engineering [84], and applying in-situ strain during deposition [93] |
| Flexibility (low elastic modulus) | Allows bending and stretching without damage | Positive | Achieved by lowering modulus through material design |
| Brittleness | Leads to cracking and device failure | Negative | Being mitigated through polymer reinforcement [3, 94] and nanostructuring [95] |
| Environmental instability | Affects performance in humid or wet conditions | Negative | Being addressed via encapsulation, polymer composites [94, 96] and hydrophobic layers [97] |
| Bulk modulus | Affects rigidity and compressibility | Composition-dependent | Being tuned by chemical substituting [98] and strain regulation [99] |

#### 3.3.2.1 Alkyl Group Additives

The use of alkyl group additives is a hot research topic addressing the mechanical limitations of perovskite materials in wearable device applications such as low fracture toughness.

One mechanism here is cross-linking and grain boundary reinforcement. Alkyl materials bearing specific functional end groups demonstrate remarkable ability to enhance mechanical stability through intermolecular interactions with perovskite structures. Compounds such as 5-ammonium valeric acid, which possesses both amino (–$NH_2$) and carboxyl (–COOH) functionalities, create robust linkages between adjacent perovskite grains. This cross-linking mechanism fundamentally alters the fracture behavior of the material, transforming it from a brittle ceramic-like response to a more ductile, failure-resistant system. The reported improvement in fracture energy from 0.33 to 1.19 J/cm$^2$ exemplifies this enhancement, while the concurrent reduction in elastic modulus from 31.8 to 2.57 GPa [84] indicates increased flexibility perfect for wearable applications.

The incorporation of alkyl ammonium additives extends beyond mechanical reinforcement to address fundamental structural deficiencies within perovskite crystals. These additives effectively passivate lattice defects that typically serve as stress concentration points and failure initiation sites. In two-dimensional tin halide perovskites, alkyl ammonium cations not only improve crystallinity but also reduce non-radiative recombination pathways that can lead to localized heating and thermal stress [100]. This dual benefit of structural stabilization and defect mitigation creates a more homogeneous stress distribution throughout the material, reducing the likelihood of premature failure under mechanical loading.

The dramatic enhancement of cohesion energy through alkyl additives has been also reported and appears to be a future research trend in this field. The incorporation of 5-aminovaleric acid (5-AVA) increases cohesion energy from 0.53 to 6.04 J/m$^2$—more than an order of magnitude improvement [101]. This enhancement stems from increased material plasticity and the promotion of crack deflection mechanisms around grain boundaries. Rather than propagating directly through the material in a catastrophic manner, cracks are forced to follow tortuous paths around reinforced grain boundaries, dramatically increasing the energy required for failure and improving overall toughness.

The alkyl chains present in polymer additives such as Poly[(*R*)-3-hydroxybutyric acid] (PHB) contribute an additional layer of mechanical stability through environmental protection. These hydrophobic chains create moisture barriers that prevent water infiltration—a critical failure mechanism in perovskite materials. By maintaining structural integrity under humid conditions and preventing hygroscopic expansion and contraction cycles, these additives ensure consistent mechanical performance across diverse environmental conditions encountered in wearable applications [102].

The combined effects of alkyl group additives create perovskite materials with mechanical properties specifically tailored for wearable electronics. The increased fracture toughness, reduced stiffness, enhanced cohesion, and environmental stability collectively enable the development of truly flexible perovskite-based devices.

#### 3.3.2.2 Polymer Reinforcement

Polymer reinforcement strategies have been also developed to address the mechanical limitations of perovskite materials in wearable applications. The incorporation of specialized elastomeric polymers represents one of the most promising fields of research in perovskite mechanical enhancement. Polyurethane elastomers containing disulfide bonds (PUDS) exemplify this approach [96], introducing dynamic covalent bonding networks that enable unprecedented self-healing capabilities in flexible perovskite solar cells. The disulfide linkages within the polymer matrix can reversibly break and reform under mechanical stress, allowing the material to autonomously repair micro-cracks and structural damage that would otherwise propagate and lead to catastrophic failure. The self-healing [96] mechanism is particularly critical in wearable applications where repeated flexing, stretching, and impact loading create continuous opportunities for damage accumulation. The elastomeric nature of these polyurethane systems simultaneously reduces the overall stiffness of the perovskite-polymer composite, distributing mechanical loads more uniformly and preventing stress concentration at grain boundaries and interfaces.

Encapsulation strategies utilizing flexible polymer matrices provide an additional layer of mechanical protection while maintaining the electronic functionality of perovskite active layers. Polydimethylsiloxane (PDMS) serves as an exemplary encapsulant, offering a unique combination of mechanical compliance, environmental barrier properties, and optical transparency [94]. The low elastic modulus of PDMS allows it to accommodate substrate deformation without transferring excessive stress to the embedded perovskite layer, effectively decoupling the mechanical and electronic functions of the device. This encapsulation approach enhances mechanical durability and also improves long-term stability by preventing moisture infiltration and chemical degradation, addressing two critical failure modes that limit the practical application of perovskite devices in wearable electronics.

The development of perovskite-polymer composite systems represents a more integrated approach to mechanical enhancement, where the polymer and perovskite phases are intimately mixed to create hybrid materials with tailored properties. Perovskite-organogel composites [3] demonstrate how polymer networks can be designed to maintain the optical and electronic properties of perovskites while dramatically improving their mechanical characteristics. The three-dimensional polymer network within these organogels provides a flexible scaffold that supports the perovskite crystals, preventing crack propagation and enabling large-strain deformation without loss of functionality. This approach is particularly effective because the polymer network can be tuned to match the thermal expansion characteristics of the perovskite phase, minimizing thermomechanical stress during temperature cycling.

Nanofiber membrane architectures incorporating perovskite-polymer composites offer another pathway for mechanical enhancement, creating materials with exceptional flexibility and processability [95]. These composite nanofiber membranes leverage the high surface-to-volume ratio of nanofiber structures to maximize the interfacial interaction between polymer and perovskite phases, resulting in enhanced load transfer and improved overall mechanical performance. The fibrous morphology inherently provides multiple pathways for stress distribution, preventing localized failure and enabling the material to maintain integrity under complex multiaxial loading conditions typical of wearable applications. The interconnected nature of the nanofiber network also contributes to improved toughness by promoting energy dissipation through fiber bridging and pull-out mechanisms.

The mechanical enhancement achieved through polymer reinforcement extends beyond simple toughening to encompass fundamental changes in failure mechanisms and deformation behavior. The viscoelastic properties of polymeric additives may introduce time-dependent mechanical responses that could potentially accommodate both rapid impact loading and sustained deformation, suggesting that perovskite devices might become more resilient to the varied mechanical environments encountered in wearable applications. This viscoelastic behavior appears to provide a degree of mechanical memory, potentially allowing the material to recover from deformation and maintain its original properties over multiple loading cycles, though further investigation is needed to fully understand these recovery mechanisms. The combination of improved flexibility, enhanced toughness, self-healing capabilities, and environmental protection achieved through polymer reinforcement shows promise for enabling the development of wearable perovskite devices, though continued research is required to optimize these systems for maintaining high performance throughout extended operational lifetimes in real-world applications.

#### 3.3.2.3 Nanostructuring

Nanostructuring offers a promising way to overcome the mechanical shortcomings of perovskite materials in wearable electronics by enabling precise control over their structure and behavior. It enhances surface-to-volume ratio, porosity, and flexibility, improving key properties like piezoelectricity, ferroelectricity, and conductivity. This approach allows perovskite systems to endure the mechanical demands of wearables while retaining their superior optoelectronic performance.

Imprinting and electrospinning at nanoscales are employed to create flexible perovskite films with improved mechanical properties. Nacre-inspired "brick-and-mortar" designs, using polymer-glued perovskite grains, mimic natural hierarchical structures to allow controlled deformation and energy dissipation. This approach enables flexible perovskite modules that retain their structural integrity and photovoltaic performance under repeated mechanical stress [103].

Electrospinning has also advanced the development of perovskite nanofibers with improved mechanical properties for wearable electronics [104]. Their one-dimensional structure and high surface area enhance strength, flexibility, and load distribution while reducing defects. These fibers offer advantages in flexibility, conductivity, and transparency, and their properties can be finely tuned through control of fiber diameter, crystallinity, and morphology.

The integration of perovskite nanocrystals within polymer matrices through nanostructuring approaches creates composite materials with synergistic mechanical properties. These nanostructured composites leverage the high surface-to-volume ratio of perovskite nanocrystals to maximize interfacial interactions with the polymer matrix, resulting in enhanced load transfer and improved overall mechanical performance. The nanoscale dimensions of the perovskite phase also reduce the critical flaw size, leading to increased fracture toughness and resistance to crack propagation. This approach is particularly effective for wearable applications where the material must maintain functionality under complex multiaxial loading conditions.

Nanofiber membrane architectures renders another significant advancement in nanostructured perovskite materials and future research trend for wearable electronics. These membranes combine the mechanical advantages of electrospun nanofibers [104] with the optoelectronic properties of perovskites, creating materials that exhibit exceptional flexibility and durability. The interconnected nature of the nanofiber network provides multiple pathways for stress distribution, preventing localized failure and enabling the material to accommodate large-strain deformation without loss of functionality. The high porosity of these membranes also contributes to their mechanical compliance, allowing them to conform to curved surfaces and undergo repeated flexing without degradation.

The crystalline orientation control achieved through nanostructuring approaches offers additional mechanical benefits for wearable perovskite devices [105]. N-CNFs provide nucleation sites as templates to induce crystalline orientation. By controlling grain orientation and size through nanostructured templates, researchers can optimize the mechanical anisotropy of perovskite films [106] to match the expected stress directions in wearable applications. This orientation control can significantly improve the mechanical properties along preferred directions while maintaining the desired optoelectronic characteristics. The use of functionalized nanofiber templates provides nucleation sites that guide perovskite crystal growth, resulting in films with reduced defect density and improved mechanical stability.

Recent advances in hierarchical nanostructuring have demonstrated significant improvements in the mechanical properties of perovskite materials, making them increasingly attractive for potential wearable applications. Structures such as nanoposts on cross-nanograting patterns not only enhance light trapping for photodetectors but also contribute to exceptional mechanical and operational stability. Devices employing these architectures maintain consistent performance under bending radii as small as 8 mm and after more than 1,000 bending cycles [107], highlighting their durability under mechanical stress.

The incorporation of biomimetic strategies further improves the mechanical robustness of perovskite films in the prospective research in the field. Inspired by natural materials like wood and bone, the integration of hierarchical multiscale structures has been shown to significantly increase both strength and toughness. For example, adding sodium hyaluronate to organometallic halide perovskite films [91] enhances elongation at break and fracture strength, producing materials that are both strong and flexible. These improvements are critical for the required mechanical resilience in dynamic, real-world conditions.

In addition to the topics given here, by reducing grain sizes and increasing the concentration of specific components, it is possible to simultaneously improve fracture toughness and nanoductility. A notable example is nanosized polycrystalline $MAPbI_3$, which demonstrates significantly enhanced nanoductility and fracture toughness due to extensive amorphization during the yielding process [108]. This transformation allows the material to better accommodate mechanical stress, improving its flexibility and mechanical resilience, critical attributes for reliable performance in dynamic, wearable electronic systems.

#### 3.3.2.4 Dynamic Hydrogen Bonding

Dynamic hydrogen bonding networks offer a novel advancement in enhancing the mechanical stability of wearable perovskite devices. The integration of poly(acrylic acid) and dopamine branch chains creates multiple dynamic hydrogen bonding sites that function as energy dissipation mechanisms during stress deformation. This network architecture effectively reduces the Young's modulus of the perovskite layer by 54.3% [109], transforming the inherently brittle perovskite into a more compliant material capable of accommodating mechanical strain without catastrophic failure. The dynamic nature of these hydrogen bonds allows for reversible bond breaking and reformation during bending cycles, enabling the material to maintain structural integrity while dissipating mechanical energy that would otherwise lead to crack propagation and device degradation.

Another strategy in this realm involves the use of hydrogen bonding within a dual-dynamic cross-linking network. By incorporating α-lipoic acid (LA), which undergoes ring-opening polymerization through dynamic disulfide and hydrogen bonds, the resulting polymer network significantly improves mechanical flexibility and stability [110]. This approach is reportedly allowing flexible perovskite solar cells to retain over 80% of their efficiency after 3000 bending cycles and recover up to 95% efficiency with mild thermal treatment.

Enhancing the mechanical stability of wearable perovskite solar cells was also achieved by reinforcing interfacial adhesion through hydrogen bonding. In this approach, polyamide-amine-based hyperbranched polymers are introduced [111] at the mechanically weakest interface, between the electron transport layer and the perovskite, where their adhesive functional groups and dynamic hydrogen bond networks significantly increase fracture energy from 1.08 to 2.13 J $m^{-2}$. These hydrogen bonds strengthen the interface and improve mechanical robustness and flexibility. As a result, the polymer-functionalized devices demonstrate high power conversion efficiencies of 25.05% for rigid and 23.86% for flexible cells, while also mitigating lead leakage via intramolecular cavities that capture $Pb^{2+}$ ions.

The hydrogen bond-assisted self-assembly of a 3D supramolecular framework (3D-SF) has been also research work on spotlight where dendritic nanopolymers were used to enhance the mechanical properties of wearable perovskite devices [112]. Acting as a crystallization template, the 3D-SF guides the formation of uniform, pinhole-free, and compact perovskite films by coordinating with perovskite grains and remaining at intergranular boundaries to crosslink them. This hydrogen-bonded network improves film flexibility and bending resistance, enabling stable resistive switching performance under mechanical stress. Devices incorporating this structure maintain high ON/OFF ratios and long data retention, even after prolonged environmental exposure and repeated bending, demonstrating the potential of hydrogen-bond-driven frameworks for mechanically resilient wearable perovskite electronics.

As the last instance in this section, interchain hydrogen bonding, regulated by cyclic linkages in copolymers, forms a hybrid cross-linked network within the perovskite thin film, significantly enhancing its mechanical resilience. This network has been reported to enable efficient energy dissipation and self-recoverability, allowing the film to endure 2000 bending cycles at a 1 mm radius without performance loss [113]. The strong hydrogen bonding also contributes to controlled crystal growth, defect passivation, and environmental stability, resulting in perovskite solar cells with high efficiency (23.25%) and extended operational lifetime under harsh conditions.

#### 3.3.2.5 Encapsulation Techniques

Encapsulation strategies are reported to be capable to address mechanical challenges for wearables by simultaneously protecting perovskites from degradation and enhancing their mechanical performance.

One of the most pronounced fields in this area is the integration of protective polymer matrices into perovskites. Polydimethylsiloxane (PDMS) and polymethyl hydrogen siloxane (PMHS) layers encapsulation [114, 115] demonstrated dual functionality, providing environmental protection from moisture and oxygen while enhancing mechanical properties. This approach transforms brittle perovskite films into flexible composites suitable for wearable applications. Also, a liquid-to-liquid self-encapsulation inkjet-printing technique using PDMS [116] has been shown to significantly improve the mechanical properties of wearable perovskite devices. In this method, perovskite inks are directly printed into a liquid PDMS precursor, forming a perovskite single-crystal-embedded PDMS structure. The space-confined environment of the PDMS slows crystallization and supports uniform crystal growth, resulting in mechanically robust and flexible composites. The inherent elasticity and sealing capability of PDMS not only enhance flexibility but also protect the perovskite crystals from ambient degradation. This approach enables the fabrication of wafer-scale, air-stable, and mechanically resilient perovskite fluorescent patterns, making it highly suitable for wearable light-emitting applications.

A bioinspired liquid-based encapsulation strategy, utilizing a polymer matrix infused with hydrophobic oil, significantly enhanced the mechanical properties and environmental stability of wearable perovskite devices [97]. Applied to halide perovskites, this approach creates a flexible, transparent, and self-healing barrier that effectively blocks water in all forms—vapor, droplets, and ice—without compromising the material's functional performance. The infused liquid fills coating defects, eliminating diffusion pathways and promoting cluster-based water transport, which drastically reduces moisture ingress. Additionally, the liquid core enables reversible water transport and allows for periodic replenishment of the hydrophobic oil, further extending device lifetime.

The use of room-temperature vulcanizing (RTV) silicone rubber as part of a bilayer encapsulation system with polymethyl methacrylate (PMMA) has been also reported to improve the environmental stability and mechanical durability of perovskite solar cells [117]. This low-cost, commercially available encapsulation package effectively prevents moisture and oxygen from entering the system without requiring additional post-processing such as heat treatment or UV curing. Under conditions of 50% humidity and room temperature, the encapsulated devices maintained over 80% of their initial efficiency after 1000 h, in contrast to severe degradation in unencapsulated cells. The RTV silicone contributes to mechanical protection by forming a flexible and water-repellent barrier, making it particularly thinkable for enhancing the long-term mechanical integrity of wearable perovskite devices exposed to harsh environmental conditions.

An ambient spray coating method using a polystyrene (PS)-4033/PS-4033-$SiO_2$ organic–inorganic composite film has been developed [118] to enhance the mechanical and environmental stability of perovskite solar cells, particularly for flexible applications. This superhydrophobic encapsulation layer, characterized by high compactness and uniformity, significantly improved moisture resistance while maintaining mechanical compliance. The combination of PS for structural compactness and $SiO_2$ for hydrophobicity enables a tenfold increase in device lifetime under humid conditions (60–70% RH, 35 °C), compared to unencapsulated cells. Unlike rigid or complex encapsulation methods, this approach offers a simple, scalable, and flexible solution suitable for large-area wearable perovskite electronics.

A polymer encapsulation strategy using hydrophobic room-temperature vulcanized silicone rubber (RTV-2) has been used to enhance the mechanical flexibility and moisture stability of methylammonium lead bromide ($MAPbBr_3$) perovskite films for wearable X-ray scintillators [119]. The $MAPbBr_3$/RTV-2 composite film, further treated with water for surface modification, demonstrates an eightfold fluorescence enhancement and substantial improvement in luminous efficiency. The flexible composite structure maintains high sensitivity under X-ray irradiation, with a detection limit of 12 $nGy_{air}\ s^{-1}$ that is far below typical diagnostic levels. The RTV-2 matrix not only protects the perovskite from moisture-induced degradation but also imparts mechanical resilience, making the film suitable for flexible, wearable radiation detection devices.

In addition to the research areas mentioned above, there are novel encapsulation methods, such as plasma-enhanced molecular layer deposition [120] that appears to be emerging now with dual offer of flexibility and barrier performance. These fields may affect the future of research in mechanical enhancement of wearable perovskites.

Figure 3.13 summarizes the approaches identified in the literature and those with potential for future research to enhance the mechanical properties of perovskite materials for wearable applications.

The mechanical properties of perovskites require significant optimization for wearable devices, as there are no universally optimal values for the critical mechanical parameters. The requirements are highly application-dependent and often involve trade-offs between competing properties.

For instance, bulk modulus presents a fundamental challenge: inorganic perovskites typically exhibit excessively high values (150–300 GPa) that compromise flexibility, while organic–inorganic hybrid perovskites display values that are often too low (10–30 GPa) for adequate structural integrity. The optimal bulk modulus depends entirely on the specific wearable application—whether the device requires high conformability (lower modulus) or structural robustness (higher modulus).

Similar optimization challenges exist for other mechanical properties. Fracture toughness must balance crack resistance with processability, elastic modulus needs to provide flexibility without sacrificing durability, and brittleness must be minimized while maintaining electronic performance. These property requirements are inherently application-specific and require careful engineering based on the intended use case.

However, environmental stability represents a universal requirement that must be maximized regardless of perovskite composition. Unlike other mechanical properties that require balanced optimization, chemical and structural stability under ambient conditions (humidity, temperature, mechanical stress) is always beneficial and should be enhanced through every available approach.

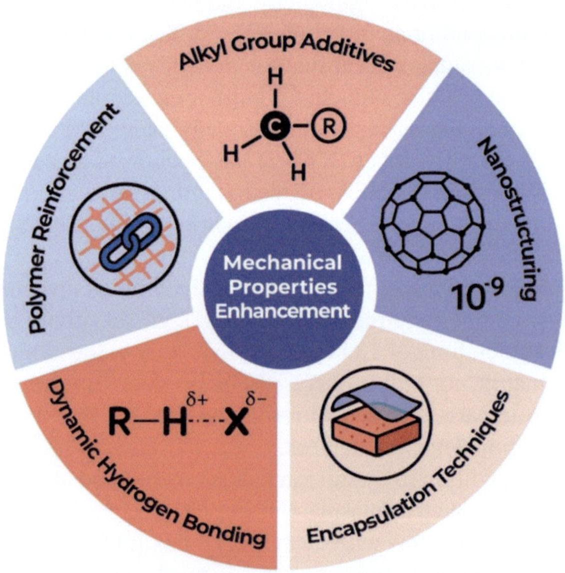


**Fig. 3.13** Main approaches in research trends to improve mechanical properties of wearable perovskites

The synthesis techniques for achieving perovskites with optimized optoelectronic and mechanical properties, as well as their integration strategies into flexible wearable substrates, are detailed in Chap. 7.

## References


1. M. Tabassum, Q. Zia, Y. Zhou, Y. Wang, M.J. Reece, L. Su, A review of recent developments in smart textiles based on perovskite materials. Textiles **2**(3), 447–463 (2022). https://doi.org/10.3390/textiles2030025
2. Y.-F. Kang, A.-R. Wang, R. Li, Y.-L. Song, Q.-F. Dong, A review: flexible perovskite solar cells towards high mechanical stability | 高机械稳定性的柔性钙钛矿太阳能电池的研究进展. Acta Polymerica Sinica **52**(8), 920–937 (2021). https://doi.org/10.11777/j.issn1000-3304.2021.21069
3. Y. Zhang et al., Homogeneous freestanding luminescent perovskite organogel with superior water stability. Adv. Mater. **31**(37) (2019). https://doi.org/10.1002/adma.201902928
4. A. Miglani, S.B. Ogale, O.S. Game, Architectural innovations in perovskite solar cells. Small **21**(15) (2025). https://doi.org/10.1002/smll.202411355
5. C.C. Li, T. Yu Huang, Y.H. Lai, Y.C. Huang, C. Shan Tan, Lead-free perovskites for flexible optoelectronics. Mater. Today Electron. **8** (2024). https://doi.org/10.1016/j.mtelec.2024.100095
6. R. Zhao, Z. Gu, P. Li, Y. Zhang, Y. Song, Flexible and wearable optoelectronic devices based on perovskites. Adv. Mater. Technol. **7**(3) (2022). https://doi.org/10.1002/admt.202101124
7. T. Wallach, L. Etgar, Highly transparent and semi-transparent perovskites and their applications. Appl. Phys. Rev. **12**(1) (2025). https://doi.org/10.1063/5.0237977
8. L. Rakocevic et al., Translucent, color-neutral and efficient perovskite thin film solar modules. J. Mater. Chem. C Mater. **6**(12), 3034–3041 (2018). https://doi.org/10.1039/c7tc05863b
9. I.T. Bello, S. Zhai, Q. He, Q. Xu, M. Ni, Scientometric review of advancements in the development of high-performance cathode for low and intermediate temperature solid oxide fuel cells: three decades in retrospect. Int. J. Hydrogen Energy **46**(52), 26518–26536 (2021). https://doi.org/10.1016/J.IJHYDENE.2021.05.134
10. E. Poorshakoor, M. Darab, Advancements in the development of nanomaterials for lithium-ion batteries: a scientometric review. J. Energy Storage **75**, 109638 (2024). https://doi.org/10.1016/J.EST.2023.109638
11. N.J. van Eck, L. Waltman, Software survey: VOSviewer, a computer program for bibliometric mapping. Scientometrics **84**(2), 523–538 (2010). https://doi.org/10.1007/s11192-009-0146-3
12. V.M. Goldschmidt, Die Gesetze der Krystallochemie. Naturwissenschaften **14**(21), 477–485 (1926). https://doi.org/10.1007/BF01507527
13. Z. Li, M. Yang, J.-S. Park, S.-H. Wei, J.J. Berry, K. Zhu, Stabilizing perovskite structures by tuning tolerance factor: formation of formamidinium and cesium lead iodide solid-state alloys. Chem. Mater. **28**(1), 284–292 (2016). https://doi.org/10.1021/acs.chemmater.5b04107
14. D. Hodges et al., Synchrotron and optical probing of hybrid organic-inorganic perovskite halides for photovoltaics, in *Conference Record of the IEEE Photovoltaic Specialists Conference* (2019), pp. 1170–1174. https://doi.org/10.1109/PVSC40753.2019.8980653
15. A.M. Glazer, The classification of tilted octahedra in perovskites. Acta Crystallogr. B **28**(11), 3384–3392 (1972). https://doi.org/10.1107/S0567740872007976
16. C. Zhao et al., Chemical vapor transport deposition of stable cubic $CsPbI_3$ optical films on the porous alumina substrate. MRS Adv. **4**(36), 1973–1979 (2019). https://doi.org/10.1557/adv.2019.171
17. M. Lai et al., Structural, optical, and electrical properties of phase-controlled cesium lead iodide nanowires. Nano Res. **10**(4), 1107–1114 (2017). https://doi.org/10.1007/s12274-016-1415-0

18. R.J. Sutton et al., Cubic or orthorhombic? Revealing the crystal structure of metastable black-phase $CsPbI_3$ by theory and experiment. ACS Energy Lett. **3**(8), 1787–1794 (2018). https://doi.org/10.1021/acsenergylett.8b00672
19. M. Yu et al., Laser-driven insulator-metal phase transitions in $CsPbI_3$ quantum dots and influence of doped metal nanowires. J. Phys. Chem. Lett. **14**(44), 10012–10018 (2023). https://doi.org/10.1021/acs.jpclett.3c02487
20. R.X. Yang, L.Z. Tan, Understanding size dependence of phase stability and band gap in $CsPbI_3$ perovskite nanocrystals. J. Chem. Phys. **152**(3) (2020). https://doi.org/10.1063/1.5128016
21. R.Q. Yang, W. Huang, M.B. Santos, Narrow bandgap photovoltaic cells. Sol. Energy Mater. Sol. Cells **238**, 111636 (2022). https://doi.org/10.1016/j.solmat.2022.111636
22. L. Wang et al., Recent progress in solar-blind photodetectors based on ultrawide bandgap semiconductors. ACS Omega **9**(24), 25429–25447 (2024). https://doi.org/10.1021/acsomega.4c02897
23. V. Raj, *Heterojunction Perovskite Solar Cells* (2018). https://doi.org/10.1016/B978-0-12-812915-9.00010-1
24. I.L. Braly et al., Hybrid perovskite films approaching the radiative limit with over 90% photoluminescence quantum efficiency. Nat. Photonics **12**(6), 355–361 (2018). https://doi.org/10.1038/s41566-018-0154-z
25. A.Y. Samsonova, P.P. Teslina, E.I. Deribina, N.I. Selivanov, C.C. Stoumpos, Y.V. Kapitonov, Low-temperature refractive index dispersion in $MAPbI_3$ halide perovskite single crystal. J. Phys. Chem. C **128**(23), 9730–9734 (2024). https://doi.org/10.1021/acs.jpcc.4c02066
26. T. Handa, H. Tahara, T. Aharen, Y. Kanemitsu, Large negative thermo-optic coefficients of a lead halide perovskite, Sci. Adv. **5**(7) (2019). https://doi.org/10.1126/sciadv.aax0786
27. F. Ramiro-Manzano et al., Optical properties of organic/inorganic perovskite microcrystals through the characterization of Fabry-Pérot resonances. Dalton Trans. **49**(36), 12798–12804 (2020). https://doi.org/10.1039/d0dt02254c
28. S. Chang, Z. Bai, H. Zhong, In situ fabricated perovskite nanocrystals: a revolution in optical materials. Adv. Opt. Mater. **6**(18) (2018). https://doi.org/10.1002/adom.201800380
29. R.K. Singh, R. Kumar, N. Jain, S.R. Dash, J. Singh, A. Srivastava, Investigation of optical and dielectric properties of $CsPbI_3$ inorganic lead iodide perovskite thin film. J. Taiwan Inst. Chem. Eng. **96**, 538–542 (2019). https://doi.org/10.1016/j.jtice.2018.11.001
30. Y.-J. Jung, S.-Y. Cho, J.-W. Jung, S.-Y. Kim, J.-H. Lee, Influence of indium-tin-oxide and emitting-layer thicknesses on light outcoupling of perovskite light-emitting diodes. Nano Converg. **6**(1) (2019). https://doi.org/10.1186/s40580-019-0196-z
31. Q. Shan et al., Perovskite light-emitting/detecting bifunctional fibres for wearable LiFi communication. Light Sci. Appl. **9**(1), 163 (2020). https://doi.org/10.1038/s41377-020-00402-8
32. L. Hu, S. Huang, R. Patterson, J.E. Halpert, Enhanced mobility in PbS quantum dot films: via PbSe quantum dot mixing for optoelectronic applications. J. Mater. Chem. C Mater. **7**(15), 4497–4502 (2019). https://doi.org/10.1039/c8tc06495d
33. J.-T. Shieh, C.-H. Liu, H.-F. Meng, S.-R. Tseng, Y.-C. Chao, S.-F. Horng, The effect of carrier mobility in organic solar cells. J. Appl. Phys. **107**(8) (2010). https://doi.org/10.1063/1.3327210
34. Y. Yuan et al., Significantly enhanced detectivity of CIGS broadband high-speed photodetectors by grain size control and ALD-$Al_2O_3$ interfacial-layer modification. ACS Appl. Mater. Interfaces **11**(22), 20157–20166 (2019). https://doi.org/10.1021/acsami.9b04248
35. Md.H. Miah, M.U. Khandaker, Md.B. Rahman, M. Nur-E-Alam, M.A. Islam, Band gap tuning of perovskite solar cells for enhancing the efficiency and stability: issues and prospects. RSC Adv. **14**(23), 15876–15906 (2024). https://doi.org/10.1039/D4RA01640H
36. T. Zhang, Y. Chen, Y. Chu, S.-J. Ding, W. Liu, X. Wu, Flexible perovskite and organic semiconductor heterojunction devices for tunable band-selective photodetection. ACS Appl. Electron. Mater. **4**(6), 2805–2814 (2022). https://doi.org/10.1021/acsaelm.2c00322
37. L. Chao, T. Niu, Y. Xia, Y. Chen, W. Huang, Ionic liquid for perovskite solar cells: an emerging solvent engineering technology. Acc. Mater. Res. **2**(11), 1059–1070 (2021). https://doi.org/10.1021/accountsmr.1c00154

38. X. Yu et al., Additive engineering for efficient wide-bandgap perovskite solar cells with low open-circuit voltage losses, Front. Chem. **12** (2024). https://doi.org/10.3389/fchem.2024.1441057
39. D. Zhang, X. Wang, Z. Fan, X. Xia, F. Li, Nondestructive post-treatment enabled by in situ generated 2D perovskites derived from multi-ammonium molecule vapor for high-performance 2D/3D bilayer perovskite solar cells. ACS Appl. Mater. Interfaces **14**(45), 51053–51065 (2022). https://doi.org/10.1021/acsami.2c17151
40. X. Hu et al., Methylammonium chloride as an interface modificator for planar-structure perovskite solar cells with a high open circuit voltage of 1.19 V. J. Power Sources 480 (2020). https://doi.org/10.1016/j.jpowsour.2020.229073
41. C.M. Sutter-Fella et al., Cation-dependent light-induced halide demixing in hybrid organic-inorganic perovskites. Nano Lett. **18**(6), 3473–3480 (2018). https://doi.org/10.1021/acs.nanolett.8b00541
42. H. Hasegawa, K. Kobayashi, Y. Takahashi, J. Harada, T. Inabe, Effective band gap tuning by foreign metal doping in hybrid tin iodide perovskites. J. Mater. Chem. C Mater. **5**(16), 4048–4052 (2017). https://doi.org/10.1039/c7tc00446j
43. D.B. Straus, R.J. Cava, Tuning the band gap in the halide perovskite $CsPbBr_3$ through Sr substitution. ACS Appl. Mater. Interfaces **14**(30), 34884–34890 (2022). https://doi.org/10.1021/acsami.2c09275
44. K. Sandeep et al., Band gap engineering of lead halide perovskite nanocrystals via room temperature bi-phasic anion exchange reactions. Mater. Today: Proc. 1274–1276 (2019). https://doi.org/10.1016/j.matpr.2020.03.697
45. Md.A. Sarker, M.M. Hasan, Md. Al Momin, A. Irfan, Md.R. Islam, A. Sharif, Band gap engineering in lead free halide cubic perovskites $GaGeX_3$ ($X = Cl$, Br, and I) based on first-principles calculations. RSC Adv. **14**(14), 9805–9818 (2024). https://doi.org/10.1039/D4RA00224E
46. R. Wei, H. Li, Z. Chen, Q. Hua, G. Shen, K. Jiang, Revolutionizing wearable technology: advanced fabrication techniques for body-conformable electronics. npj Flex. Electron. **8**(1), 83 (2024). https://doi.org/10.1038/s41528-024-00370-8
47. S.X. Tao, X. Cao, P.A. Bobbert, Accurate and efficient band gap predictions of metal halide perovskites using the DFT-1/2 method: GW accuracy with DFT expense. Sci. Rep. **7**(1), 14386 (2017). https://doi.org/10.1038/s41598-017-14435-4
48. C. Yu, Y. Ren, Z. Chen, K. Shum, First-principles study of structural phase transitions in $CsSnI_3$. J. Appl. Phys. **114**(16) (2013). https://doi.org/10.1063/1.4826068
49. R. Kashikar, M. Gupta, B.R.K. Nanda, Defining the topological influencers and predictive principles to engineer the band structure of halide perovskites. Phys. Rev. B **101**(15), 155102 (2020). https://doi.org/10.1103/PhysRevB.101.155102
50. Y. Xie et al., Dimensionality-controlled confinement effects for tunable optoelectronic properties in quasi-1D hybrid perovskites. ACS Nano **19**(13), 12895–12909 (2025). https://doi.org/10.1021/acsnano.4c16359
51. C. Katan, N. Mercier, J. Even, Quantum and dielectric confinement effects in lower-dimensional hybrid perovskite semiconductors. Chem. Rev. **119**(5), 3140–3192 (2019). https://doi.org/10.1021/acs.chemrev.8b00417
52. L.E. Brus, Electron–electron and electron–hole interactions in small semiconductor crystallites: the size dependence of the lowest excited electronic state. J. Chem. Phys. **80**(9), 4403–4409 (1984). https://doi.org/10.1063/1.447218
53. M. Ezzeldien et al., Electronic and optical properties of bulk and surface of $CsPbBr_3$ inorganic halide perovskite a first principles DFT 1/2 approach. Sci. Rep. **11**(1), 20622 (2021). https://doi.org/10.1038/s41598-021-99551-y
54. J. Butkus et al., The evolution of quantum confinement in $CsPbBr_3$ perovskite nanocrystals. Chem. Mater. **29**(8), 3644–3652 (2017). https://doi.org/10.1021/acs.chemmater.7b00478
55. H. Zhang et al., Bright and stable quantum confined $CsPbBr_3$ nanocrystals through post-synthetic chemical cutting and in-situ encapsulation. Nano Res. **18**(2), 94907131 (2025). https://doi.org/10.26599/NR.2025.94907131

56. B. Ren et al., Full-color fiber light-emitting diodes based on perovskite quantum wires. Sci. Adv. **10**(20) (2024). https://doi.org/10.1126/sciadv.adn1095
57. E.B. Kim, M.S. Akhtar, H.S. Shin, S. Ameen, M.K. Nazeeruddin, *A Review on Two-Dimensional (2D) and 2D-3D Multidimensional Perovskite Solar Cells: Perovskites Structures, Stability, and Photovoltaic Performances* (Elsevier B.V., 2021). https://doi.org/10.1016/j.jphotochemrev.2021.100405
58. G. Grancini, M.K. Nazeeruddin, Dimensional tailoring of hybrid perovskites for photovoltaics. Nat. Rev. Mater. **4**(1), 4–22 (2018). https://doi.org/10.1038/s41578-018-0065-0
59. S. Ye et al., Expanding the low-dimensional interface engineering toolbox for efficient perovskite solar cells. Nat. Energy **8**(3), 284–293 (2023). https://doi.org/10.1038/s41560-023-01204-z
60. J. Dou, Q. Chen, Interfacial engineering for improved stability of flexible perovskite solar cells. Energy Mater. Adv. **2022** (2022). https://doi.org/10.34133/energymatadv.0002
61. J. Wang et al., Bilayer interface engineering through 2D/3D perovskite and surface dipole for inverted perovskite solar modules. eScience **4**(6), 100308 (2024). https://doi.org/10.1016/j.esci.2024.100308
62. L. Huang, D. Zhang, S. Bu, R. Peng, Q. Wei, Z. Ge, Synergistic interface energy band alignment optimization and defect passivation toward efficient and simple-structured perovskite solar cell. Adv. Sci. **7**(6) (2020). https://doi.org/10.1002/advs.201902656
63. Z. Iqbal et al., Interface modification for energy level alignment and charge extraction in $CsPbI_3$ perovskite solar cells. ACS Energy Lett. **8**(10), 4304–4314 (2023). https://doi.org/10.1021/acsenergylett.3c01522
64. S. Palei, G. Murali, C.-H. Kim, I. In, S.-Y. Lee, S.-J. Park, A review on interface engineering of MXenes for perovskite solar cells. Nanomicro Lett. **15**(1), 123 (2023). https://doi.org/10.1007/s40820-023-01083-9
65. X. Shen et al., Two-dimensional materials for highly efficient and stable perovskite solar cells. Nanomicro Lett. **16**(1), 201 (2024). https://doi.org/10.1007/s40820-024-01417-1
66. G. Wu, R. Liang, M. Ge, G. Sun, Y. Zhang, G. Xing, Surface passivation using 2D perovskites toward efficient and stable perovskite solar cells. Adv. Mater. **34**(8) (2022). https://doi.org/10.1002/adma.202105635
67. M.A. Mahmud et al., Double-sided surface passivation of 3D perovskite film for high-efficiency mixed-dimensional perovskite solar cells. Adv. Funct. Mater **30**(7) (2020). https://doi.org/10.1002/adfm.201907962
68. Y. Shen et al., Rational interface engineering for efficient flexible perovskite light-emitting diodes. ACS Nano **14**(5), 6107–6116 (2020). https://doi.org/10.1021/acsnano.0c01908
69. Z. Yang et al., Self-healing and efficient flexible perovskite solar cells enabled by host–guest interaction and a 2D/3D heterostructure. J. Mater. Chem. A Mater. **10**(42), 22445–22452 (2022). https://doi.org/10.1039/D2TA05968A
70. Z. Wang et al., An embedding 2D/3D heterostructure enables high-performance FA-alloyed flexible perovskite solar cells with efficiency over 20%. Adv. Sci. **8**(22) (2021). https://doi.org/10.1002/advs.202101856
71. D. Li, Z. Xing, X. Meng, X. Hu, T. Hu, Y. Chen, Selection of functional spacer cations for efficient 2D/3D perovskite solar cells. CCS Chem. **5**(4), 781–801 (2023). https://doi.org/10.31635/ccschem.023.202202409
72. L. Ning et al., Enhanced molecular interaction by polymer additive for efficient and stable flexible perovskite solar cells. J. Mater. Sci. **57**(44), 20654–20671 (2022). https://doi.org/10.1007/s10853-022-07930-1
73. G. Yan et al., Synergistic passivation via Lewis coordination and electrostatic interaction for efficient perovskite solar cells. ACS Appl. Energy Mater. **6**(13), 7014–7024 (2023). https://doi.org/10.1021/acsaem.3c00573
74. H. Zhou et al., Surface passivation using 2-methoxyphenethylammonium iodide for efficient inverted perovskite solar cells. J. Power Sources **629** (2025). https://doi.org/10.1016/j.jpowsour.2024.236011

75. M. Zhang et al., Ultralong photoluminescence lifetime enables efficient tin halide perovskite solar cells. Next Mater. **6** (2025). https://doi.org/10.1016/j.nxmate.2024.100425
76. N.-R. Jiang et al., Enhanced efficiency and mechanical robustness of flexible perovskite solar cells by using $HPbI_3$ additive. Sol. RRL **5**(4) (2021). https://doi.org/10.1002/solr.202000821
77. Q. Lou et al., Synergistic effect of Lewis base polymers and graphene in enhancing the efficiency of perovskite solar cells. ACS Appl. Energy Mater. **4**(4), 3928–3936 (2021). https://doi.org/10.1021/acsaem.1c00299
78. A. Hassan et al., Recent defect passivation drifts and role of additive engineering in perovskite photovoltaics. Nano Energy **101**, 107579 (2022). https://doi.org/10.1016/j.nanoen.2022.107579
79. Z. Wang, H. Gao, D. Wu, J. Meng, J. Deng, M. Cui, Defects and defect passivation in perovskite solar cells. Molecules **29**(9), 2104 (2024). https://doi.org/10.3390/molecules29092104
80. F. Wang, S. Bai, W. Tress, A. Hagfeldt, F. Gao, Defects engineering for high-performance perovskite solar cells. npj Flex. Electron. **2**(1), 22 (2018). https://doi.org/10.1038/s41528-018-0035-z
81. B. Kang et al., Grain boundary passivation by alkylammonium salt for highly stable perovskite solar cells. J. Ind. Eng. Chem. **135**, 397–405 (2024). https://doi.org/10.1016/j.jiec.2024.01.051
82. S. Abicho, B. Hailegnaw, G.A. Workneh, T. Yohannes, Role of additives and surface passivation on the performance of perovskite solar cells. Mater. Renew. Sustain. Energy **11**(1), 47–70 (2022). https://doi.org/10.1007/s40243-021-00206-9
83. S. Rathore, W.L. Leong, A. Singh, Mechanical properties estimation of 2D–3D mixed organic-inorganic perovskites based on methylammonium and phenylethyl-ammonium system using a combined experimental and first-principles approach. J. Alloys Compd. **936** (2023). https://doi.org/10.1016/j.jallcom.2022.168328
84. K. Fan et al., Enhanced mechanical stability of perovskite film by modulating the toughness of grain boundary. Org. Electron. **117** (2023). https://doi.org/10.1016/j.orgel.2023.106778
85. N. Rolston et al., Mechanical integrity of solution-processed perovskite solar cells. Extreme Mech. Lett. **9**, 353–358 (2016). https://doi.org/10.1016/j.eml.2016.06.006
86. F. Di Giacomo, A. Fakharuddin, R. Jose, T.M. Brown, Progress, challenges and perspectives in flexible perovskite solar cells. Energy Environ. Sci. **9**(10), 3007–3035 (2016). https://doi.org/10.1039/C6EE01137C
87. C.C. Boyd, R. Cheacharoen, T. Leijtens, M.D. McGehee, Understanding degradation mechanisms and improving stability of perovskite photovoltaics. Chem. Rev. **119**(5), 3418–3451 (2019). https://doi.org/10.1021/acs.chemrev.8b00336
88. W. Yao et al., In situ microscopic observation of humidity-induced degradation in all-inorganic perovskite films. ACS Appl. Energy Mater. **5**(7), 8092–8102 (2022). https://doi.org/10.1021/acsaem.2c00576
89. G.R. McAndrews et al., Moisture uptake relaxes stress in metal halide perovskites at the expense of stability. ACS Energy Lett. **9**(8), 4153–4161 (2024). https://doi.org/10.1021/acsenergylett.4c01817
90. Y.B. Kim et al., Clarifying the degradation process of luminescent inorganic perovskite nanocrystals. RSC Adv. **14**(52), 38378–38384 (2024). https://doi.org/10.1039/d4ra07548j
91. Z. Lu et al., Grain-slip derived network topology to remarkable strength–toughness combination of perovskite film for flexible solar cells. Adv. Energy Mater. **12**(39) (2022). https://doi.org/10.1002/aenm.202202298
92. A. Giuri et al., Additive engineering: a route towards flexible and robust perovskite solar cells, in *2023 Photonics and Electromagnetics Research Symposium Piers 2023 Proceedings* (2023), pp. 54–64. https://doi.org/10.1109/PIERS59004.2023.10221237
93. D.B. Kim, Y.S. Cho, J.W. Lee, Anisotropic in situ strain-engineered halide perovskites for high mechanical flexibility. Adv. Funct. Mater. **31**(4) (2021). https://doi.org/10.1002/adfm.202007131
94. J. Wu et al., Low-temperature encapsulation with silicone grease enhances efficiency and stability of perovskite solar cells via $Pb^0$ defect passivation. Adv. Funct. Mater. (2025). https://doi.org/10.1002/adfm.202425979

95. L. Ning et al., Flexible hybrid perovskite nanofiber for all-inorganic perovskite solar cells. Mater. Res. Bull. **149** (2022). https://doi.org/10.1016/j.materresbull.2022.111747
96. Y. Lan et al., Thermally driven self-healing efficient flexible perovskite solar cells. Nano Energy **100** (2022). https://doi.org/10.1016/j.nanoen.2022.107523
97. B. Lemaire et al., Flexible fluid-based encapsulation platform for water-sensitive materials. Proc. Natl. Acad. Sci. U. S. A. **120**(34) (2023). https://doi.org/10.1073/pnas.2308804120
98. A.S. Verma, A. Kumar, Bulk modulus of cubic perovskites. J. Alloys Compd. **541**, 210–214 (2012). https://doi.org/10.1016/j.jallcom.2012.07.027
99. H.-S. Kim, N.-G. Park, Importance of tailoring lattice strain in halide perovskite crystals. NPG Asia Mater. **12**(1), 78 (2020). https://doi.org/10.1038/s41427-020-00265-w
100. H. Kim et al., Alkylammonium passivation for 2D tin halide perovskite field-effect transistors. J. Mater. Chem. C Mater. **13**(13), 6806–6815 (2025). https://doi.org/10.1039/d4tc05307a
101. M. Gutwald et al., Perspectives on intrinsic toughening strategies and passivation of perovskite films with organic additives. Sol. Energy Mater. Sol. Cells **209** (2020). https://doi.org/10.1016/j.solmat.2020.110433
102. Y. Qi et al., Perovskite films passivated by poly[(R)-3-hydroxybutyric acid] for improved photovoltaic performance. Org. Electron. **104** (2022). https://doi.org/10.1016/j.orgel.2022.106487
103. Y. Zhan et al., Nacre inspired robust self-encapsulating flexible perovskite photodetector. Nano Energy **98** (2022). https://doi.org/10.1016/j.nanoen.2022.107254
104. Y. Chen, Z.Y. Zhang, G.P. Wang, Water erosion highly recoverable and flexible photodetectors based on electrospun, waterproof perovskite-polymer fiber membranes. ACS Appl. Polym. Mater. **5**(8), 6124–6132 (2023). https://doi.org/10.1021/acsapm.3c00800
105. F. Song et al., Mechanical durability and flexibility in perovskite photovoltaics: advancements and applications. Adv. Mater. **36**(18) (2024). https://doi.org/10.1002/adma.202312041
106. G. Nazir et al., Advancements in flexible perovskite solar cells and their integration into self-powered wearable optoelectronic systems. Adv. Powder Mater. 100304 (2025). https://doi.org/10.1016/j.apmate.2025.100304
107. Y.H. Lee et al., Boosting the performance of flexible perovskite photodetectors using hierarchical plasmonic nanostructures. Small Struct. **5**(7) (2024). https://doi.org/10.1002/sstr.202300546
108. J. Yu, M. Wang, S. Lin, Probing the soft and nanoductile mechanical nature of single and polycrystalline organic-inorganic hybrid perovskites for flexible functional devices. ACS Nano **10**(12), 11044–11057 (2016). https://doi.org/10.1021/acsnano.6b05913
109. S. Zhu et al., Multiple dynamic hydrogen bonding networks boost the mechanical stability of flexible perovskite solar cells. Adv. Funct. Mater. **34**(48) (2024). https://doi.org/10.1002/adfm.202408487
110. J. Yang et al., Synergistic toughening and self-healing strategy for highly efficient and stable flexible perovskite solar cells. Adv. Funct. Mater. **33**(23) (2023). https://doi.org/10.1002/adfm.202214984
111. Z. Li et al., Hyperbranched polymer functionalized flexible perovskite solar cells with mechanical robustness and reduced lead leakage. Nat. Commun. **14**(1) (2023). https://doi.org/10.1038/s41467-023-41931-1
112. X. Ma, J. Zhou, Y. Liu, S. Xu, S. Cao, Supramolecular framework constructed by dendritic nanopolymer for stable flexible perovskite resistive random-access memory. Small **19**(11) (2023). https://doi.org/10.1002/smll.202206852
113. T.-H. Han et al., Spontaneous hybrid cross-linked network induced by multifunctional copolymer toward mechanically resilient perovskite solar cells. Adv. Funct. Mater. **32**(40) (2022). https://doi.org/10.1002/adfm.202207142
114. S. Tanaka, Y. Chao, S. Araki, Y. Miyake, Pervaporation characteristics of pore-filling PDMS/PMHS membranes for recovery of ethylacetate from aqueous solution. J. Memb. Sci. **348**(1–2), 383–388 (2010). https://doi.org/10.1016/j.memsci.2009.11.033
115. Z. Luo et al., Evaluation of the underwater stability of encapsulated perovskite solar cells. Sol. Energy Mater. Sol. Cells **262** (2023). https://doi.org/10.1016/j.solmat.2023.112557

116. Z. Gu et al., In situ inkjet printing of the perovskite single-crystal array-embedded polydimethylsiloxane film for wearable light-emitting devices. ACS Appl. Mater. Interfaces **12**(19), 22157–22162 (2020). https://doi.org/10.1021/acsami.0c04131
117. M. Samadpour, P. Parand, M. Heydari, N. Taghavinia, M. Mohammadi, Water repellent room temperature vulcanized silicone for enhancing the long-term stability of perovskite solar cells. Sol. Energy **218**, 28–34 (2021). https://doi.org/10.1016/j.solener.2021.02.035
118. Z. Luo, C. Zhang, L. Yang, J. Zhang, Ambient spray coating of organic-inorganic composite thin films for perovskite solar cell encapsulation. ChemSusChem **15**(3) (2022). https://doi.org/10.1002/cssc.202102008
119. Y. Li et al., Water-assistant ultrahigh fluorescence enhancement in perovskite polymer-encapsulated film for flexible X-ray scintillators. Chem. Eng. J. **452** (2023). https://doi.org/10.1016/j.cej.2022.139132
120. Z. Wang et al., Crosslinking and densification by plasma-enhanced molecular layer deposition for hermetic seal of flexible perovskite solar cells. Nano Energy **109** (2023). https://doi.org/10.1016/j.nanoen.2023.108232

# Chapter 4
# Characterization Techniques

This chapter presents essential characterization techniques for evaluating wearable perovskite materials and devices. The discussion covers fundamental physicochemical methods including X-ray diffraction, scanning and transmission electron microscopy for structural and morphological analysis. Optoelectronic characterization techniques specific to perovskite device performance are explored, followed by mechanical testing approaches critical for wearable applications, including nanoindentation and bending tests that assess flexibility and durability under stress.

The chapter addresses washability testing protocols essential for textile-integrated devices, thermal analysis for temperature-dependent behavior, and interface analysis methods for understanding critical material interactions. While acknowledging standard characterization techniques commonly used in materials science, the focus remains on wearable-specific evaluation approaches.

## 4.1 Introduction

The characterization of perovskites for wearable technologies builds upon the foundation of conventional perovskite comprehensive analysis while expanding into specialized domains unique to wearable applications. Standard characterization protocols including but not limited to X-ray diffraction (XRD), scanning electron microscopy (SEM), band gap measurements, electrical transport and conductivity analyses, and various optical spectroscopies, remain essential for understanding 'building blocks' properties. However, wearable perovskite systems demand an extended analytical framework that encompasses mechanical deformation studies, environmental resilience testing, and substrate compatibility assessments.

The characterization workflow for wearable perovskites (schematically shown in Fig. 4.1) follows a systematic multiphase approach beginning with synthesis of



M. Darab et al., *Perovskites for Next-Generation Wearable Technology*, Engineering Materials, https://doi.org/10.1007/978-3-032-03931-6_4

the perovskite materials, followed by comprehensive physicochemical characterization to establish baseline material properties. The characterized materials that were successfully screened out of several synthesized batches, then undergo integration into flexible substrates or wearable device architectures and possibly encapsulation, after which device performance evaluation encompasses optical, electrical, and connectivity testing to validate functional capabilities. Subsequently, the integrated devices are subjected to stress protocols that physically simulate real-world usage conditions, culminating in post-stress performance re-evaluation to assess durability and identify plausible performance degradation.

This comprehensive workflow ensures that perovskite materials maintain their functional properties throughout the entire development production line, from basic material synthesis to final wearable device implementation.

Given the time and resource constraints inherent in research projects, stress protocols typically employ standardized accelerated testing methods that mimic real-world scenarios, providing reliable projections of long-term performance [1, 2]. These protocols may include bending tests, washability evaluations, thermal cycling, and humidity exposure, followed by comprehensive re-assessment of all device performance metrics to quantify performance retention and identify specific degradation mechanisms. For example, a smartwatch incorporating perovskite layers with specific optical absorption or electrical conductivity properties must retain these optoelectronic performances, or at minimum, a substantial percentage thereof, after

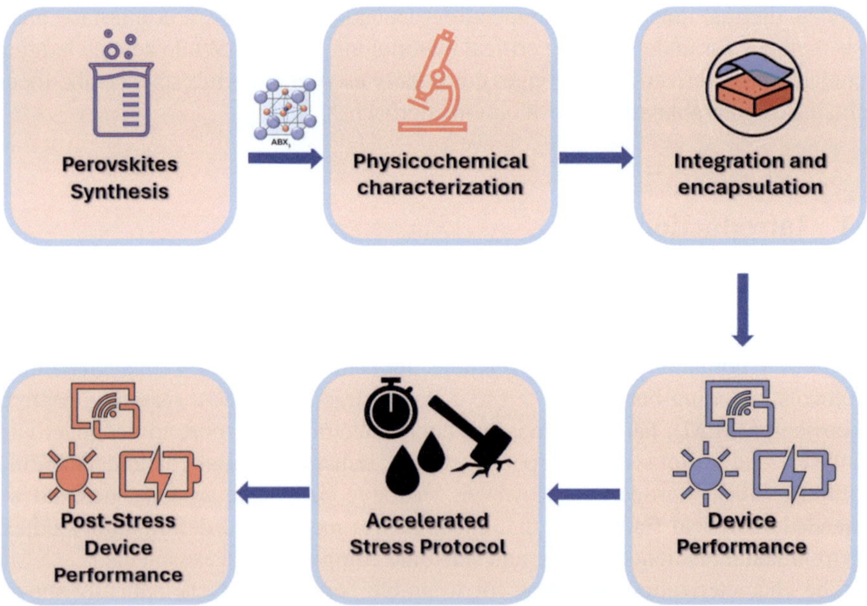


**Fig. 4.1** Schematic characterization workflow for wearable perovskites

undergoing a certain number of bending cycles that simulate actual human wear patterns.

Beyond conventional analyses, wearable perovskite characterization requires enhanced microstructural investigation to ensure optimal flexibility and substrate compatibility. This includes detailed examination of adhesion mechanisms with diverse flexible substrates—polymer films, paper matrices, cellulose composites, and textile fibers—each presenting unique interfacial challenges. Understanding these substrate-perovskite interactions is crucial for developing electrically robust and mechanically stable wearable devices.

The characterization must also address the unique operational demands of body-worn electronics: demonstrating biocompatibility for human contact, validating long-term durability under physiological conditions, and ensuring reliable performance across maintenance scenarios ranging from self-powered operation to complete washability.

This chapter presents an account of the specialized techniques essential for wearable applications of perovskites. The discussion emphasizes how traditional analytical approaches must be adapted and supplemented to provide comprehensive insights into wearability set of traits perovskite systems.

## 4.2 Physicochemical Characterization

### *4.2.1 X-Ray Diffraction (XRD)*

X-ray diffraction is arguably the first structural analysis technique for any material scientist working with ordered crystalline structures, particularly perovskite crystals [3–5]. XRD serves multiple critical functions in perovskite characterization: confirming phase identification and crystal structure determination, estimating crystallite size through peak broadening analysis, assessing overall crystallinity and phase purity, and detecting secondary phases or structural impurities that could compromise device performance.

The connection between XRD studies and wearable applications extends far beyond conventional perovskite characterization, representing a critical analytical foundation that addresses the challenges of flexible electronic devices. In addition to traditional XRD information, wearable applications impose two additional fundamental requirements that can only be evaluated through careful X-ray diffraction studies. First, materials must demonstrate exceptional mechanical resilience to withstand the repetitive stresses that wearable devices inevitably encounter during daily use, including bending, stretching, twisting, and compression cycles that can reach thousands of iterations over the device's lifetime. Second, these materials must maintain their desired electronic and optical properties not only in their pristine state but also after experiencing mechanical deformation, ensuring that device performance remains stable throughout the demanding operational conditions of wearable electronics.

The atomic-level structural information provided by XRD helps to address both requirements through its ability to detect subtle changes in crystal lattice parameters, identify stress-induced phase transitions, monitor the evolution of crystallographic defects, and correlate structural modifications with performance degradation. This comprehensive characterization enables researchers to predict mechanical failure modes, optimize material compositions for enhanced flexibility, and design protective strategies that preserve optoelectronic functionality under mechanical stress.

A critical crystallographic consideration for wearable perovskite devices involves the complex structure–property relationships that govern the trade-off between functional performance and mechanical resilience. Cubic perovskite phases typically exhibit superior optoelectronic properties essential for device functionality, including reduced optical anisotropy, enhanced light absorption efficiency, and improved charge carrier mobility due to their high structural symmetry and optimal band structure characteristics. Conversely, alternative perovskite polymorphs—including tetragonal and orthorhombic phases—often demonstrate enhanced mechanical properties crucial for flexible applications, such as improved mechanical resilience, better adhesion to flexible substrates, and superior tolerance to mechanical deformation.

Successfully balancing this trade-off between optimal optoelectronic performance and mechanical robustness requires careful compositional engineering and processing optimization guided by comprehensive XRD analysis. Through systematic approaches, X-ray diffraction studies provide the fundamental structure–property relationships necessary to develop perovskite-based devices that meet the stringent durability and performance requirements of flexible electronics.

To demonstrate the practical application of these XRD-based design principles, the following analysis examines the temporal stability of $CsPbI_3$ perovskite films reported by Zhao et al. [6] under different environmental and chemical conditions, providing crucial insights into the degradation mechanisms that must be understood and controlled for successful wearable device implementation.

Figure 4.2 presents the systematic evolution of XRD patterns for $CsPbI_3$ perovskite films processed with and without diphenylphosphinic chloride (DPPOCl) additive, comparing their structural stability under air and nitrogen atmospheres across different exposure time intervals.

It can be seen that all samples initially exhibit sharp, well-defined diffraction peaks characteristic of the pure $\gamma$-phase $CsPbI_3$ crystal structure. However, the subsequent evolution dramatically diverges depending on the experimental conditions. The control sample (panel a) shows rapid deterioration of peak sharpness and the emergence of additional peaks indicating the presence of both $\gamma$ and $\delta$ phases within just 1 h. In contrast, the sample with DPPOCl under air (panel b) maintains better structural integrity for extended periods, though eventual degradation becomes apparent. Most remarkably, the sample under nitrogen atmosphere (panel c) demonstrates exceptional stability, with XRD patterns remaining virtually unchanged even after 1000 h of exposure, containing only $\gamma$-phase $CsPbI_3$ crystals.

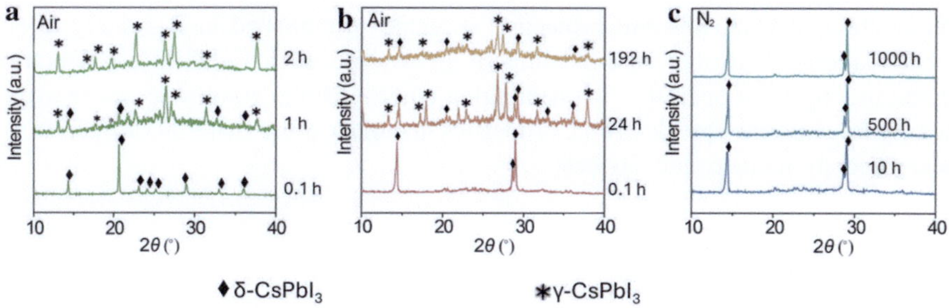


**Fig. 4.2** XRD pattern evolution of $CsPbI_3$ perovskite films over time: **a** without DPPOCl additive under air, **b** with DPPOCl under air exposure, and **c** with DPPOCl under nitrogen protection reproduced from [6] published by Springer Nature. Licensed under Creative Commons Attribution 4.0

The sharp, characteristic peaks observed in the first measurement of each series (0.1 h for panels a and b, 10 h for panel c) indicate high crystallinity and phase purity, confirming that the black-phase perovskite formation is not significantly affected by these processing variables during the initial synthesis stage. The degradation patterns, however, reveal particular kinetic behaviors that highlight the protective mechanisms at play. For the control sample without DPPOCl (panel a), degradation proceeds rapidly with significant structural changes evident within 1 h. The XRD pattern becomes increasingly complex, showing peak broadening, intensity reduction, and the appearance of additional reflections corresponding to the δ-phase $CsPbI_3$ and partially amorphous content. By 2 h, the pattern exhibits severe degradation with substantial loss of crystalline order, indicating nearly complete transformation from the desired black phase to a mixture of yellow δ-phase and amorphous material. This type of structure is typically not favorable for optoelectronic applications and thus the synthesis route leading to such crystallization can be neglected for the fabrication of wearable electronics.

The mechanistic implications of these observations point to multiple degradation pathways that can be systematically addressed through appropriate materials engineering. The rapid degradation in air atmosphere indicates that moisture and oxygen serve as primary catalysts for phase transformation, likely through protonation of the perovskite framework and oxidation of iodide ions. The protective effect of DPPOCl suggests that this phosphorus-containing additive forms coordination bonds with lead centers, creating a more stable local environment that resists structural rearrangement. The nitrogen atmosphere eliminates reactive species from the environment, preventing the initiation of degradation cascades that would otherwise lead to phase transformation.

These findings have profound implications for the practical application of $CsPbI_3$ perovskites in wearable health monitoring devices. The demonstration that stability can be extended from hours to potentially thousands of hours through careful materials design and atmospheric control provides a clear pathway toward commercially viable perovskite-based optoelectronic devices. The systematic understanding of degradation mechanisms and protective strategies established in this study offers a foundation for further optimization of perovskite device stability, potentially enabling long-term operation in challenging environmental conditions while maintaining the superior optoelectronic properties that make these materials attractive for next-generation electronic devices.

#### 4.2.1.1 Wide Angle X-Ray Scattering

While conventional X-ray diffraction provides essential bulk structural information, Grazing Incidence Wide Angle X-ray Scattering (GIWAXS) [7] offers critical complementary insights specifically tailored for thin film perovskite applications in wearable devices. GIWAXS employs a shallow-angle X-ray beam geometry that maximizes signal from the thin film surface while minimizing substrate interference, making it uniquely suited for characterizing the device-relevant perovskite layers typically only hundreds of nanometers thick. Additionally, GIWAXS provides detailed texture and crystallographic orientation analysis, revealing how grain boundaries and preferred orientations affect both mechanical resilience and charge transport properties—critical parameters that conventional XRD cannot adequately assess. Figure 4.3 shows Giwaxs results in Zhao et al.'s work as a complementary analysis to the reports X-ray patterns under different atmospheres and residence time regimes.

As shown in the figure Giwas results show the excellent phase stability for the perovskite films prepared with DPPOCl additive at the humidity of 40% humidity and 25 °C. This is an indication of the mechanical compatibility of the sample synthesized with DPPOCl additive for potential wearable applications.

The technique's compatibility with operando measurements allows researchers to observe degradation processes, phase transitions, and structural evolution under simulated device operating conditions, providing the structural understanding necessary to optimize perovskite films for the demanding mechanical and environmental conditions of wearable electronics.

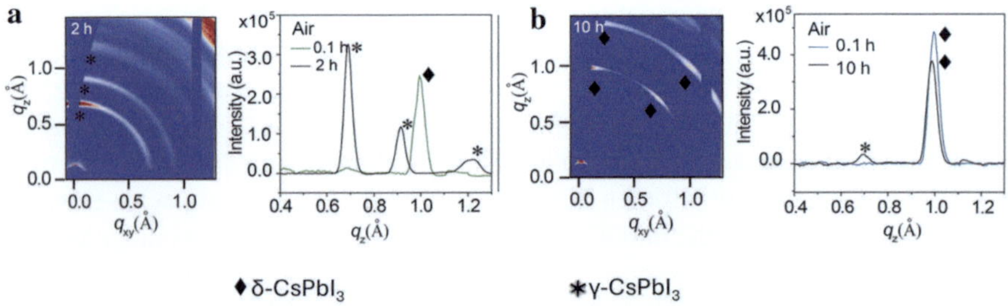


**Fig. 4.3** Giwaxs results of $CsPbI_3$ perovskite films **a** without DPPOCl, and **b** with DPPOCl additive both under air atmosphere, reproduced from [6] published by Springer Nature. Licensed under Creative Commons Attribution 4.0

#### 4.2.1.2 In-Situ XRD

In addition to standard XRD and Giwaxs, In-situ XRD is also a critical analytical tool for developing perovskite materials in wearable applications, used for real-time monitoring of structural changes in perovskite materials under various conditions, such as high pressure and temperature [8, 9] which is essential for understanding their stability and performance in flexible devices. For instance, lead halide perovskites are utilized in wearable ionizing radiation dosimeters integrated into textiles [10]. In-situ XRD has been reported as a tool to monitor the crystallinity and phase purity of these perovskite films during fabrication, ensuring high performance and reliability [11].

### *4.2.2 Scanning Electron Microscopy (SEM)*

SEM provides high-resolution structural and compositional information at the micro- and nanoscale, typically complementing the bulk analysis capabilities of XRD [12].

SEM analysis reveals the surface morphology, grain structure, and film quality of perovskite materials. High-resolution imaging capabilities enable visualization of grain boundaries, surface roughness, and defect structures that directly influence device performance. For wearable applications, SEM is essential for assessing the integrity of perovskite films after mechanical deformation. In the case of $CsPbI_3$ perovskite for wearable health monitoring application reported by Zhao et al., the morphology of the perovskite materials synthesized with and without DPPOCl is visibly presented in the SEM images as shown in Fig. 4.4 [6].

Figure 4.5 demonstrates high quality SEM images of sample with deposition of composite nanofibers, with polyvinylidene fluoride (PVDF) and polyethylene oxide (PEO) prepared with different amount of perovskite precursor after an annealing step. The exceptional image quality enables precise morphological analysis, allowing for accurate measurement of fiber diameter distributions and perovskite particle sizes,

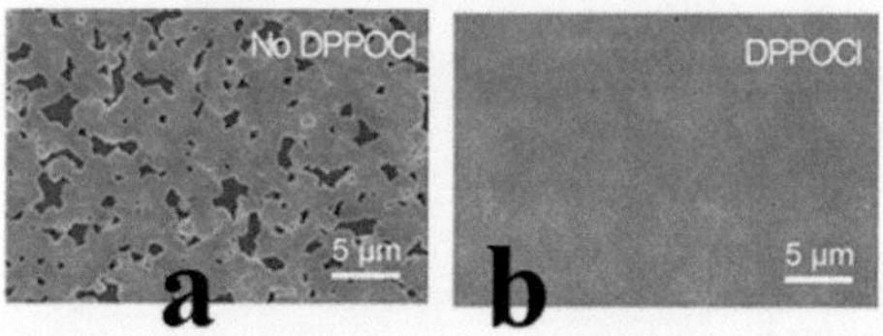


**Fig. 4.4** SEM images of $CsPbI_3$ perovskite films without DPPOCl (**a**) and with DPPOCl additive (**b**), reproduced from [6] published by Springer Nature. Licensed under Creative Commons Attribution 4.0

as quantified in the accompanying histograms in the figure. This level of microscopic detail facilitates comprehensive quantitative characterization, providing critical insights for optimizing perovskite-based materials in wearable device applications. The clear visualization of nanofiber morphology and perovskite particle decoration demonstrates the successful integration of the perovskite phase within the polymer matrix, essential for understanding structure–property relationships in flexible optoelectronic devices.

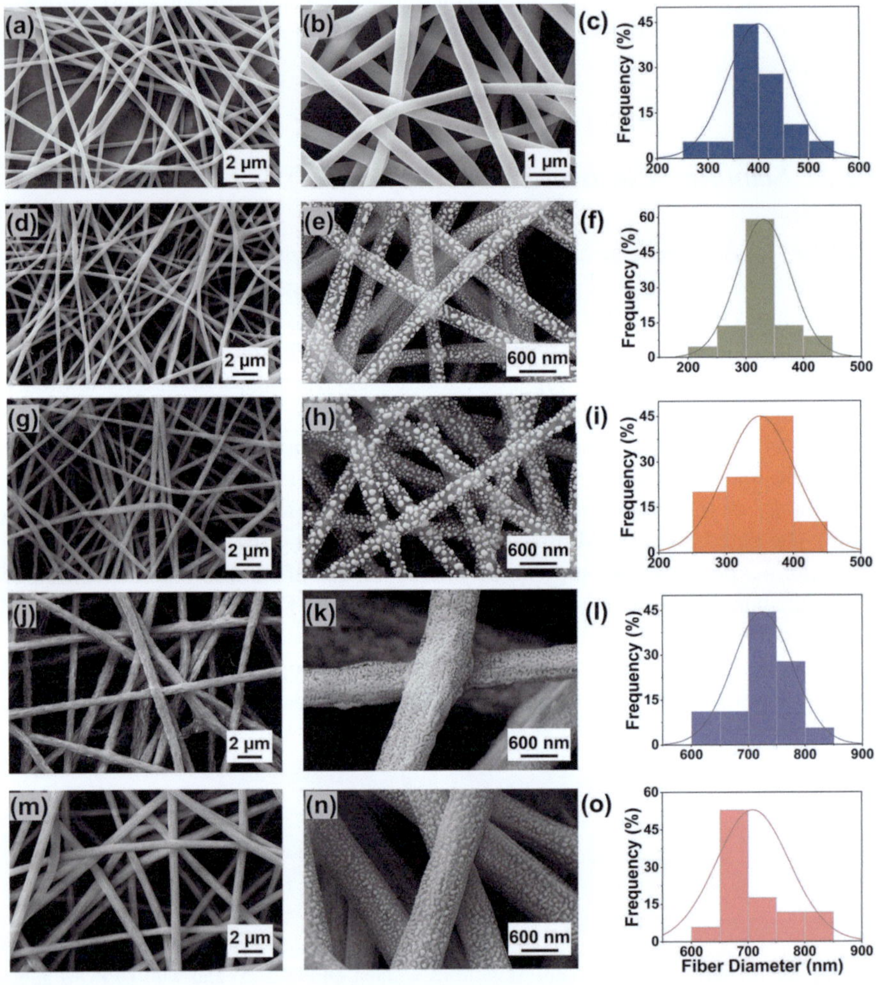


**Fig. 4.5** SEM images and corresponding histograms of diameter distribution of **a–c** pristine PVDF-PEO (mean diameter 399 nm), **d–f** BR_0.2 (mean diameter 331 nm), **g–i** HI_0.2 (mean diameter 358 nm), **j–l** BR_1.5 (mean diameter 725 nm), and **m–o** HI_1.5 (mean diameter 708 nm) nanofibers after annealing. Reproduced from [13] Copyright © 2023 The Authors. Published by American Chemical Society. Licensed under Creative Commons Attribution 4.0

In general, high-resolution, detailed electron microscopy provides concrete evidence of particle configurations within the microstructure, supporting theoretical explanations for the observed mechanical properties. These insights can be linked to specific synthesis parameters or operating conditions, as corroborated by other structural analysis methods, ultimately guiding the design of perovskite materials optimized for targeted wearable applications.

#### 4.2.2.1 Cross-sectional SEM

In addition to standard top-view SEM imaging, which captures surface morphology, cross-sectional SEM has become an increasingly important tool for characterizing perovskite structures. This is especially relevant for applications such as wearables, where perovskite layers need to adhere to other functional layers—typically a flexible substrate. In these multilayer configurations, the interface quality, layer uniformity, and potential delamination or voids become critical factors that influence mechanical integrity and charge transport. Cross-sectional imaging enables direct observation of layer thicknesses, interface continuity, and morphological defects that can compromise performance. Figure 4.6 shows a cross-sectional SEM image of a graded $CsPbI_{3-x}Br_x$ perovskite layer integrated with a polymer substrate, designed for use in energy harvesting applications [14].

### *4.2.3 Transmission Electron Microscopy (TEM)*

TEM offers atomic-resolution imaging capabilities essential for understanding the detailed crystal structure and defect chemistry of perovskite materials in high magnification (such as $CsPbBr_3$ QDs in Fig. 4.7) as well as a clear overview of the detailed surface configurations in low magnification (such as the one shown in

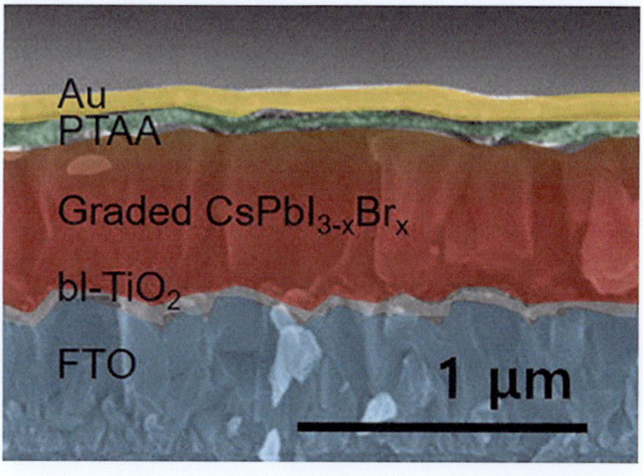


**Fig. 4.6** The SEM cross-sectional view of graded CsPbI3-xBrx reproduced with permission from [15] © 2020 Elsevier

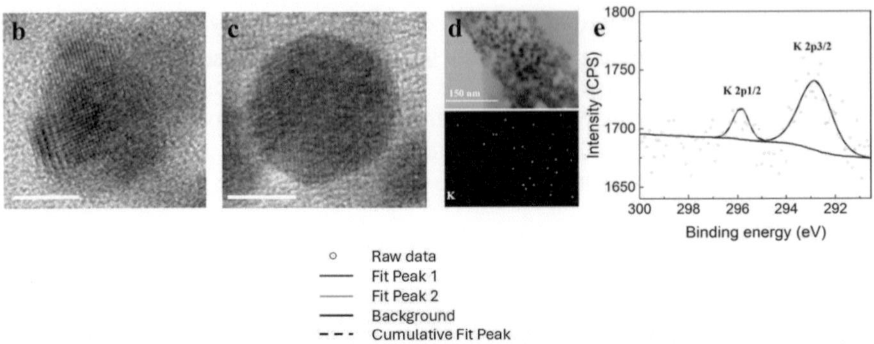


**Fig. 4.7** TEM images (scale bars show 5 nm) of pristine $CsPbBr_3$ QDs (**b**) and 0.025 CsBr/KBr doped $CsPbBr_3$ QDs (**c**). TEM images at low magnification (**d**) and its elemental mapping (K). High-resolution XPS spectra of K 2p core level of CsBr/KBr (**e**) assisted QDs thin film. Reproduced from [17] published by Wiley–VCH GmbH. Licensed under Creative Commons Attribution 4.0

Fig. 4.7d). High-resolution TEM (HRTEM) can directly visualize crystal lattices, grain boundaries, and structural defects such as dislocations or stacking faults.

Nanoscale features of the synthesized perovskites as well as nanostructures such as quantum dots are best viewed with TEM technology for a wide range of applications including wearables [16].

#### 4.2.3.1 Selected Area Electron Diffraction (SAED)

SAED patterns are often an option in TEM systems providing local crystallographic information, enabling identification of crystal phases and orientation relationships at the nanoscale. This capability is particularly important for understanding the formation of secondary phases or the behavior of grain boundaries in polycrystalline perovskite films.

For wearable perovskite devices, SAED is particularly valuable in assessing crystal quality and orientation when perovskites are deposited on flexible substrates. There have been reports of the device uses a flexible quasi-two-dimensional perovskite solar cell module that provides ample power under outdoor and indoor illumination conditions [18] and SAED is used in such studies to ensure that the perovskite maintains its desired crystallographic properties despite substrate flexibility.

#### 4.2.3.2 Scanning Transmission Electron Microscopy (STEM)

STEM combines the high spatial resolution of TEM with analytical capabilities, enabling simultaneous structural and compositional analysis at the atomic scale.

High-angle annular dark-field (HAADF) STEM imaging provides Z-contrast information, allowing direct visualization of different atomic species within the perovskite structure.

The technique is invaluable for studying compositional variations, such as halide segregation or cation redistribution, which can significantly affect the optoelectronic properties of mixed-composition perovskites commonly used in wearable applications.

## 4.3 Optoelectronic Characterization

Optoelectronic characterization is the set of techniques evaluating the photophysical properties of perovskite materials in wearable device applications, where mechanical flexibility must be balanced with optimal electronic performance. The unique operational environment of wearable systems necessitates comprehensive understanding of charge carrier dynamics and optical behavior under both static and stressed conditions.

Key optoelectronic techniques including photoluminescence (PL) spectroscopy [19], UV-visible absorption spectroscopy [20], and external quantum efficiency (EQE) measurements [21] provide quantitative assessment of critical parameters such as optical bandgap, carrier lifetime, diffusion length, and radiative recombination efficiency. These measurements directly correlate with device performance metrics for different applications.

For wearable perovskite systems, optoelectronic characterization must additionally address the impact of substrate curvature, cyclic bending, and environmental factors on the fundamental photophysical processes, ensuring that device functionality is maintained throughout the operational lifetime under realistic wearing conditions.

Photoluminescence (PL) spectroscopy is a powerful technique for probing the optoelectronic properties of perovskite materials [22]. It provides valuable insights into charge carrier recombination dynamics, transport behavior, and the influence of structural or compositional modifications on the electronic structure [23, 24]. Such information is critical for the rational design and optimization of perovskite-based devices, particularly in wearable applications where material performance must be maintained under mechanical deformation and varying operational conditions.

A proper example of this technique can be seen in the optoelectrical characterization of device performance of the fiber-compatible photodetector (FCPD) as reported Chen et al. [19]. The setup for the evaluation of the photoresponse of the FCPD (Fig. 4.8a) is also a typical system for this class of optoelectrical characterization for wearable perovskites: a light source, the coupler between the specimen and the source, a powermeter to accurately measure the power of the light that is incident on the FCPD, a voltage source that controls the bias voltage and a computerized logging and monitoring system.

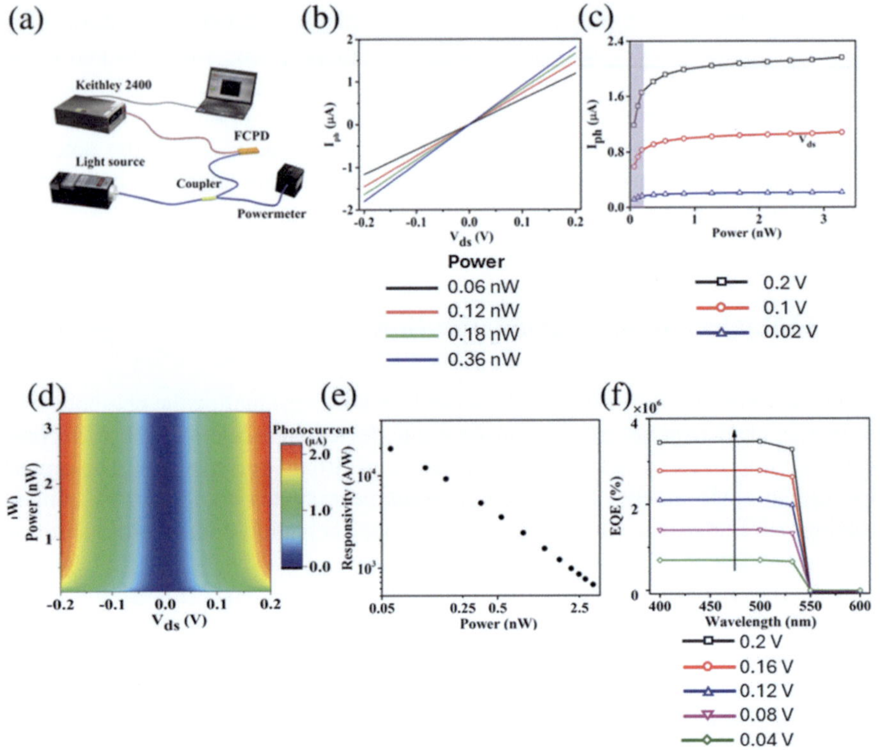


**Fig. 4.8** Electrooptical characterization of fiber-compatible photodetector (FCPD). Schematic of the photoresponse evaluation setup (**a**), Generated photocurrent on incident light power vs. applied bias voltage (**b**), Photocurrent of the FCPD at different light power and applied voltages (**c**), map of photocurrent spectra related to the incident light power and applied voltages (**d**), Responsivity of the FCPD to illuminating light power at a given wavelength and bias (**e**), and EQE of the FCPD as a function of spectral wavelength at different bias voltage. Reproduced from [19] published by Optica Publishing Group. Licensed under Creative Commons Attribution 4.0

In this typical study, the device performance is assessed through systematic photoelectric measurements. The photocurrent response is analyzed as a function of incident light power at a certain wavelength and varying bias voltages, revealing a quasi-linear relationship in a specific illumination range which is critical for reliable signal detection. The linear I–V characteristics in dark conditions further confirm the low leakage and stable baseline of the device (Fig. 4.8b). As an additional electrical test connected to the optical behavior, photocurrent of the device at different light power values and a range of applied voltages is recorded (Fig. 4.8c). The region market with violet in the figure is the quasi-linear light power response for the device. Photocurrent mapping across different optical powers and voltages enables visualization of the device's sensitivity landscape as shown in Fig. 4.8d, while responsivity measurements demonstrates strong performance at low operating voltages (e.g., 0.2 V). Additionally, external quantum efficiency (EQE) spectra collected across varying wavelengths and bias conditions offered insight into the device's spectral selectivity and charge collection efficiency. These comprehensive optoelectronic analyses are essential for validating the practical viability of perovskite-based photodetectors in flexible and wearable platforms, where consistent photoresponse under low power and mechanical constraints is paramount.

The fiber-based graphene-perovskite photodetectors demonstrated exceptional performance characteristics suitable for wearable applications. The devices exhibited high external responsivity across a broad spectral range that aligned well with the perovskite absorption spectrum, operating effectively at low bias voltages. Response times were characterized by relatively fast rise and fall dynamics, making them suitable for dynamic light detection applications. Mechanical robustness was confirmed through extensive bending tests, where the photodetectors maintained their performance through multiple flexing cycles with minimal degradation in photocurrent. The devices also showed good washability when properly encapsulated, retaining most of their photoresponse after standard washing procedures. Optical characterization revealed significant photoluminescence quenching and reduced lifetime in the perovskite layer, confirming efficient charge transfer to the graphene electrode. The combination of high responsivity, adequate response times, mechanical flexibility, washing durability, and low operating voltage requirements makes these photodetectors particularly attractive for integration into wearable electronic systems and textile-based optoelectronic applications.

Another example of such comprehensive optoelectrical characterization can be found in the work of Akhavan et al. [25] for their fiber-based graphene-perovskite photodetectors that demonstrated exceptional performance characteristics suitable for wearable applications. The devices exhibited high external responsivity across a broad spectral range that aligned well with the perovskite absorption spectrum, operating effectively at low bias voltages. Response times were characterized by relatively fast rise and fall dynamics, making them suitable for dynamic light detection applications.

Being concerned with the wearability, mechanical robustness was also confirmed through extensive bending tests in this work [25], where the photodetectors maintained their performance through multiple flexing cycles with minimal degradation in photocurrent. The devices also showed good washability when properly encapsulated, retaining most of their photoresponse after standard washing procedures. These are part of the stress tests mentioned in the beginning of the chapter and also in Fig. 4.1 for the characterization workflow of wearable perovskite, where the optoelectrical characterizations are required to be repeated post-stress protocol as well. Optical characterization revealed significant photoluminescence quenching and reduced lifetime in the perovskite layer for this device, confirming efficient charge transfer to the graphene electrode. The combination of high responsivity, adequate response times, mechanical flexibility, washing durability, and low operating voltage requirements makes these photodetectors particularly attractive for integration into wearable electronic systems and textile-based optoelectronic applications.

## 4.4 Mechanical Testing

Evaluating the mechanical properties of perovskites is essential for predicting their performance and lifespan in flexible and wearable electronics. A range of techniques is employed to probe deformation, strength, and durability under conditions simulating practical use, as shown in Fig. 4.9.

### 4.4.1 Nanoindentation

Nanoindentation is a relatively novel technique for quantitatively assessing the mechanical properties of perovskite materials, particularly relevant for thin films and microstructures employed in next-generation wearable technologies. This method enables the determination of parameters such as hardness, elastic modulus, and time-dependent deformation (viscoelasticity) at micro- and nanoscale resolutions. A key advantage of nanoindentation is its precise spatial resolution, facilitating the characterization of localized mechanical properties across thin films and multilayer heterostructures commonly found in wearable systems. This capability is vital for identifying and mapping mechanical heterogeneity arising from variations in composition, crystallinity, or fabrication processes. Additionally, nanoindentation can be conducted under variable environmental conditions, allowing for assessment of mechanical stability in response to factors such as humidity, temperature fluctuations, and simulated physiological environments, which is particularly important given the environmental sensitivity of many hybrid halide perovskites [26].

Recent studies employing nanoindentation have provided significant insights into the deformation mechanisms, fracture toughness, and viscoelastic behavior of

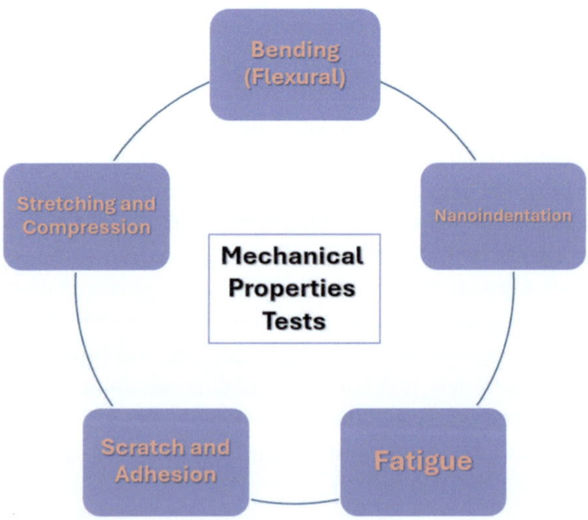


**Fig. 4.9** Mechanical techniques for the characterization of perovskite wearables

perovskite materials. Mechanical properties determined via nanoindentation have been shown to critically influence device performance and operational stability, especially under cyclic loading or bending conditions that mimic service environments for wearable devices. Cyclic or fatigue nanoindentation experiments further enable the assessment of the long-term mechanical durability of perovskite films and layers, facilitating correlations with the projected lifespan and reliability of wearable devices. However, challenges persist in interpreting nanoindentation data for perovskites due to their intrinsic softness, susceptibility to ion migration, and prominent viscoelastic responses. Consequently, advanced data analysis methods and in situ measurement techniques—such as simultaneous mechanical, electrical, or optical probing—are increasingly utilized to gain a more comprehensive understanding of failure mechanisms and the degradation of functional properties under mechanical stress [27].

### 4.4.2 Bending (Flexural) Tests

Bending, or flexural, tests constitute a fundamental approach for assessing the mechanical response of perovskite materials, particularly relevant given their increasing adoption in flexible and wearable devices. Unlike conventional uniaxial tensile or compression tests, flexural tests subject a material to a combination of tensile and compressive stresses, closely simulating the loading conditions encountered during practical use in wearable technology. Flexural testing typically involves applying a controlled bending force, often in a three-point or four-point configuration, to a specimen. This allows for the direct measurement of flexural strength, flexural modulus, and the material's capacity to withstand deformation prior to fracture. For perovskites, commonly deployed as thin films, multilayer stacks, or composite structures, flexural tests provide critical insights into flexibility, ductility, and resilience to repeated bending cycles [28].

The mechanical flexibility of wearable perovskite devices is quantitatively assessed through systematic bending tests that evaluate device performance under various curvature conditions. Figure 4.10 illustrates the experimental setup for bendability measurements, showing the key parameters used to characterize device bending behavior. The bending radius (R) serves as the primary parameter for quantifying the degree of mechanical deformation, with smaller radii representing more severe bending conditions. As demonstrated in Fig. 4.10a, b, devices are subjected to different bending radii (R = 3.5 mm and R = 2.5 mm, respectively) to evaluate their mechanical limits and performance retention under varying stress levels. The optical image in Fig. 4.10c shows the practical implementation of the bending measurement setup, providing visual confirmation of the device's ability to conform to curved surfaces typical of wearable applications. These bending characterizations are essential for determining the operational limits of wearable perovskite devices and ensuring their reliability during real-world use where continuous flexing and deformation are expected.

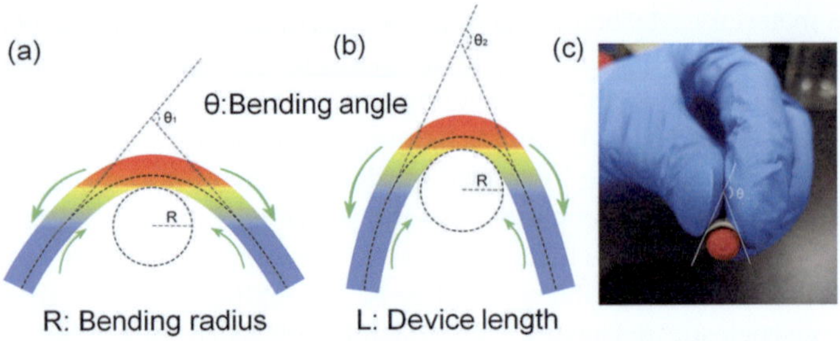


**Fig. 4.10** Schematic representation of the device bending parameters. **a** R = 3.5 mm; **b** R = 2.5 mm; **c** Optical image of the bending measurement. Reproduced from with permission from [13] © 2021 Elsevier

The electrical performance of wearable devices can also be comprehensively evaluated before and after mechanical deformation to assess the impact of bending on device functionality. Figure 4.11 demonstrates one example of this approach through piezoelectric energy generator (PENG) devices, showing output voltage and current measurements under various mechanical configurations and stress conditions. As illustrated in Fig. 4.11a–d, electrical characteristics are monitored during different mechanical manipulations including twisting, knotting, braiding, and fabric interlacing, demonstrating how device geometry and mechanical state directly influence electrical output. This methodology of comparing pre- and post-bending electrical performance provides crucial insights into the mechanical robustness of wearable electronic devices and helps establish operational guidelines for maintaining device functionality under real-world wearing conditions. Such characterization approaches are equally applicable to wearable perovskite devices, where understanding the relationship between mechanical deformation and electrical performance is essential for reliable operation in body-worn applications.

Flexural testing of perovskite-based films, often deposited on flexible substrates like polymers, evaluates their ability to withstand dynamic mechanical strain without initiating cracks or experiencing performance degradation. These tests enable the quantification of key parameters such as the strain-at-break, endurance to cyclic bending (fatigue), and the interfacial adhesion between perovskite layers and their substrates—factors that critically influence device longevity and user comfort [30].

Recent research demonstrates the essential role of flexural testing in understanding failure mechanisms in perovskite layers, including crack initiation, propagation, and delamination. Flexural tests can also be conducted in situ, under variable environmental conditions (e.g., humidity or temperature), or coupled with real-time electrical and optical measurements to directly correlate mechanical deformation with changes in electronic performance. This integrated approach is particularly valuable for hybrid organic–inorganic perovskites, which can exhibit time-dependent or environment-sensitive mechanical behavior. Despite these benefits, challenges remain in preparing standardized perovskite specimens, especially at the thin-film scale, and ensuring consistent test conditions. Moreover, the intrinsic brittleness

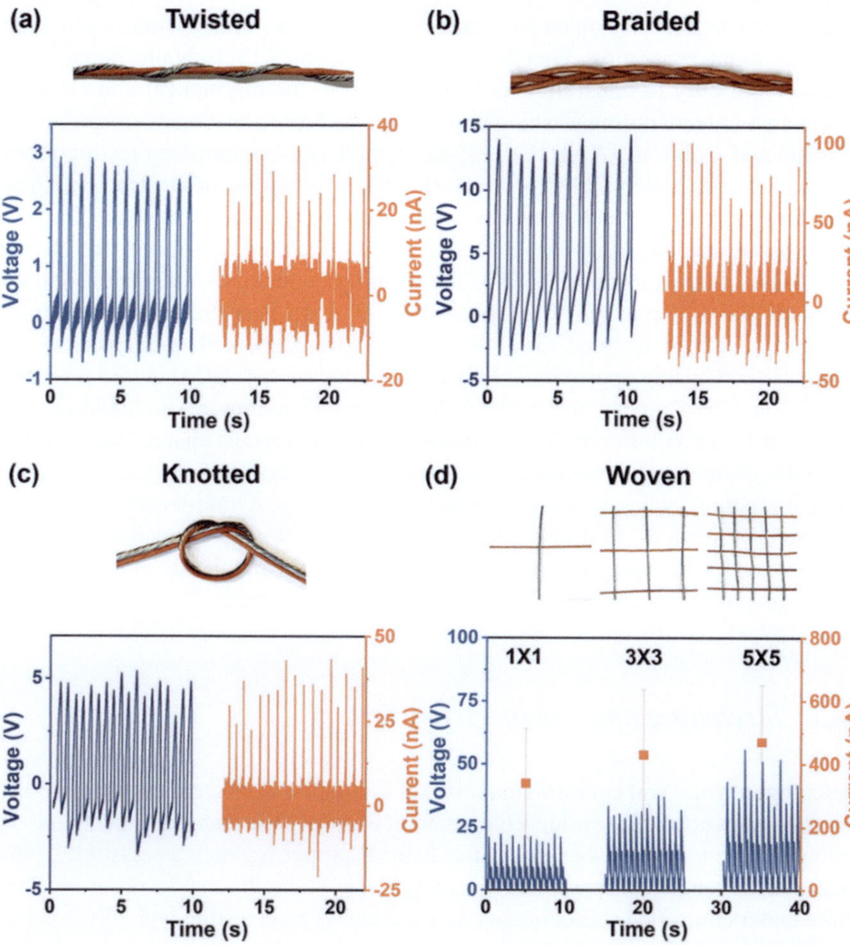


**Fig. 4.11** Generated output voltage and current from **a** a twisted PENG yarn responding to mild finger touching; **b** the response of a knotted piezoelectric yarn and stainless-steel yarn to hand pressing; **c** 3 braided piezoelectric yarns responding to hand pressing; **d** PENG fabrics of different interlacing points responding to actuator pressing with a rectifier. Reproduced from [29] Copyright © 2023 The Authors. Published by American Chemical Society. Licensed under Creative Commons Attribution 4.0

and environmental sensitivity of many perovskites can complicate the interpretation of flexural test data. Continued development of flexible device architectures and improved encapsulation strategies, informed by robust flexural testing, are thus critical for advancing perovskite-based wearable electronics [31].

A less well-known but critically important characterization protocol for wearable materials, including wearable perovskites, is washability testing, which can be interpreted as cyclic bending and deformation under harsh chemical and mechanical conditions. Washability represents a critical requirement for wearable perovskite devices, as these materials must withstand regular cleaning cycles while maintaining their functional properties. Wearable electronic textiles [32] should demonstrate resistance not only to moisture exposure but also to the mechanical and chemical stresses encountered during washing processes. The American Association of Textile Chemists and Colorists (AATCC) [33] has established standardized testing protocols that serve as benchmarks for evaluating the washability of electronic textiles. According to AATCC testing methods, one laboratory cycle is equivalent to five home washing cycles, achieved through accelerated testing conditions that include the addition of steel or rubber balls to the washing container, with specific agitator speeds of 179 $\pm$ 2 rpm and spin speeds of 645 $\pm$ 15 rpm. Standard washability tests are typically performed for a minimum of 20 cycles, with each cycle lasting 45 min at 40 °C using approximately 0.37% volume of AATCC standard reference detergent and 10 stainless steel balls (6 mm diameter) or appropriate load ballast to simulate abrasion conditions. These rigorous testing protocols ensure that wearable perovskite devices can maintain their electrical and mechanical integrity throughout repeated washing cycles, making them suitable for practical long-term use in textile-integrated applications. In studies such as the reported perovskite-based wearable PD by Akhavan [25], this protocol was used as the stress test and the set of performance characterization were performed before and after the washing.

### *4.4.3 Stretching and Compression Tests*

Stretching (tensile) and compression tests are fundamental mechanical characterization techniques utilized to evaluate the behavior of perovskite materials under uniaxial loading. These tests assess essential mechanical properties such as elasticity, plasticity, yield strength, ultimate tensile/compressive strength, and strain-at-break or failure under controlled stretching or compression. Tensile testing typically involves applying a uniaxial force to elongate a sample while measuring its response until fracture or a predetermined strain limit. This allows for the determination of key mechanical parameters crucial for understanding the material's response to tensile stress [34]. Compression testing similarly applies a compressive force to a material to evaluate its resistance to deformation under squeezing [35]. Both tensile and compression data provide critical insights into the material's ability to withstand uniaxial stresses encountered in various device configurations and operational modes within wearable systems.

### *4.4.4 Scratch and Adhesion Tests*

Scratch and adhesion tests are essential for characterizing the surface robustness, cohesion, and interfacial integrity of perovskite materials and their stacks, particularly relevant for wearable technology applications. Wearable devices are subjected to continuous mechanical stress from flexing, abrasion, and contact with skin or clothing, necessitating reliable surface and interfacial properties for long-term durability and function. Scratch tests typically employ a diamond or hard-tipped stylus drawn across the material surface under a progressively increasing load. This method assesses the material's resistance to scratching, providing insight into surface hardness and fracture behavior under localized stress [36]. Adhesion tests, such as tape tests or peel tests, quantify the bonding strength between different layers in a device stack or between the perovskite film and its substrate, which is critical for preventing delamination under mechanical stress or environmental exposure [37].

### *4.4.5 Fatigue Testing*

Fatigue testing represents a crucial method in the mechanical characterization of perovskite materials, particularly as they are increasingly considered for use in next-generation wearable technologies. Fatigue refers to the progressive, localized structural damage that accumulates when a material is subjected to cyclic loading, such as the repeated bending, stretching, or compression experienced during the regular operation of wearable devices [38].

Fatigue testing typically involves subjecting perovskite films, multilayers, or device stacks to repeated mechanical cycles at controlled amplitudes and frequencies. The primary objective is to evaluate the material's resistance to crack initiation, propagation, and eventual failure under conditions closely simulating the stresses encountered in practical wearable applications. Mechanical endurance under cyclic loading is determined by monitoring changes in key material properties—such as modulus, fracture strain, or crack density—over a large number of cycles. Important benchmarks derived from these tests include the number of cycles to failure (fatigue life) and the endurance limit (the maximum cyclic stress the material can withstand over a specified number of cycles without failure). For wearable devices, perovskites must demonstrate mechanical resilience under frequent flexing, stretching, and other repetitive motions imposed by human movement and daily activity. Fatigue testing enables the assessment of long-term mechanical durability and provides direct insight into the functional reliability and expected service life of perovskite-based wearables. Changes in electrical, optical, or piezoelectric performance during or after fatigue cycles are commonly measured concurrently to establish the impact of mechanical degradation on device functionality. Furthermore, fatigue tests can reveal critical aspects of failure mechanisms in perovskite films, such as the onset of microcracks, delamination between layers, and interfacial degradation. The results inform improvements in material formulations (e.g., incorporating flexible polymers), device architecture, and encapsulation strategies aimed at enhancing durability. Fatigue testing of perovskites presents unique challenges due to their material softness, viscoelasticity, and sensitivity to environmental conditions (e.g., humidity, temperature). Specialized test geometries and environmental controls are often required to ensure meaningful results. Moreover, the intrinsic nature of hybrid perovskites, which may undergo ion migration and possess dynamic crystal structures, adds complexity to the interpretation of fatigue data. Recent developments include in situ fatigue experiments, where electrical or optical measurements are performed concurrently with mechanical cycling. These experiments enable real-time monitoring of device degradation and offer a more comprehensive understanding of the relationship between mechanical loading and functional performance loss [39, 40].

## 4.5 Thermal Analysis

Thermal analysis comprises a suite of techniques essential for investigating the thermal stability, phase transition behavior, and degradation pathways of perovskite materials. Given that wearable electronic devices are frequently subjected to fluctuating temperatures resulting from environmental conditions, body heat, or internal power dissipation, a comprehensive understanding of the thermal response of perovskite components is critical for ensuring long-term performance and device longevity [41]. The primary thermal analysis techniques applied to perovskite materials are detailed below.

Thermogravimetric Analysis (TGA): TGA is employed to quantify the mass change of a perovskite sample as a function of temperature or time under a controlled atmosphere. This technique is instrumental in determining thermal decomposition temperatures, assessing the intrinsic compositional stability, and evaluating the impact of device encapsulation or incorporated additives on the thermal degradation profile of perovskite films [42].

Differential Scanning Calorimetry (DSC): DSC measures the heat flow associated with thermal transitions, including crystallization, melting, glass transition phenomena, and solid-state phase transformations in perovskite materials. Characterization of these transitions is pivotal for establishing material processing windows, predicting upper operational temperature limits, and ensuring the requisite phase stability of active perovskite layers throughout the intended service life of wearable devices [43].

Thermomechanical Analysis (TMA): TMA assesses the dimensional changes (e.g., expansion, shrinkage, or warpage) of a material as a function of temperature. This enables the determination of the coefficient of thermal expansion (CTE) for perovskite films or multilayer architectures. The CTE is a particularly relevant parameter for wearable applications, where mechanical flexibility and thermal compatibility with disparate substrate materials are required to prevent interfacial delamination or mechanical failure induced by temperature-driven differential expansion during use [44].

For perovskite materials integrated into wearable devices, maintaining structural integrity, functional performance, and adhesion to flexible substrates despite exposure to temperature fluctuations is paramount [45]. Thermal analysis facilitates:

- Precise determination of suitable operational and storage temperature ranges.

- Identification of critical thermal failure points, such as decomposition temperatures or adverse phase transitions that compromise device functionality.
- Informing strategies for thermal management, optimal device encapsulation design, and the judicious selection of compatible flexible substrates or interlayer materials.

Furthermore, comprehensive thermal characterization is indispensable for identifying and mitigating degradation pathways, particularly those linked to moisture or oxygen ingress, which can be significantly accelerated at elevated temperatures—a critical concern for wearables exposed to sweat and body heat. Hybrid organic–inorganic perovskites present unique challenges for thermal analysis, including often complex multi-step degradation processes, pronounced sensitivity to atmospheric conditions, and the potential emission of volatile byproducts during heating. Consequently, controlled testing environments and careful interpretation of data from multicomponent device systems are frequently required for accurate evaluation. Advanced thermal analysis approaches often involve the integration of real-time functional measurements (e.g., simultaneous monitoring of electrical conductivity or photoluminescence during heating) to directly correlate thermal events with device-level performance changes, thereby yielding practical insights for the design of robust wearable electronics [46].

## 4.6 Interface and Surface Analysis

Interface and surface characteristics represent indispensable aspects in the comprehensive characterization of perovskite materials, particularly concerning their prospective implementation in next-generation wearable technology. The operational efficiency, mechanical resilience, and environmental durability of perovskite-based devices are profoundly influenced by phenomena occurring at surfaces and internal interfaces. Consequently, thorough analytical characterization of these regions is paramount for the rational design and development of robust wearable devices. In the context of wearable applications, perovskite materials are typically integrated as thin films within complex multilayer architectures. The functional performance, mechanical flexibility, and long-term device reliability are critically dependent upon [47]:

- The uniformity, surface roughness, and precise chemical composition of the perovskite surface.
- The quality and stability of interfaces formed between perovskite layers and adjacent functional or protective materials, such as electrodes, encapsulation layers, and flexible polymer substrates.
- The presence and distribution of defects, contaminants, or inhomogeneities at surfaces and interfaces, which can serve as preferential pathways for material degradation or sites for mechanical failure initiation.

Conducting interface and surface analysis on perovskites can present significant challenges due to their intrinsic sensitivity to ambient conditions (e.g., air, moisture, light) and the potential for facile ion migration. To obtain artifact-free and reliable characterization data, stringent environmental controls, such as inert or vacuum atmospheres, and meticulous sample handling procedures are often required. Furthermore, the dynamic, multi-layered nature inherent to wearable device architectures implies that surface and interface properties can evolve substantially during device operation, underscoring the value of in situ or operando analytical approaches. Insights derived from detailed surface and interface analyses are fundamental to guiding material and process optimization strategies. This includes targeted interface engineering, surface passivation techniques, design of effective encapsulation schemes, and the incorporation of adhesion-promoting interlayers. Such advancements directly contribute to enhancing the mechanical resilience, environmental stability, and ultimately, the functional performance of perovskite components, which are pivotal for meeting the rigorous demands of wearable technology applications [48].

Key analytical techniques routinely employed for the characterization of perovskite surfaces and interfaces include the following.

Atomic Force Microscopy (AFM)

AFM provides high-resolution nanoscale topographical mapping of perovskite surfaces. This capability enables quantitative assessment of surface roughness, detailed analysis of grain morphology, and the identification of surface defects such as pinholes or cracks. Advanced AFM modes, such as PeakForce Quantitative Nanomechanical Mapping, can be utilized to assess local mechanical properties, including modulus and adhesion, at the nanoscale. Such localized mechanical information is particularly critical for understanding the behavior of flexible and stretchable perovskite-based devices [49].

Electron Microscopy at Interface

Scanning Electron Microscopy (SEM) offers high-resolution imaging of the surface morphology of perovskite films and devices, providing information on grain size, distribution, and overall film coverage. Transmission Electron Microscopy (TEM), particularly when applied in cross-sectional imaging modes, affords even higher resolution and is essential for visualizing the internal grain structure of perovskite crystals, confirming the continuity and integrity of deposited layers, and assessing the quality of interfaces between different materials within a device stack. Electron microscopy techniques are also invaluable for identifying and characterizing morphological changes induced by mechanical stress, environmental exposure, or device degradation.

Spectrometry at Interface

Spectroscopic techniques are vital for obtaining elemental and chemical information from perovskite surfaces and interfaces as previously explained. X-ray Photoelectron Spectroscopy (XPS) provides surface-sensitive elemental composition and chemical state information, typically probing depths of a few nanometers [50, 51]. XPS is widely used to detect surface impurities, characterize the chemical termination of perovskite layers, and monitor changes in oxidation states or elemental ratios indicative of degradation [52]. Time-of-Flight Secondary Ion Mass Spectrometry (ToF–SIMS) offers high surface sensitivity and the ability to perform depth profiling, allowing for detailed elemental and molecular analysis of surfaces and interfaces, including the tracking of compositional gradients and the identification of degradation products [51]. Furthermore, Ultraviolet Photoelectron Spectroscopy (UPS) is a powerful technique for mapping the electronic structure near the surface, providing crucial information about surface potentials and work functions. This information directly relates interfacial energetics to the efficiency of charge injection or extraction processes in perovskite-based electronic and optoelectronic devices [52]. These interfacial characterization techniques are particularly important for wearable perovskite devices, where understanding surface chemistry and interface stability becomes critical for maintaining device performance under mechanical stress, environmental exposure, and long-term operational conditions typical of body-worn applications.

## 4.7 Standard Characterization Techniques

Electrical transport measurements and comprehensive material characterization techniques represent fundamental analytical approaches widely employed in materials science and semiconductor research. Electrical transport measurements encompass conductivity analysis, carrier mobility determination through Hall effect studies, temperature-dependent transport behavior, impedance spectroscopy, and carrier concentration profiling. Similarly, material characterization methods including Energy Dispersive X-ray Spectroscopy (EDS), X-ray Photoelectron Spectroscopy (XPS), Secondary Ion Mass Spectrometry (SIMS), Nuclear Magnetic Resonance (NMR) Spectroscopy, Fourier Transform Infrared Spectroscopy (FTIR) and Raman Spectroscopy provide comprehensive insights into elemental composition, chemical bonding states, surface chemistry, ion dynamics, and structural properties across various length scales.

For perovskite materials, both electrical transport measurements and standard material characterization techniques follow established protocols and methodologies that are universally applied across different perovskite applications, whether in traditional solar cells, light-emitting devices, or other optoelectronic systems. The fundamental principles of charge transport characterization and analytical techniques remain consistent regardless of the specific device architecture or application target. Given that these characterization methods represent standard procedures for perovskite systems and have been extensively documented in the literature for various perovskite applications, detailed discussion of these techniques is not included in this chapter. Instead, the focus was directed toward characterization aspects that are specifically relevant to wearable applications, such as mechanical flexibility, environmental stability, and integration challenges unique to textile-based and body-worn electronic systems, and the workflow for the characterization of such system including the physicochemical characterization of the synthesized materials, device testing after integration, stress protocol and post-stress performance device testing.

## References


1. Q. Jiang et al., Towards linking lab and field lifetimes of perovskite solar cells. Nature **623**(7986), 313–318 (2023). https://doi.org/10.1038/s41586-023-06610-7
2. Q.C. Burlingame, Y.-L. Loo, E.A. Katz, Accelerated ageing of organic and perovskite photovoltaics. Nat. Energy **8**(12), 1300–1302 (2023). https://doi.org/10.1038/s41560-023-01330-8
3. P. Hosseinzadeh Kouhi, M. Darab, R. Ashiri, Efficient and cost-effective synthesis pathway for obtaining porous La1-xCaxFeO3 material through a modified sol-gel route. Ceram Int. **50**(15), 26938–26948 (2024). https://doi.org/10.1016/j.ceramint.2024.04.425
4. M.S. Toprak, M. Darab, G.E. Syvertsen, M. Muhammed, Synthesis of nanostructured BSCF by oxalate co-precipitation—as potential cathode material for solid oxide fuels cells. Int. J. Hydrog. Energy **35**(17), 9448–9454 (2010). https://doi.org/10.1016/j.ijhydene.2010.03.121
5. H. Sohrabpoor, M. Elyasi, M. Aldosari, N.E. Gorji, Modeling the PbI2 formation in perovskite solar cells using XRD/XPS patterns. Superlattices Microstruct. **97**, 556–561 (2016). https://doi.org/10.1016/j.spmi.2016.07.026
6. Y. Zhao et al., Low-temperature fabrication of stable black-phase CsPbI3 perovskite flexible photodetectors toward wearable health monitoring. Nano Micro Lett. **17**(1) (2025). https://doi.org/10.1007/s40820-024-01565-4
7. J.A. Steele et al., How to GIWAXS: grazing incidence wide angle X-Ray scattering applied to metal halide perovskite thin films. Adv. Energy Mater. **13**(27) (2023). https://doi.org/10.1002/aenm.202300760
8. M. Li et al., Synthesis of edge-shared octahedral $MAPbBr_3$ via Pressure- and temperature-induced multiple-stage processes. Chem. Mater. **35**(3), 1177–1185 (2023). https://doi.org/10.1021/acs.chemmater.2c03172
9. L. Liu, Z. Dong, Characterization of lead halide perovskites using synchrotron x-ray techniques. **303** (2020). https://doi.org/10.1007/978-981-15-6637-0_6
10. L. Possanzini, L. Basiricò, A. Ciavatti, M. Tessarolo, B. Fraboni, Fully textile X-Ray detectors based on fabric-embedded perovskite crystals. Adv. Mater Interfaces **9**(15) (2022). https://doi.org/10.1002/admi.202101417
11. B. Dou et al., Roll-to-roll printing of perovskite solar cells. ACS Energy Lett. **3**(10), 2558–2565 (2018). https://doi.org/10.1021/acsenergylett.8b01556
12. T. Li et al., Rational interface design toward mechanically durable flexible perovskite solar cells. Small (2025). https://doi.org/10.1002/smll.202503109
13. S. Wu et al., Cesium lead halide perovskite decorated polyvinylidene fluoride nanofibers for wearable piezoelectric nanogenerator yarns. ACS Nano **17**(2), 1022–1035 (2023). https://doi.org/10.1021/acsnano.2c07320
14. T.I. Alanazi, Current spray-coating approaches to manufacture perovskite solar cells. Results Phys. **44**, 106144 (2023). https://doi.org/10.1016/j.rinp.2022.106144
15. J.H. Heo et al., Efficient and stable graded CsPbI3−xBrx perovskite solar cells and submodules by orthogonal processable spray coating. Joule **5**(2), 481–494 (2021). https://doi.org/10.1016/j.joule.2020.12.010
16. L. Hao, S. Cai, Progress in *in situ* TEM investigations of halide perovskites. Microstructures **5**(1) (2025). https://doi.org/10.20517/microstructures.2024.10
17. K. Shen et al., Flexible and self-powered photodetector arrays based on all-inorganic $CsPbBr_3$ quantum dots. Adv. Mater. **32**(22) (2020). https://doi.org/10.1002/adma.202000004
18. J. Min et al., An autonomous wearable biosensor powered by a perovskite solar cell. Nat. Electron. **6**(8), 630–641 (2023). https://doi.org/10.1038/s41928-023-00996-y

19. J. Chen, Q. Jing, F. Xu, Z. Lu, Y. Lu, High-sensitivity optical-fiber-compatible photodetector with an integrated CsPbBr_3–graphene hybrid structure. Optica **4**(8), 835 (2017). https://doi.org/10.1364/OPTICA.4.000835
20. V.N. Kuznetsov et al., Absorption edge fine structure in the $PyPbI_3$ perovskite revealed by variable-temperature UV–Vis diffuse reflectance spectroscopy. J. Phys. Chem. C **127**(34), 17085–17095 (2023). https://doi.org/10.1021/acs.jpcc.3c02828
21. L. Kong et al., Efficient and stable hybrid perovskite-organic light-emitting diodes with external quantum efficiency exceeding 40 per cent. Light Sci. Appl. **13**(1), 138 (2024). https://doi.org/10.1038/s41377-024-01500-7
22. N.S. Mahon et al., Photoluminescence kinetics for monitoring photoinduced processes in perovskite solar cells. Sol. Energy **195**, 114–120 (2020). https://doi.org/10.1016/j.solener.2019.11.050
23. M. Ghafarkani, N. Safari, R.T. Ghahrizjani, E. Mohajerani, M. Ameri, Harnessing prefabrication techniques for accurate and efficient optoelectronic characterization of perovskite thin films. Sol. Energy Mater. Sol. Cells **287** (2025). https://doi.org/10.1016/j.solmat.2025.113596
24. T.P.A. van der Pol, K. Datta, M.M. Wienk, R.A.J. Janssen, The intrinsic photoluminescence spectrum of perovskite films. Adv. Opt. Mater. **10**(8) (2022). https://doi.org/10.1002/adom.202102557
25. S. Akhavan et al., Graphene-perovskite fibre photodetectors. Adv. Mater. **36**(35) (2024). https://doi.org/10.1002/adma.202400703
26. S. Sun, Y. Fang, G. Kieslich, T.J. White, A.K. Cheetham, Mechanical properties of organic–inorganic halide perovskites, $CH_3NH_3PbX_3$ (X = I, Br and Cl), by nanoindentation. J. Mater. Chem. A Mater. **3**(36), 18450–18455 (2015). https://doi.org/10.1039/C5TA03331D
27. S. Seong, Y. Liu, X. Gong, Mechanical study of perovskite solar cells: opportunities and challenges for wearable power source. Opt. Mater. Express **12**(2), 772 (2022). https://doi.org/10.1364/OME.450214
28. W. Wang et al., Structural, optical and flexible properties of CH3NH3PbI3 perovskite films deposited on paper substrates. Opt. Mater. (Amst) **114**, 110926 (2021). https://doi.org/10.1016/j.optmat.2021.110926
29. J. Yu, M. Wang, S. Lin, Probing the soft and nanoductile mechanical nature of single and polycrystalline organic-inorganic hybrid perovskites for flexible functional devices. ACS Nano **10**(12), 11044–11057 (2016). https://doi.org/10.1021/acsnano.6b05913
30. M. Zhang, Z. Li, Z. Gong, Z. Li, C. Zhang, Perspectives on the mechanical robustness of flexible perovskite solar cells. Energy Adv. **2**(3), 355–364 (2023). https://doi.org/10.1039/D2YA00303A
31. B. Li, H. Wu, Y. Zheng, Mechanical stability analysis of flexible perovskite solar cells via opto-electro-mechanical simulation. Sol. Energy **287**, 113211 (2025). https://doi.org/10.1016/j.solener.2024.113211
32. S. Rotzler, M. von Krshiwoblozki, M. Schneider-Ramelow, Washability of e-textiles: current testing practices and the need for standardization. Text. Res. J. **91**(19–20), 2401–2417 (2021). https://doi.org/10.1177/0040517521996727
33. Association of Textile Chemists and Colorists (AATCC) AATCC61a, Test Method for Colorfastness to Laundering: Accelerated. (2017)
34. E. Ercan et al., Stretchable and ambient stable perovskite/polymer luminous hybrid nanofibers of multicolor fiber mats and their white LED applications. ACS Appl. Mater. Interfaces **11**(26), 23605–23615 (2019). https://doi.org/10.1021/acsami.9b05527
35. M. Zhang, Y. Qiang, Z. Li, Z. Li, C. Zhang, Mechanism and regulation of tensile-induced degradation of flexible perovskite solar cells. Energy Adv. **3**(6), 1431–1438 (2024). https://doi.org/10.1039/D4YA00086B
36. H. Liang et al., Strain effects on flexible perovskite solar cells. Adv. Sci. **10**(35) (2023). https://doi.org/10.1002/advs.202304733
37. J.C. Sadighian, C.Y. Wong, Just scratching the surface. *In situ* and surface-specific characterization of perovskite nanocrystal growth. J. Phys. Chem. C **125**(38), 20772–20782 (2021). https://doi.org/10.1021/acs.jpcc.1c06582

38. F. Huang et al., Fatigue behavior of planar CH3NH3PbI3 perovskite solar cells revealed by light on/off diurnal cycling. Nano Energy **27**, 509–514 (2016). https://doi.org/10.1016/j.nanoen.2016.07.033
39. M. Chen et al., Stress engineering for mitigating thermal cycling fatigue in perovskite photovoltaics. ACS Energy Lett. **9**(6), 2582–2589 (2024). https://doi.org/10.1021/acsenergylett.4c00988
40. K. Poorkazem, D. Liu, T.L. Kelly, Fatigue resistance of a flexible, efficient, and metal oxide-free perovskite solar cell. J. Mater. Chem. A Mater. **3**(17), 9241–9248 (2015). https://doi.org/10.1039/C5TA00084J
41. M.T. Mbumba, D.M. Malouangou, J.M. Tsiba, L. Bai, Y. Yang, M. Guli, Degradation mechanism and addressing techniques of thermal instability in halide perovskite solar cells. Sol. Energy **230**, 954–978 (2021). https://doi.org/10.1016/j.solener.2021.10.070
42. D. Fan, C. Chen, T. Xiao, Q. Li, Flexible thermochromic foil derived from fine perovskite nanoparticles with excellent energy-saving performance. Int. J. Energy Res. **43**(9), 4546–4553 (2019). https://doi.org/10.1002/er.4582
43. Y. Wang et al., Halide perovskite inducing anomalous nonvolatile polarization in poly(vinylidene fluoride)-based flexible nanocomposites. Nat. Commun. **15**(1), 3943 (2024). https://doi.org/10.1038/s41467-024-48348-4
44. M. Gonzalez et al., Thermo-mechanical analysis of flexible and stretchable systems, in *2010 11th International Thermal, Mechanical & Multi-Physics Simulation, and Experiments in Microelectronics and Microsystems (EuroSimE)* (IEEE, 2010), pp. 1–7. https://doi.org/10.1109/ESIME.2010.5464566
45. Md.H. Miah et al., Key degradation mechanisms of perovskite solar cells and strategies for enhanced stability: issues and prospects. RSC Adv. **15**(1), 628–654 (2025). https://doi.org/10.1039/D4RA07942F
46. A. Mavlonov et al., Thermal stability test on flexible perovskite solar cell modules to estimate activation energy of degradation on temperature. Sol. Energy Mater. Sol. Cells **277**, 113148 (2024). https://doi.org/10.1016/j.solmat.2024.113148
47. L.K. Ono, Y. Qi, Surface and interface aspects of organometal halide perovskite materials and solar cells. J. Phys. Chem. Lett. **7**(22), 4764–4794 (2016). https://doi.org/10.1021/acs.jpclett.6b01951
48. F. Matteocci et al., Interface and composition analysis on perovskite solar cells. ACS Appl. Mater. Interfaces **7**(47), 26176–26183 (2015). https://doi.org/10.1021/acsami.5b08038
49. C. Shakher Pathak, Application of atomic force microscopy in organic and perovskite photovoltaics, in *Recent Developments in Atomic Force Microscopy and Raman Spectroscopy for Materials Characterization* (IntechOpen, 2022). https://doi.org/10.5772/intechopen.98478
50. C.-H. Hou et al., Validated analysis of component distribution inside perovskite solar cells and its utility in unveiling factors of device performance and degradation. ACS Appl. Mater. Interfaces **12**(20), 22730–22740 (2020). https://doi.org/10.1021/acsami.9b22492
51. M. Duan et al., Mechanically stable screen-printed flexible perovskite solar cells via selective self-assembled siloxane coupling agents. npj Flex. Electron. **9**(1), 30 (2025). https://doi.org/10.1038/s41528-025-00407-6
52. X. Li et al., Highly efficient flexible perovskite solar cells with vacuum-assisted low-temperature annealed SnO2 electron transport layer. J. Energy Chem. **67**, 1–7 (2022). https://doi.org/10.1016/j.jechem.2021.09.021

# Chapter 5
# Perovskite-Based Wearable Sensors

The increasing complexity of modern life demands the development of advanced wearable sensors that can monitor food safety, pollution, health, and illness management in real time. Conventional detection methods are frequently insufficient because of their high cost and requirement for specialized equipment. This calls for a switch to innovative sensing technologies. Because perovskite-based wearable sensors leverage the unique properties of perovskite materials—such as their affordability, high performance, and flexibility—to improve wearable electronics, they constitute a significant advancement. These sensors are incorporated into smart textiles and have the potential to work in harmony with the body to track physiological indicators like pulse rate, muscle contraction, and respiration as well as bio-fluids. Additionally, these sensors could detect toxic gases in poisonous environments.

## 5.1 Introduction

Contemporary lifestyles bring forth several challenges, including the need for devices to assess the condition of food safety, environmental contamination, health surveillance, and disease management. In addition, the collecting and interpretation of biochemical and biophysical signals might provide further information about one's health and previously unknown illnesses. Conversely, traditional approaches are not well-suited for real-time or on-site detection owing to their high cost, reliance on sophisticated equipment, need for highly skilled professionals, limited sensitivity and selectivity, instability, and the necessity for sample preparation [1, 2]. The good news is that there is a promising solution for overcoming these obstacles with the advent of innovative sensing techniques including physical, electrochemical and optical sensors. These sensors provide practical, secure, efficient, on-site, and quick real-time detection, offering both qualitative and quantitative information in a straightforward and cost-effective manner [1–3]. A sensor is a device that recognizes and quantifies



M. Darab et al., *Perovskites for Next-Generation Wearable Technology*, Engineering Materials, https://doi.org/10.1007/978-3-032-03931-6_5

the concentration of analytes in a sample by responding to changes in its structural material's characteristics. In an advanced sensor, there is always at least one layer that can detect and react to stimuli, which is positioned between two functional layers. The sensor consists of a transducer, i.e., a readout system, and a detecting system referred to as a receptor. For example, when the analyte undergoes chemical interactions that alter the physical characteristics of the sensor materials, chemosensors are able to transform the chemical information into a measurable signal [4]. Chemical sensing applications also benefit from optical sensors' remarkable sensitivity to environmental changes like temperature, humidity, and analytes, which change the color or fluorescence intensity of their structural material in response to chemical compounds [5].

Three key factors are propelling the market for newly-emerged sensors: miniaturization, wearability, and multifunctionality. These sensors improve sensing and communication, fit different structural forms including those of the human body, are very flexible, light-weight, and readily connected with other devices [6]. Wearable electronics are able to conduct targeted analysis of biofluids detected by sensors as they come into close proximity with the human body. There are already a significant number of such wearable systems in the lab and on the market. Among the prominent gadgets using this technology are smart contact lenses, smart skin, Apple watches, Google Glass, and sport monitors tracking the oxygen content in muscle blood. Fortunately, everything that can be worn, including clothing, footwear, accessories, and other objects, has the ability to transform into wearable electronics. Although wearable devices may not be as accurate as more conventional, high-tech clinical analysis tools, they do provide great benefits in terms of personal care and overall health assessment. Customers' enthusiasm for wearable health and fitness trackers remains strong, despite the fact that the problem of creating robust, intimate, and precise skin interfaces using stretchy components has not yet been solved [7].

Textiles possess a range of exceptional characteristics that make them highly suitable for wearable applications. These include their lightweight nature, soft texture, inherent microstructure, comfort, and ability to allow air to pass through. Individuals in the academic and business communities are particularly interested in e-textiles made of flexible materials. This is because they have the potential to serve as a fabric that is both comfortable and versatile, capable of accommodating individuals of various sizes and shapes. Unlike regular textiles, smart (or intelligent) textiles can sense and react to their surrounding stimuli depending on their mechanical, electrical, magnetic, or thermal properties. It is common practice to classify intelligent fabrics as either passive or active. Applying an external stimulus can change the condition of passive smart textiles. Smart textiles that can actively respond to their environments are equipped with sensors and actuators. The simultaneous use of smart textiles and wearable electronics opens up new possibilities for user interface design. Integrating touch-sensitive materials, interactive displays, or haptic feedback systems into smart textiles can greatly enhance the user experience and expand the range of interaction options [8–10].

Sensors may be grouped based on the signals they are designed to detect as well as the functions they perform. There are two types of sensors that employ different detection techniques: chemical sensors and physical (or mechanical) sensors [11, 12]. However, for sensors to detect physical or chemical values, signal acquisition and signal transduction are generally critical procedures. Signal acquisition utilizes less complicated approaches to establishing a relationship between the target biological parameter and another, often assessed by less complex means (e.g., volume, resistance, and capacity). Signal transduction works on a fundamental basis to transform raw input into useful information. Once a precise relationship between input and output is established, transduction can be achieved through simple calculations [7]. The sensors, which can be stretched or bent, are positioned in various parts of the body, such as the fingers, wrists, elbows, arms, and neck. The devices are specifically engineered to identify and measure various physiological signals, including wrist pulse, muscle movement, human pulse wave, body posture and movement, phonation, and breathing. Physical sensors commonly measure motion, pressure, and temperature. The heart rate monitor is widely demanded in the market, whether it is measured by a strain or an electrophysiology (voltage difference) sensor. Kinematic sensors, which encompass speed, acceleration, and vibration sensors, are gaining significant interest for applications such as virtual reality (VR), augmented reality (AR), and gaming. They are being developed for gait recognition [13], joint angle estimation [14], VR games applications [15], fall risk assessment, activity recognition, and consumer sport application [16]. To monitor blood pressure and other bodily pressures, pressure sensor arrays are now used in both invasive and noninvasive methods. A tactile sensor is another common type of physical sensor capable of detecting data generated by physical contact. The biological sense of cutaneous touch is used by tactile sensors to detect temperature and mechanical stimuli. In modern computer hardware, especially touchscreens and robots, tactile sensors have become widespread [17].

Light, gases, and humidity play crucial roles in monitoring both internal and external conditions. One important area where wearable sensing technology might greatly improve customized healthcare is the detection of ultraviolet radiation, which is a major risk factor for skin cancer. A self-powered, flexible photodetector-based wearable UV sensor can fulfill the requirements for real-time UV monitoring, in contrast to the inflexible and expensive technologies currently used to fabricate UV sensors, which limit their use in cutting-edge electronics [18]. Furthermore, industrial emissions of hazardous gases contribute significantly to climate change, cause a variety of health problems, and are detrimental to all forms of life on Earth. As a result, there has been a significant investment in creating reusable gas sensors that are inexpensive and very sensitive. The basic idea behind gas sensors is that when exposed to the target gas, they undergo changes in their optical, electrical, charge storage, crystal structure, and physicochemical characteristics [19]. Combining neural networks with non-contact, lead-free, humidity-sensitive sensors has recently made it possible to accurately measure labor intensity and monitor breathing in real time. Their design allows air to flow through them and sticks to the skin, which makes them more comfortable for users and provides continuous, accurate data for better labor management [20]. Biofluids, such as saliva, interstitial fluid, and sweat, include substances that may aid medical professionals in diagnosing illnesses and identifying individuals' health problems. Biosensors, a biochemical sensor, can be used to study and detect chemical molecules involved in biological processes in cells, enzymes, tissues, bacteria, and amino acids. Biosensors for DNA, microbes, and metastatic cancer cells are among their many applications in the fields of biotechnology, agriculture, food technology, biomedicine, and health monitoring. A state-of-the-art biosensor that can detect and isolate diseased cells has been developed as a result of recent advances in next-generation medications, including personalized medicine and targeted drug delivery [21, 22]. Biosensors may utilize nanomaterials in their design to enhance sensitivity, accelerate reactions, increase the possibility of reacting at low potentials, and facilitate rapid electron transfer from the active reaction center to the electrode surface (in electrochemical nanosensors). The incorporation of nanomaterials in biosensor construction eliminates the need for chemical intermediates in electron transfer, a critical factor in the development of advanced biosensors. In addition, nanoparticles have the advantage of encouraging biomolecule adsorption because of their unique properties, such as a large surface area and high free surface energy [23, 24]. Nanostructure, stability, toxicity, environmental cost, and quantum yield are some of the criteria that affect the sensory uses of nanomaterials, making it vital to construct nanostructures for specific applications [23].

The "Internet of Medical Things" (IoMT) is one way the Internet of Things (IoT) is changing the landscape of wearable technology in healthcare. The IoMT allows for the continuous monitoring of patients' vital signs via the use of sensors and wearable devices, which improves communication between physicians and patients and ultimately leads to improved treatment outcomes [25]. Watches, bands, contact lenses, and electronic skin patches that look like tattoos are just a few examples of the accessories that researchers are working to integrate with multi-modal wearable electronics. A constant source of energy is necessary for all four main functions of these wearable systems: data collection, communication, processing, and user delivery [26, 27]. Researchers are exploring sustainable, lightweight, flexible, stretchy, and twistable wearable power sources, such as intelligent nanogenerators (NGs) and Energy Autonomous Wearable Sensors (EAWS). These technologies provide reliable bioelectric signals and precise measurements, enabling continuous monitoring for disease prevention and promoting lifelong optimal health, thereby addressing the limitations of conventional wearable electronics, i.e., rigidity and high weight [28, 29]. For example, the thermoelectric sensor is a component that can power wearable technology utilizing the wearer's body [30, 31]. In another circumstance, a photoactive substance could be utilized in the design of optoelectronic sensors, as it possesses the capability to produce charge carriers when exposed to light. However, only perovskite and silicon have shown practical efficiency in converting light into electricity. Researchers are now focusing on perovskite materials for their affordability and high performance in optoelectronics. Organic–inorganic hybrid perovskites have shown potential for developing self-powered wearable devices, which can be manufactured using low-temperature solutions [6]. The emergence of portable perovskite sensors is a major turning point in the field of health and environmental monitoring. These sensors fix the problems with prior technologies by providing easy-to-use, real-time detection that is also cheap. Because perovskite materials make them work better, they can be used for a lot of things, like finding environmental hazards and keeping track of vital signs. As the market for wearable technology grows, it is important to make progress in developing sensors that are autonomous, flexible, and energy-efficient. This is important for the future of personal health care and overall health. Wearable perovskite sensors are making it possible for health monitoring to be a normal part of daily life because they can provide accurate and continuous data. This makes them leaders in the management of healthcare.

## 5.2 Wearable Perovskite Sensor

Perovskites, as mentioned above, are among the most promising materials for wearable sensors due to their defect tolerance and outstanding optoelectronic capabilities [32]. In crystal structure, the standard formula for perovskites is $ABX_3$, where X can be a halogen or an oxygen ion, and A is a large cation that occupies the cubooctahedral position with twelve X ions. In an octahedral configuration, a small cation named B sits alongside the six X anions, Fig. 5.1 [23, 33, 34]. Solar cell, LED, and photodetector manufacturers are increasingly interested in perovskite materials due to their exceptional intrinsic photoelectric properties, such as their high photoluminescence quantum yield, balanced carrier diffusion length, high defect tolerance, and easily adjustable bandgap [32, 34]. Through careful manipulation of their chemical composition, perovskites have the ability to alter their band gaps, enabling them to absorb a wider range of photons. When considering photovoltaic power sources for wearable sensor modules, it is crucial to have the capability to fine-tune the band gap. Additionally, sensors, actuators, and memory devices benefit from the magnetic, superconducting, ferroelectricity, and piezoelectric capabilities shown by some perovskite materials [35]. The stability and luminosity of three-dimensional perovskites, which are excellent for light harvesting and emission, are being improved by researchers investigating lower-dimensional metal halide perovskite derivatives [36]. They require economically and practically feasible fabrication processes involving low temperatures, spin-coating, vapor phase, and/or the roll-to-roll techniques [37]. Improved accessibility and efficiency of health diagnostics could be achieved via the use of perovskite materials in smartphone sensors. These scalable, cost-manageable, and adaptable materials are ideal for mass-produced biosensors, reducing maintenance requirements and monitoring metabolic or enzymatic activities [38].

Non-contact monitoring has become increasingly important in order to address the challenges associated with traditional sensors, which require physical attachment to the body for data collection. When it comes to health and safety, using non-contact photovoltaic devices can significantly reduce the risk of disease transmission. Certain sensors utilize exceptionally sensitive and perceptive 1D perovskite photodetectors to monitor the movements of an athlete in real time, eliminating the need for physical contact. Through the elimination of the need for physical contact with the monitoring

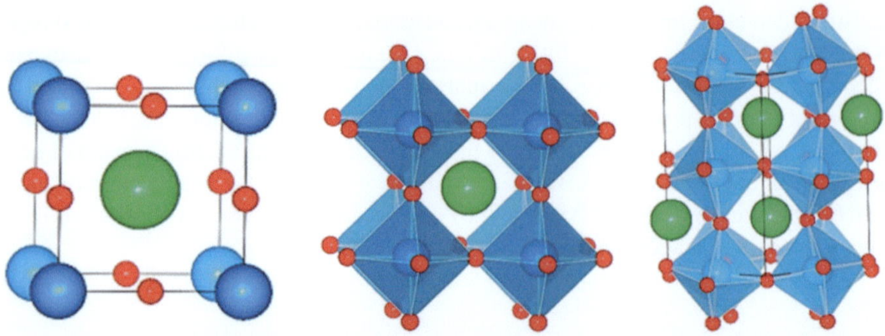

**Fig. 5.1** The ideal cubic perovskite unit cell. **a** The lattice corners are occupied with A cations (blue), the interstitial sites are filled with B cations (green), and the lattice faces are packed with X anions (red). **b** The B-X bonds determine the electrical properties, so another way to look at it is that B cations make $BX_6$ octahedra by coming together around X anions. **c** The BX6 octahedra tilt because of size effects that aren't perfect and other things, which puts stress on the B–X bonds. Reproduced from Ref. [39] published by Nano Today. Licensed under Creative Commons Attribution 4.0

equipment, this approach enhances the user's comfort and overall enjoyment while participating in sports [40, 41]. Wearable photoplethysmography sensors, capable of detecting volumetric changes in blood in peripheral circulation, along with flexible image sensor applications, have been made possible by the integration of perovskite materials into flexible photodetectors. These materials improve the performance of the photodetectors by minimizing dark current, preventing ion migration, and ensuring long-term operational stability [42, 43]. Wearable X-ray detectors have become more accessible with the creation of a flexible X-ray detector that utilizes a biocompatible, metal-free perovskite membrane. This detector demonstrates exceptional sensitivity, minimal detection limits, and the ability to produce X-ray images at reduced dose rates [44]. Additionally, investigations into perovskite-based electrochemical sensors have also yielded encouraging results in the non-enzymatic detection of bio-analytes; these sensors could find use in areas like drug delivery and medical diagnostics where faster and more accurate biomarker quantification is desired [45]. Wearable sensors based on perovskites have the potential to revolutionize healthcare monitoring and human–computer interaction. The key aspects of perovskite-based wearable sensors will be discussed in the following sections.

## 5.3 Perovskite-Based Wearable Photodetectors

Photodetectors are widely used in various fields, but their complex manufacturing process under vacuum conditions hinders their development. Polymer and quantum dot photodetectors offer low-cost solutions and broad detected spectrum, but their low light absorption capacity and poor electric performance hinder practical applications [46]. Different kinds of photodetectors react to different wavelength bands of light, which may be anything from X-rays to ultraviolet (UV) spectrum to visible and infrared light. One sort of photodetector is a narrowband photodetector, while another is a broadband (multiple wavelength bands) photodetector. Photodetectors rely on semiconductor materials, and traditional inorganic semiconductor photodetectors such as Si and GaN have found extensive applications in deep space detection and ultraviolet detection. They demonstrate exceptional performance when it comes to ultraviolet radiation within the 210–362 nm wavelength range, although their capabilities are limited in visible light and near-infrared. Besides, their cost is high because of intricate manufacturing methods and their quantum efficiencies are low [47, 48]. Perovskites, with over 22% energy conversion efficiency, are promising for solar cells, light-emitting diodes, photodetectors, and lasers due to their high quantum efficiency, low dark current, and efficient charge transport. Research in photodetectors focuses on detecting light intensity, but there is growing interest in multifunctional photodetectors that can detect polarization, spectrum, and incidence angle simultaneously. These multifunctional photodetectors have applications in imaging, environmental monitoring, quantum communication, autonomous driving, and robot vision, Fig. 5.2 [49].

Metal-halide perovskites (MHP) have emerged as a promising option for future devices. Perovskite photodetectors show potential for broad spectral response from UV to near-IR. As a result of increased crystal-structure deformation, electron–phonon interaction induces trap states in perovskites. Intrinsic and extrinsic doping both influence conductivity and device function, and controlling these defects and traps may improve device performance [50]. A flexible broadband photodetector employing organolead halide perovskite has enhanced sensitivity throughout the ultraviolet and visible spectrum, exhibiting photo-responsivities of 3.49 and 0.0367 A $W^{-1}$, with external quantum efficiencies of $1.19 \times 10^3\%$ at 365 nm and 5.84% at 780 nm, respectively [51]. Flexible perovskite photodetectors (FPDs) are a great material for use in optical communication, image sensing, health monitoring, and many other areas because they are flexible and easy to move. Self-powered FPDs are a long-lasting and maintenance-free alternative to traditional photodetectors. They use energy sources inside or outside the device, so they don't need batteries, which are hard to recycle and harmful for the environment. These devices will be very important for future smart systems and wearable electronics, such as smart homes, bioimaging, and better sensing platforms [48]. In this section the above-mentioned aspects of wearable perovskite photodetectors will be discussed.

### *5.3.1 Operation Mechanisms of Perovskite Photodetectors*

Photodetectors capture optical signals and convert them into electrical signals instantly. To function properly, photodetectors must produce photoconductive gain through charge-carrier photogeneration and charge transport. Improving perovskite photodetector (PD) performance may be achieved by better understanding these

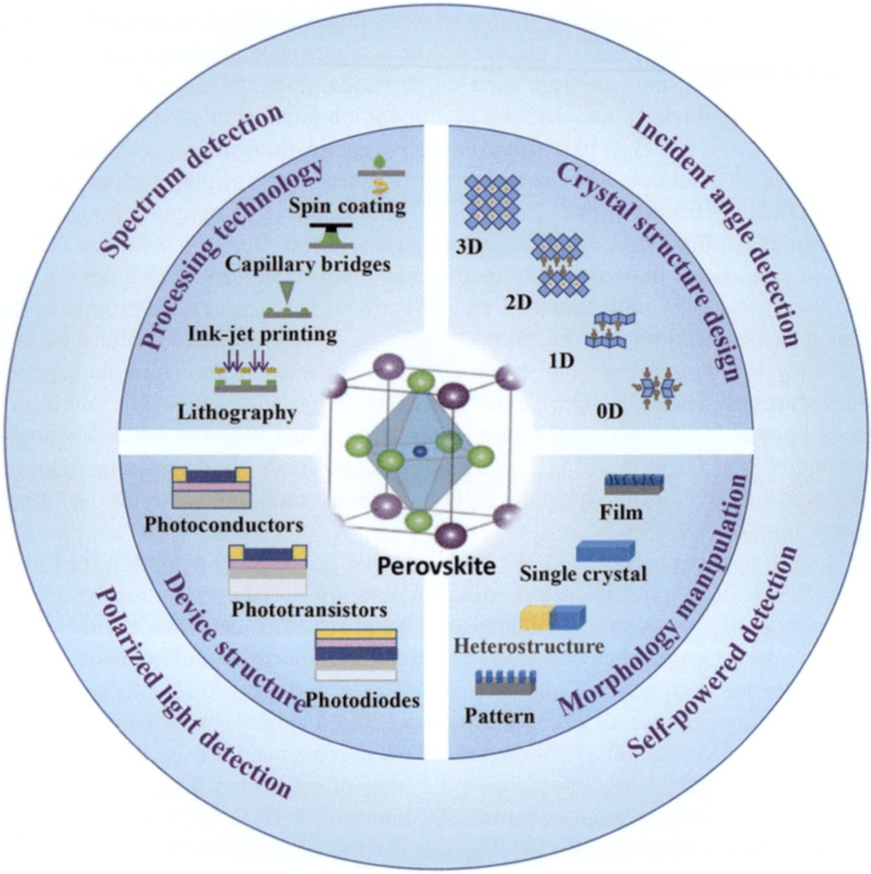


**Fig. 5.2** Comprehensive overview of the progress and obstacles in multifunctional photodetectors using perovskite materials, highlighting its essential contribution to the improvement of optical communication and artificial intelligence applications. The picture delineates the structural benefits, fabrication techniques, and operational mechanisms of these photodetectors, including their capacities to detect intensity, wavelength, polarization, and angle of incident light. Reproduced from Ref. [49] published by Nano-Micro Letters. Licensed under Creative Commons Attribution 4.0

processes and implementing strategies to increase the efficiency of exciton separation and selective charge trapping [52]. External quantum efficiency (EQE), also known as incident photon-to-current efficiency (IPCE), is a parameter used to assess the conversion of photons into charges. It represents the ratio of photo-induced electrons/holes to light photons and is proportional to responsivity. The term "gain" is used when EQE is larger than 100%, while PD lacks gain if EQE is less than 100%. Photoconductive gain is defined as the number of charges flowing through an external circuit per incident photon. High-gain photodetectors enhance sensitivity and are suitable for applications like optical fiber communications, high-resolution imaging, and laser microscopy. They maximize photocurrent spectral responsivity without combining photovoltaic power generation [53, 54]. Photocurrent increases significantly in reverse bias, and photodiode output is linear with illumination intensity. Excessive reverse-bias voltage could make the device work less effectively. Understanding these mechanisms is essential for optimizing devices. PDs' key figure-of-merit parameters include responsivity (R), detectivity (D*), noise equivalent power (NEP), linear dynamic range (LDR), and response speed. LDR is crucial for PDs to have an identical responsivity over a wide range of light intensity [55].

External quantum efficiency and gain play a crucial role in determining device performance. PDs are typically operated under a reverse bias, which generates electrical current signal even in the absence of light. This electrical current generated in the absence of light illumination is termed as the dark current. Dark current significantly impacts photodetectors' performance, affecting detectivity, responsivity, and signal-to-noise ratio. Strategies to reduce it include improving film quality, passivating grain boundaries, and reducing structural defects [56]. Photodetectors' response speeds are affected by a number of important factors, such as the active layer thickness, electrical bias, and interfacial charge transport materials. PD's specific detectivity is affected by their noise equivalent power (NEP) and low dark current under reverse bias. In order to attain high specific detectivity, photodetectors need to be able to detect light intensities that are very low, i.e., equal to the noise equivalent power [57]. Photodetector noise, a fluctuation in output current and voltage signals during photoelectric conversion, affects weak light detection accuracy. It can be classified into external and internal noises, including shot noise, generation-recombination noise, photon noise, thermal noise, $1/f$ noise, temperature noise, and amplifier noise. Shot noise is random generation of photoelectrons, generation-recombination noise results from photogenerated carriers, thermal noise occurs from charge carrier motion, and $1/f$ noise is caused by trace impurities during microspark discharge [58]. Perovskite photodetectors present a significant advancement in the field of high-gain photodetection, showcasing their potential across a wide spectral range from UV to IR light. Their ability to achieve high specific detectivity, coupled with low-dimensional structures that enhance charge-carrier mobility, positions them as ideal candidates for various applications, including environmental monitoring and surveillance. However, to fully realize their capabilities, further advancements in buffer-layer engineering and noise suppression are necessary.

### *5.3.2 Perovskite Photodetector Type*

Perovskite photodetectors are developed based on factors such as crystal structure, operating mechanism, device design, and detection wavelength range. They can be categorized into polycrystalline thin films, single crystals, and low-dimensional nanostructures. PDs can also be classified into visible light, NIR, and broadband types based on the detected wavelength spectrum. PDs are primarily divided into photoconductive, photodiode, and phototransistor structures, Fig. 5.3. Photodiodes are two-terminal devices with vertical and lateral configurations, while phototransistors are three-terminal devices with drain, source, and gate electrodes. Photodiode and photoconductive devices have compact electrode spacing and shorter carrier transport distances, while phototransistor-type PDs require reducing response time for high photocurrent [59, 60].

***Photoconductor***—Perovskite photoconductive devices are characterized by their ability to amplify photocurrent through multiple carrier recirculation processes. These devices use metal electrodes to form a metal–semiconductor-metal (MSM) coplanar structure, enhancing conductivity by producing free charge carriers due to light. The electric field directly impacts the active material, causing charge carriers to move in opposite directions. When photon energy exceeds the band gap width, electrons form an electron–hole pair, collected by metal electrodes. This process continues until the holes reach the other pole, enabling an incident photon to stimulate multiple electrons and generate current. These devices can achieve efficiencies exceeding 100%, but require a balance between gain and response time. Perovskite photoconductors are divided into lateral and vertical types, with lateral photoconductors requiring higher operating voltages due to defects in the film, and vertical photoconductors achieving high gain at lower voltages. Compared to photodiodes, lateral photoconductors with electrodes contacted with photoactive films in symmetrical or asymmetrical Ohmic states can be prepared and integrated easily [61–63]. Perovskite photoconductors offer simplicity, convenience, and large photocurrent, but their large distance between electrodes and metal electrode area results in slow response speed, low sensitivity, and specific detectivity [64].

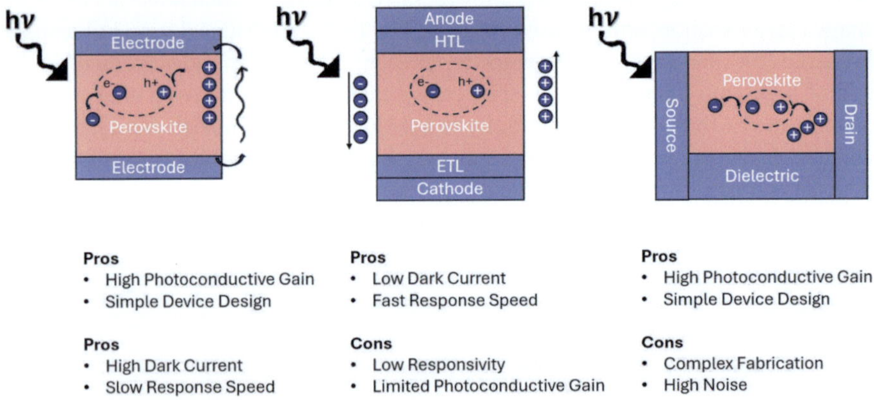


**Fig. 5.3** Exploring the adaptability of perovskite photodetectors: A comparative analysis of device architectures—Photoconductor, Photodiode, and Phototransistor—each possessing distinct advantages and limitations in conjunction with an innovative flexible skin-mounted design exhibiting multifunctional capabilities for advanced applications in optical communication and health monitoring

***Photodiodes***—Photodiodes are photovoltaic devices that generate photocurrent using built-in potentials in junctions. They generate only one electron–hole pair and have built-in electric fields to separate and drift the pairs towards electrodes. Perovskite photodiodes are p-i-n structures with a wide intrinsic semiconductor layer, offering enhanced sensitivity and a larger working area. They have better rectification characteristics, faster response speed, and lower dark current compared to photoconductors. However, they have lower responsivity and photocurrent. The response speed of photodiodes is determined by the time carriers take to traverse the depletion layers, resulting in superior performance. However, they have a low external quantum efficiency (EQE) below 100% due to the absence of charge injections under reverse bias voltages with blocking or rectifying contacts [61, 64–66].

***Phototransistors***—Perovskite phototransistors, mainly Bipolar Junction Transistors (BJTs) and Field-Effect Transistors (FETs), utilize Perovskite to enhance or inhibit charge transfer, achieving external quantum efficiencies above 100%. These three-end structures absorb light to generate photogenerated carriers, simplifying electrical control and synchronization. They adjust carrier transport by applying a gate voltage, creating an electron-conducting channel between the source and drain. This accumulates electrons generated by light, reducing dark current and extending carrier lifetime, making the device more responsive. However, these phototransistors have high material quality requirements, multiple production procedures, and high manufacturing costs, especially when fabricated into large-area devices [61, 64, 65].

### *5.3.3 Processing for Flexible Photodetectors (FDPs)*

Textile-based photodetectors are increasingly being used in various applications, including video imaging, bioinspired sensing, optical communication, and biomedical imaging. The efficiency of wearable PDs depends on factors like substrate morphology and performance retention after repeated bending, stretching, or folding. Low-temperature solution-processing techniques can be used to synthesize provskites, which can be used in making wearable PDs. Flexible photodetectors consist of substrates, electrodes, perovskite films, and encapsulation layers. These thin perovskite films can be fabricated using solution-based or vapor-based methods. Flexible substrates like polymers, carbon cloth, fibers, and paper provide mechanical flexibility, while soft electrodes made of metal networks, carbon materials, and 2D conductive materials offer conductivity and transparency. Polymer/perovskite composite nanofibers have shown potential for stretchable and wearable PDs. Utilizing low-dimensional perovskites and conformal encapsulation are strategies for enhancing performance, as conformal encapsulation layers safeguard the perovskite from moisture and oxygen [67, 68].

***Substrate***—Photodetectors' flexibility is primarily determined by their flexible substrates, which are essential for transparent devices. Common substrates include polymers, carbon cloth, fibers, and paper. Light transmittance and elastic modulus are crucial factors for substrate selection. Polymers like as polyester, polyimide (PI), polyetherimide (PEI), polyethylenenaphthalate (PEN), polyethylene terephthalate (PET), and polyvinylidene fluoride (PVDF) are commonly used. These materials offer flexibility, scalability, lightweight, and low-cost advantages. Elastomers like polydimethylsiloxane, Ecoflex, thermoplastic urethane, and styrene-ethylene-butylenestyrene are also used as substrates for stretchable wearable sensors. Fabrics made of natural or synthetic polymers, such as cotton, silk, Nomex, Kevlar, and nylon, are also used as substrates for wearable sensors. Carbon cloth and fibers offer superior flexibility and temperature tolerance. Fibrous yarn bundles are widely used as templates for fabricating perovskite films with controlled morphology and dimensionality due to their high flexibility and ease of knitting into various patterns. Additionally, silicon, traditionally used in rigid microelectronics, can be used as a substrate for flexible electronics if sufficiently thin. Besides, research revealed that laser-structured polymers can act as powerful, flexible substrates [69, 70].

***Electrode***—Electrodes and interconnectors for flexible wearable electronics are commonly made of solid metals, conducting polymers, carbons, and liquid metals-based conductors. Traditional metallic materials like silver, gold, and aluminum are commonly used as electrodes in photodetectors due to their easy access and compatibility with substrate materials. However, these materials are brittle and cause high Schottky barriers between the perovskite film and electrodes, reducing device performance. To overcome this, stretchable structures have been designed to decorate these materials, such as spiral snake-like, island-bridge, and wrinkled structures. Conducting polymers can also be used, but they are inherently rigid and require modifications to gain flexibility. Carbons and metal nano/micro particles are often dispersed in elastomers to create flexible conductive traces. Liquid metals, when incorporated with polymers or elastomers, are highly conductive and fluidic, making them ideal for high-performance flexible conductors. These materials can be used to create conductive inks for low-cost printing methods [69–71]. These structures are typically fabricated on a planner surface to enhance their stretching ability. Metal-based conductive nanomaterials such as metal nanowires (NWs) are increasingly being utilized as flexible electrodes in Flexible Photovoltaic Devices (FPDs). These nanomaterials combine the high electrical conductivity of metals with the flexibility and ductility of nanomaterials, making them an ideal choice. Moreover, carbon-based conductive materials, and newly emerging conductive materials like 2D Maxens have been studied to balance light transmittance, conductivity, and mechanical flexibility [72–74].

***Conformal Encapsulation***—Conformal encapsulation is crucial for the long-term operation of perovskite-based flexible photodetectors, as it protects the perovskite film from moisture, oxygen, and thermal damage, and addresses lead leakage. Materials used for encapsulation must not damage the underlayer perovskite and have excellent moisture/oxygen resistance, high transparency, and mechanical flexibility. Methods for conformal encapsulation include single-layer and multilayer encapsulation, with polymers and inorganic metal oxides being commonly used. Single-layer encapsulation can cause surface permeation due to defects, while multilayer stacked encapsulation can address issues like edge permeation, intrinsic defects, and film density [59, 69].

***Perovskite-Based Flexible Photoactive Layer***—In addition to the substrates, electrodes, and encapsulation layers, the flexibility of the photoactive films does have a distinct and decisive role in the performance of flexible photodetector. Traditional inorganic photodetectors, such as silicon-based materials, have poor flexibility and complex film-forming processes, making them unsuitable for FPDs. Organic and perovskite materials are now the star materials in FPD. Research on perovskite FPDs is in its early stages, with two main types: polycrystalline perovskite film-based FPDs and low-dimensional perovskite-based FPDs. Polycrystalline perovskite films are widely used in optoelectronic devices due to their simple preparation process and excellent optoelectronic properties. They were first prepared using the drop-coating method and achieved broad-spectrum photodetection from 310 to 780 nm. In a study, $MAPbI_3$ active layer perovskite FPDs were combined with triboelectric nanogenerators (TENGs) to present self-powered FPDs with high detectivity and voltage response [75].

Perovskite film fabrication methods include solution-based and vapor-based techniques. Solution-based methods include spin coating, spray coating, slot-die coating, blade coating, and ink-jet printing. Vapor-based methods include flashing evaporation, co-evaporation, sequential evaporation, and chemical vapor deposition. Spin coating, either one-step or two-step, is primarily used for lab-scale fabrication of wearable and flexible smart optoelectronic devices with small areas. It requires a low temperature post-annealing treatment to increase phase purity and crystallinity of perovskites [76]. Inkjet printing is the most commonly used non-contact method for patterned device fabrication. It uses fine nozzles to release precursor solution droplets, generating high-resolution patterns on various substrates. However, it has disadvantages like discontinuous perovskite films and unavoidable waste of inkjet material [77, 78]. Slot-die coating is a scalable printing method for fabricating perovskite active layers in optoelectronic devices, allowing for large-area deposition with little precursor waste [79]. Spray coating is a technique for producing high-quality thin films over large substrates. It involves dividing a solution into a fine mist, directed towards the substrate, and passing the spray head over the substrate to achieve the desired thickness. The film's morphology closely relies on head speed, droplet volume, solution concentration, and substrate temperature [80, 81]. Blade coating is a cost-effective and efficient method for synthesizing thin films in photovoltaic and optoelectronic devices. It involves placing a sharp blade above the perovskite solution, moving linearly across the substrate, leaving a wet perovskite film [82]. Thermal evaporation is a lab-scale method for creating smooth and uniform perovskite thin films. It converts perovskite precursor materials into the vapor phase through heating in a vacuum chamber, resulting in a uniform perovskite thin film. Flash evaporation and co-evaporation are vacuum-based methods for film fabrication. Chemical vapor deposition is used for synthesizing 2D perovskites, reducing cost and simplifying preparation technology [68].

Perovskite polycrystalline films, known for their high photovoltaic performance, face challenges like large grain boundaries, and performance loss due to cracks under operation (bending, folding, stretching, etc.) or temperature changes. Polycrystalline metal halide perovskites also suffer from high bulk defects and ion migration, affecting their detection limit and long-term operations. Researchers are exploring low-dimensional perovskite structures as a potential replacement for perovskite FPDs, which have shown promising progress in device development [83, 84]. Low-dimensional perovskites mainly consist of 0D particles (e.g., quantum dots and nanocrystals), 1D nanostructures (e.g., nanowires, nanotubes, nanorods), and 2D nanostructures (e.g., nanosheets, nanonets) enhance flexibility and performance of photodetectors [69]. These materials offer advantages like longer carrier diffusion and an adjustable band gap, but they are unstable in the air and affected by temperature and humidity. Researchers are now focusing on these materials due to their high exciton binding energy, enhanced radiative recombination, rich structure, and better stability. Potential research methods include substituting A-site/B-site cations, enhancing electronic coupling, replacing A-site organic cations with polar cations, and improving optical properties [64]. Perovskite devices' functionality is closely linked to their nanometer-scale morphology, which allows for the integration of novel properties into a single material, facilitating the development of ultracompact multi-functional devices. Low-dimensional perovskites possess the advantages of single-crystalline perovskites (low-density bulk defects and grain boundaries) and additionally salient mechanical flexibility, simultaneously. Besides these, low-dimensional perovskites have also shown better stability than polycrystalline ones owing to their phase/environmental stability and suppressed ion migration. The development of processing technology is crucial for perovskite morphology manipulation, which involves precise control of perovskite crystallization to optimize film morphology, including grain size, surface coverage, and thickness. Despite extensive research and development of photodetectors, the large number of defects within polycrystalline films and non-tunable nanoscale morphology hinder device miniaturization, integration, and multi-functionalization [85, 86].

### *5.3.4 Applications of Flexible Photodetectors*

Perovskites-based photodetectors outperform non-perovskite-based ones in responsivity, detectivity, LDR, and response time. Flexible photodetectors are emerging in IoT, sensors, and health applications.

#### 5.3.4.1 Optical Communication

During the last few years, mobile devices and wireless services have witnessed exponential growth, resulting in huge demand for radio frequency-based technologies. The Internet of Things is expected to revolutionize computing, requiring more resources from devices and network infrastructures. Indoor positioning is a significant research area due to its applications in industrial production and daily life. Traditional wireless technologies like Wi Fi and Bluetooth struggle with high-precision indoor positioning due to electromagnetic interference and the multipath effect. Optical communication, divided into visible and ultraviolet light communication, consists of LEDs, photodetectors, and periphery circuits. Modulated white LEDs can transmit location information, and data is converted into a data stream and transferred to a master controller via wired or wireless connection. However, optical communication applications face limitations due to light signal quality loss and ambient light interference. Three-electrode receivers and filters are necessary for signal attenuation and interference reduction [87, 88]. The newly developed asymmetric 2D–3D–2D perovskite photodetector represents a significant advancement in addressing signal loss and interference in free-space communication. Its ability to achieve a frequency-selective photoresponse, coupled with an ultrafast response time of 18 ns, enables reliable transmission of data even in challenging lighting conditions. This innovation not only enhances the performance of photodetectors but also opens a bright horizon for accurate transmission of character and video data under strong light interference, Fig. 5.4 [89]. Another study highlights the promising capabilities of $MAPbI_3$ perovskite photodetectors for optical communications, demonstrating their high performance through innovative structural enhancements and material modifications. The use of Au nanoparticles and n-propylammonium iodide effectively reduces defects and enhances stability, resulting in impressive responsivity, rapid response times, and a significant on/off current ratio. Furthermore, the long-term environmental stability of these PDs under ambient conditions positions them as viable candidates for integration into visible light communication systems. Overall, the findings indicate a strong potential for the advancement of high-speed and reliable metal halide perovskites in optical communications [90].

#### 5.3.4.2 Image Sensing

The imaging system involves an object fixed on an X–Y scanning platform between the laser and the perovskite photodetector, controlled by computer software, moving in horizontal and vertical directions. The photodetector is connected to a phase-locked amplifier, collecting current signals for position and light intensity. Data is then transferred to a computer for image presentation. Image sensors convert optical images into electrical signals using semiconductor materials. Commercial digital sensors, like charge-coupled devices (CCDs) and complementary metal–oxide–semiconductor (CMOS), are based on amorphous or crystalline silicon. However, weak light absorption and poor mechanical flexibility make flexible sensors difficult to use. Perovskite has gained attention for high-performance flexible image sensors due to its excellent optoelectronic properties. Perovskite-based image sensors are arrays with individual photodetectors as pixels, used for high-definition images. There is hope for 3D and curved imaging using flexible sensors developed on complicated substrates. Photodetector arrays were fabricated on polymer substrates and transferred onto various surfaces like glass tubes, wavy objects, and soft lenses [91]. A high-resolution color camera, inspired by human visual cells, uses narrowband red, green, blue, and broadband white perovskite photodetectors to mimic cone cells for color imaging and rod cells for weak-light imaging, producing high-resolution color images up to 256 × 256 pixels, Fig. 5.5 [92]. In another study, electrochemical eye mimicked human retina photoreceptor using perovskite nanowires, showing potential in high-resolution imaging [93].

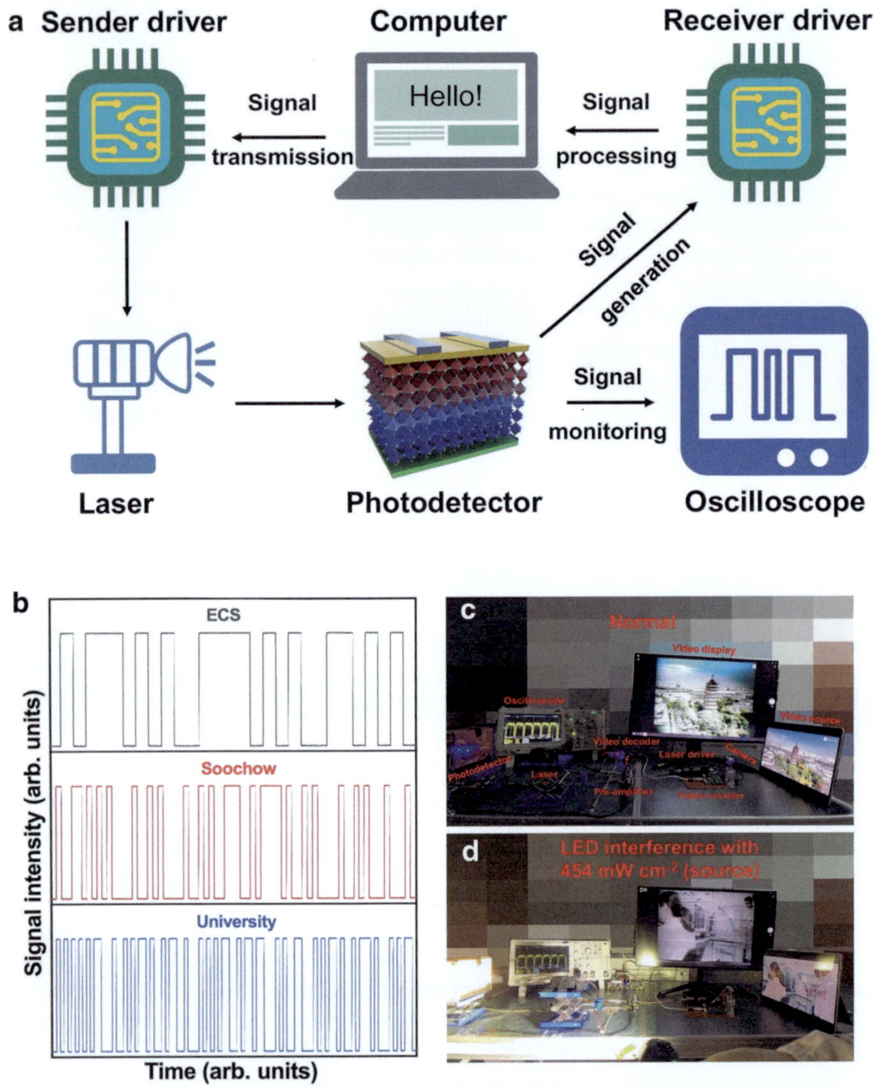


**Fig. 5.4** **a** Scheme of the optical communications infrastructure in free space. **b** During character transmission laser driving signals are used. A photograph of the video data sent in **c** a room with artificial lighting and **d** a scene with strong light interference from LEDs. The phase alternating line (PAL) is used as the video transmission encoding technique, and the photodetector is exposed to a laser intensity of 5.84 mW $cm^{-2}$. Reproduced from Ref. [89] published by Nature Communications. Licensed under Creative Commons Attribution 4.0

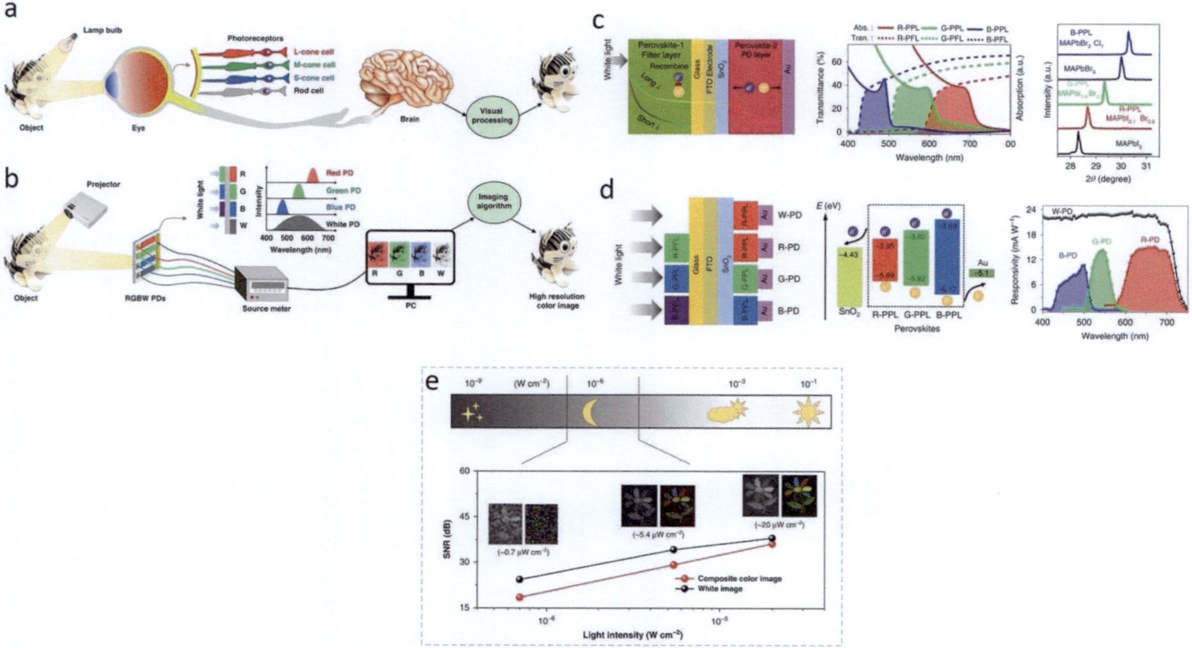


**Fig. 5.5** **a** Diagram of the human visual system illustrating L-, M-, S-cone, and rod photoreceptor cells. **b** The suggested imaging system incorporates four kinds of perovskite photodetectors (RGBW PDs) that function as L, M, S-cone, and rod cells, establishing one-to-one correspondences with **a**. **c** Conceptual illustration of the narrowband perovskite photodetectors featuring perovskite filter layer (PFL) and a perovskite PD layer (PPL), alongside experimental transmittance spectra of PFLs and absorption spectra of PPLs, as well as experimental X-ray diffraction (XRD) patterns on a narrow scale for $MAPbX_3$ (X = Cl, Br, I, or mixed) perovskites to analyze the trend in peak position variation. **d** The device architecture of the perovskite-based RGBW photodetectors (conceptual picture), depiction of the energy band of the RGB PPLs, and experimental wavelength-dependent responsivity curves of the RGBW photodetectors. **e** A comprehensive display summarizing the imaging exhibition of perovskite-based color camera across varying light intensities. Reproduced from Ref. [92] published by Light: Science & Applications. Licensed under Creative Commons Attribution 4.0

#### 5.3.4.3 Health Monitoring

Perovskite photodetectors can detect human health conditions on wearable electronic devices, particularly on the skin surface. The tissue penetration depth of photons in the NIR-I (700–1000 nm) and NIR-II (1000–1800 nm) spectral ranges can reach 1–20 mm, allowing photodetectors to be responsive to biomolecules and fluorescent information on tissue structure. Miniaturizing medical systems to be portable or wearable is a growing goal. Active and passive light detection are both important mechanisms of light signal receiving of flexible photodetectors. Active light detecting models measure heart rate, blood oxygenation, and diagnose tissue or joint [64, 69]. For advanced healthcare systems in a variety of settings, like space stations and oceans, the development of remote wearable health monitoring technology that uses flexible photodiodes is very important. Flexible photodiodes make these systems much better. Traditional inorganic photodetectors are limited by their rigidity and cost, while mixed Sn–Pb perovskite PPDs have struggled to find a balance between external quantum efficiency (EQE) and detectivity in the near-infrared (NIR) spectrum. Recent advances have led to the development of a functional passivating antioxidant (FPA) strategy that makes MSP perovskite films more stable, faster to respond, and better at crystallizing. This FPA strategy has made it possible to make a solution-processed mixed Sn–Pb PPD that works very well, with an EQE of 75.4% at 840 nm and a detectivity of $1.8 \times 10^{12}$ Jones, as well as very fast response times, Fig. 5.6 [94]. Scientists have developed waterproof wearable health monitors with high-sensitivity, detectable photodetectors based on layered perovskites, a novel class of two-dimensional (2D) layered materials, coated with natural beeswax. This innovative technology reduces the need for multiple components in electronics and increases device lifespan. Beeswax's hydrophobic nature, biocompatibility, and self-cleaning properties make it suitable for healthcare applications like photoplethysmography and heartbeat monitoring, contributing to sustainable electronics [95]. Photoplethysmography (PPG) is a low-cost, noninvasive method used to measure heart rate by detecting scattered light signals from blood volume. It is increasingly used for diagnosing cardiovascular diseases like cardiomyopathy, heart failure, and arrhythmia. However, complex analysis requires understanding noise sources (including individual variations, physiology, and external factors), which can limit PPG's application [96]. Scientists are suggesting an electrical field modulation technique for several perovskite compositions ($MAPbI_3$, $MAPbBr_3$, and $FA_{0.92}Cs_{0.04}MA_{0.04}PbI_3$) in order to enhance the performance of these photodetectors. This approach minimizes signal drift, guarantees stability over a prolonged period, and lessens dark current [43]. A reliable evaluation of heart rate and blood oxygen saturation may be achieved using the near-infrared photodetector in a pulse oximetry device. A research produced triple cation Sn–Pb perovskite films of excellent quality with a low bandgap and an extended light absorption edge of 961 nm. The study used double-side interfacial modification with polymer to create a near-infrared photodetector with enhanced photoresponse and ultra-fast response speed, providing accurate, non-invasive assessment in pulse oximetry systems [97]. A bilayer perovskite-CdSe quantum dot diode is produced for voltage-tunable green/red emission along with the photodetection. It can simultaneously emit and detect light, acting as a photoconductor when a positive bias exceeds the built-in voltage. This multifunctional diode is used in a reflective pulse oximeter system, simplifying system design and enabling future miniaturized pulse oximetry [98].

Since Wilhelm Röntgen discovered X-rays, they have been used in various applications like crystallography, medical inspection, space exploration, railway flaw detection, security screening, industrial product testing, and food inspection. X-rays are ionized by objects, producing output signals that are converted into digital images by thin-film photodetectors. X-ray detectors can be direct or indirect, with direct detectors used for soft detection like mammography. An indirect X-ray detector, scintillator, is a device that converts high-energy photons like X-rays and gamma rays into low-energy ultraviolet or visible light, which are detected by photoelectric diodes and further converted into recognizable signals. Indirect detectors, like perovskite materials, are cheaper and more stable but require high sensitivity, resolution, and fast scan rates to minimize radiation exposure [99–101]. Direct X-ray detection requires materials with large stopping power ability, high mobility-lifetime, and large resistance. Current semiconductor films for direct detection include silicon (Si), amorphous selenium (a-Se), mercury (II) iodide ($HgI_2$), and cadmium zinc telluride (CZT). However, these materials have limitations such as small stopping power and limited detection efficiency due to their small atomic number. a-Se, the most mature semiconductor in X-ray detectors, has a 10-time-larger stopping power but lower mobility-lifetime products. Applied bias voltage can increase mobility but sacrifices detectable sensitivity due to noise. $HgI_2$ is toxic and has severe leakage current, while CZT requires high temperature for preparation. To overcome these issues, it is urgent to develop novel semiconductors for direct detectors [100]. Halide perovskites, with formulas $ABX_3$ or $A_2M_IM_{III}X_6$, have relatively high atomic number elements, moderate density, and tunable band gap, which are superior to conventional materials like Si. Halide perovskites are predicted to be a promising alternative for X-ray detectors, leading to advanced work [100]. While portable X-ray imaging is essential for early active detection and treatment, there are still obstacles to overcome in terms of equipment downsizing, imagine visualization, and the interaction of external light signals. A lead-free perovskite scintillator ($Cs_2ZrCl_6$: $Lu^{3+}$) has been utilized for portable X-ray imaging under sunlight, demonstrating its feasibility. The ultrabright radioluminescence (347 lm $m^{-2}$) is attributed to the activation of triplet excitons (dark state) of the scintillator, which is unresponsive to UV and visible photons. The brightening of dark-state excitons is accelerated through lanthanide ion manipulation. These results emphasize the importance of harvesting and engineering dark-state excitons, urging future research in low-cost X-ray detection for portable, fast-response radiography [102]. A self-powered flexible all-perovskite X-ray detector has been developed, combining $CsPbBr_3$ perovskite nanocrystals as X-ray scintillator with $CH_3NH_3PbI_3$ perovskite photodetector for high sensitivity and fast radiation-to-current response time [99]. Besides, organic semiconductor photodetectors have gained attention due to their low temperature processing capabilities, flexibility, and low cost. However, their low X-ray attenuation coefficient still hinders practical application. A new organic–inorganic hybrid strategy is proposed to improve the

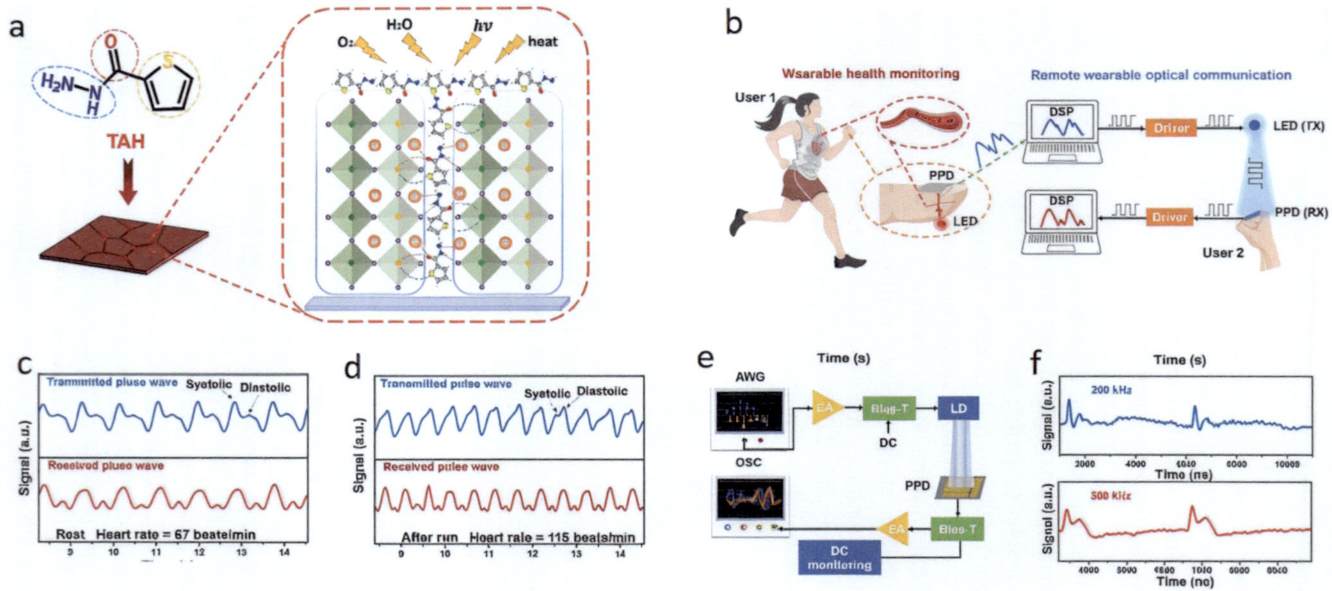


**Fig. 5.6** Demonstration of self-powered flexible mixed Sn–Pb (FMSP) perovskite photodetectors (PPDs) for the combination of wearable biomedicine and optical communications. **a** The chemical structure of thiophene-2-carbohydrazide (TAH)-passivated mixed Sn–Pb perovskite. **b** Schematic representation of the use of FMSP PPD in wearable remote health monitoring. (TX: transmitter; RX: receiver). **c, d** The pulse signal obtained from the FMSP PPD (transmitted pulse signal) and the received pulse signal via optical communication under both resting and post-exercise situations. **e** The configurations of high-frequency optical communication. **f** Transient light response of FMSP PPD at light modulation frequencies of 200 and 300 kHz without additional bias. Reproduced from Ref. [94] published by Advanced Science. Licensed under Creative Commons Attribution 4.0

X-ray sensitivity of organic photodetectors (OPDs). An X-ray-sensitive hybrid OPD is created by embedding $CsPbBr_3$ quantum dots into a P3HT:PC61BM photodiode, enabling indirect X-ray detection. The optimum OPD formula achieves a sensitivity of 229.6 e $nGy^{-1}$ $mm^{-2}$ and a dark current of 23.3 pA $cm^{-2}$ at a low operating voltage for 20–80 kV "soft" X-rays, indicating great potential for next-generation low-cost, portable, and highly sensitive X-ray detectors [103].

Ultraviolet light monitoring systems can achieve anti-radiation healthcare by continuously monitoring the surrounding ultraviolet light wearable UV wristbands are being developed using a self-powered flexible perovskite photodetector for real-time UV monitoring. The device, made of $CH_3NH_3PbCl_3$ film, offers excellent UV detection performance, mechanical robustness, and operational stability. It can detect indoor and outdoor UV radiation, display environmental information on a smartphone, and prevent prolonged exposure, enabling round-the-clock monitoring and personalized health applications [18]. A $CsPbCl_3$ film-based tunneling UV sensor (CTUS) has been developed for personal healthcare applications. The CTUS uses a photo-induced tunneling mechanism to detect UV light, a major cause of skin-related cancers. It achieves high sensitivity and portability, with a low dark current of 100 pA, high responsivity of 285 mA $W^{-1}$, and a broad linear dynamic range of 95 dB. The CTUS is combined with a self-designed UV wristband, allowing real-time monitoring of sun radiation to prevent diseases [104].

#### 5.3.4.4 Self-powered Photodetector (SPPD)

Self-powered photodetectors are potential, sustainable, and wireless nanodevices that convert light energy into electric energy through solar cells, friction electrification, piezoelectric effect, bioenergy, and thermoelectric effect. They can be classified into three types: triboelectric generator (TENG), piezoelectric NG (PENG), and pyroelectric NG (PYENG). Despite challenges in sensor device integration, further development of NG-based self-powered sensors is attractive. Two design routes for self-powered photodiodes include integrating them with miniaturized energy-harvesting/storage components like solar cells, supercapacitors, and nanogenerators, or developing self-powered photodiodes based on the photovoltaic effect. In another word, self-powered photodetectors include integrated systems and PV effect-based PDs (Schottky junction and heterojunction). These devices operate without external bias, using built-in electric fields to separate photogenerated electron–hole pairs. They offer rapid response time, high photosensitivity, and low-cost fabrication, making them suitable for future applications. The configuration of photodetectors is crucial for their detection capabilities. Perovskite-based self-powered photodetectors have been developed, but there are still challenges to overcome in transitioning from laboratory to industrial mass production and application. Optimizing intrinsic properties and device fabrication schemes can improve performance, including maintaining stable photocurrent and dark current under standard conditions [48, 67, 105]. A self-powered, flexible, and omnidirectional perovskite photodetector is developed using $CH_3NH_3PBI_3$ perovskite. This PD can operate with irregular motion, such as human finger tapping, without a bulky external power source. The PD has an impressive detectivity of $1.22 \times 10^{13}$ Jones and a large responsivity of up to 79.4 V $mW^{-1}$ $cm^{-2}$ in the self-powered voltage detection mode [106]. Self-powered, stable, and nontoxic bismuth halide perovskites show more promise for commercialization. Researchers developed photodetectors based on full-inorganic $Cs_3Bi_2I_{9-x}Br_x$ perovskites and modulated their performance by varying x in the composition. The $Cs_3Bi_2I_6Br_3$-based photodetectors showed the best performance, with excellent photosensitivity, responsivity, and detectivity. The full-inorganic perovskite photodetectors also showed excellent stability in the ambient environment, maintaining over 96% of their initial value after 100 days [107]. A novel self-powered flexible fiber-shaped photodetector is designed and fabricated using double-twisted perovskite–$TiO_2$–carbon fiber and CuO–$Cu_2O$–Cu wire. The device shows an ultrahigh detectivity of $2.15 \times 10^{13}$ Jones under 800 nm light at zero bias. The CuO–$Cu_2O$ electron block bilayer extends the response range of perovskite from 850 to 1050 nm and suppresses dark current down to $10^{-11}$ A. The fast response speed of less than 200 ms is nearly invariable after dozens of cycles of bending at the extremely 90 bending angle, demonstrating excellent flexibility and bending stability. These results suggest a promising strategy for designing photodetectors with integrated functions of self-power, flexibility, and broadband response [108]. In a study, researchers present a flexible and self-powered photodetectors demonstrating partial discharge detection in a practical environment. The device uses poly (vinylidene fluoride) (PVDF) as a transparent substrate with anti-humidity properties. A detector using mixed-halide perovskite $(FAPbI_3)_{1-x}(MAPbBr_3)_x$ as the photoactive material is constructed vertically on the PVDF substrate. The fabricated device shows fast response speed and high detectivity under 254 nm UV illumination. The device also exhibits superior mechanical flexibility at various bending angles. The device's air-exposure stability and reproducibility show almost the original performance after 6 weeks of storage. The study also demonstrates a sensitive detection for UVC leakage from germicidal lamps and simulates a partial discharge system without energy consumption. This new approach may be useful for detecting partial discharge and other practical applications [109].

As mentioned here, perovskite photodetectors have shown promise in various configurations, including photovoltaic, photoconductor, and phototransistor setups. However, there are still challenges that need to be addressed in order to fully utilize the potential of perovskite materials in optoelectronic devices. These challenges include understanding and mitigating ion migration, improving long-term stability, and addressing defects. Synthetic chemistry and interface engineering can play a role in enhancing stability, while improving crystal growth and passivating defects can improve device performance. Additionally, the development of perovskite photodetectors with different response wavelengths and understanding light-induced degradation mechanisms are important for future advancements. Overall, perovskite materials have the potential to be used in a range of optoelectronic devices, but further research is needed to overcome the current challenges and improve their performance and stability.

## 5.4 Electrophysiological Perovskite-Based Wearable Sensor

Electrophysiology is a scientific investigation of the surface-detectable electrical potentials generated by living organisms. According to the nature of generation, this area could be categorized into electrocardiograms (ECGs), electroencephalograms (EEGs), electromyograms (EMGs), and electroretinograms (ERGs). The measurement of these surface potentials depends on factors such as amplitude, spatial resolution, and frequency bandwidth. Advances in telehealth and telerehabilitation have the potential to greatly benefit the field of electrophysiology, particularly in the context of neurological disorders. Wireless communication technologies are being used to acquire data and interface with remote systems, making monitoring of electrophysiological signals more convenient and accessible. Electrophysiological measurements have a wide range of applications in healthcare. They can be used for monitoring chronic conditions, aiding in the rehabilitation of patients post-trauma, and facilitating assistive technology for individuals with disabilities. For example, surface EMG can provide valuable clinical information for neuromuscular conditions and track changes in disease progression. In rehabilitation, surface EMG can detect isometric muscle signals and promote correct patterns of activity. In the field of assistive technology, residual muscle signals can be used to control prostheses and facilitate human–computer interaction [110–112]. Wearable electrocardiogram systems are innovative tools for tracking physical activity and health metrics, particularly for cardiovascular disease management, facilitating early identification and intervention for heart issues. Current systems have drawbacks such as cumbersome wires, skin irritation, high production costs, and security concerns. Wearable sensing devices aim to detect biological signals continuously while being attached or implanted. Wireless technologies are crucial for remote power and data communication. There are two approaches to developing wireless sensor systems: integrating wireless circuits into conventional sensors or using sensors that measure radiofrequency [113]. The implementation of wearable ECG sensors involves electrodes and signal conditioning circuits. Skin electrodes measure electrophysiological graphs by making direct contact with the skin surface, measuring tiny potential differences generated by the electrical activity of the organ cycle [114]. The electrode can vary depending on the type of sensor, but the signal conditioning circuit is crucial for processing and transmitting the ECG signal without distortion. The circuit compensates for the skin electrode contact impedance and reduces external noise and common mode voltage error. The performance of wearable ECG sensors is determined by factors such as power consumption, sensor gain, and noise levels, Fig. 5.7 [115, 116].

In order to provide reliable signal monitoring with wearable electrophysiological devices, it is necessary to improve adhesion procedures and electrode–skin contact by taking electrode size, thickness, and material properties into account. Common skin electrode adhesion methods include medical tapes, e-tattoos, self-adhesive solutions, and noncontact capacitive electrodes [117]. There are two main strategies for achieving conformal skin contact with wearable sensors for biopotential. The first strategy is using van der Waals force-adhesion, which is used at the microscale

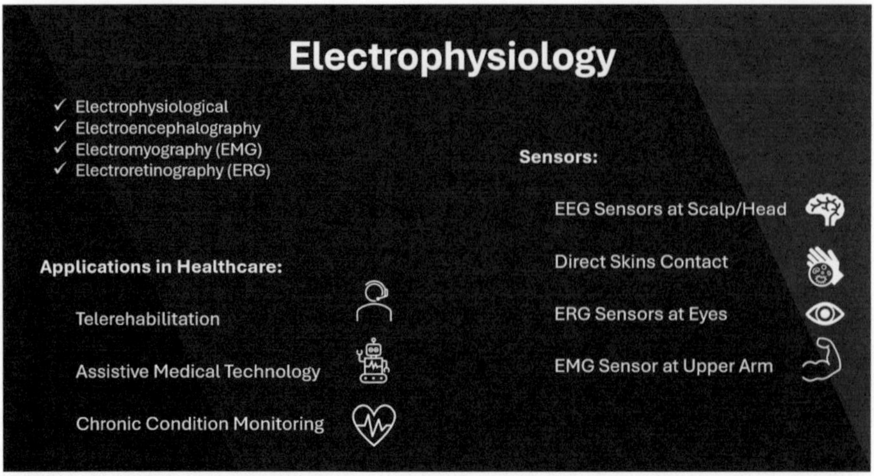


**Fig. 5.7** Electrophysiological monitoring and the advancement of wearable sensors

and nanoscale levels. This can be achieved with low modulus silicone elastomers and sacrificial polymers. Materials like nanoparticle-mixed polymers such as polydimethylsiloxane (PDMS) have been used to improve adhesion between the electrode and the skin. Another strategy is pressure sensitive adhesives (PSA), which create conformal contact between the skin and the device through electrostatic forces. Electrostatic forces have a larger value and longer dominant distance compared to van der Waals forces. It is important to note that even when PSA tape is used, van der Waals forces still play a role when the surfaces are close enough. Inadequate pressure can lead to bonding faults such as bubbles or detachment, which can affect adhesion. Adhesion and flexibility are important for wearable devices to maintain conformal contact even under strain [118]. The most common type of sensor uses the wet silver-silver chloride (Ag/AgCl) transducing electrode, but it has limitations such as baseline drift and skin irritation. Wet electrodes use electrolytic half-cell potentials and resistance between the skin and the electrode to measure electrophysiological activity. They require an electrolyte gel for contact. Dry conducting electrodes, on the other hand, offer a user-friendly approach but still face challenges like contact resistance and movement artifacts. Insulated electrode sensors show promise for healthcare applications. Dry electrodes eliminate the need for gel by using sweat as a substitute electrolyte. Researchers have developed noncontact ECG electrodes for improved monitoring of athletes' musculoneural activity during exercise and sports. These flexible noncontact electrodes, based on capacitive coupling principles, can adapt to the body's contours and minimize noise interference, providing high-quality electrophysiological signals suitable for daily-life and long-term healthcare applications [116, 117, 119, 120]. Wearable electrophysiological devices require flexible electrodes with specific properties, such as flexibility, conductivity, biocompatibility, and durability. These attributes are crucial for accurate and reliable signal monitoring, especially in dynamic environments and prolonged use. Flexible electrodes need to comfortably conform to the skin and have low skin-contact impedance. The breathability, moisture management (to prevent sweat resistance), and gas permeability could be introduced as crucial features of flexible electrodes. They also need to be nontoxic and hypoallergenic to prevent adverse reactions. Durability and stability under various conditions are also important factors to consider. Flexible ECG electrodes, made from materials like gold nanomembrane [121], carbon nanomaterials [122], silver nanowires/nanocomposites [123], and perovskite materials [124], provide precise electrophysiological signal monitoring and are designed for comfort and practicality, allowing athletes to wear skin-like devices without distractions.

In addition to electrode advancements, wearable electrophysiological sensors have also benefited from the use of smart fabrics and knitted integrated sensors (KISs). Smart fabrics, made from conductive yarn, improve comfort levels of the sensor, while KISs integrate electronic components seamlessly into the fabric, allowing for a more flexible and comfortable sensor. Textile-based sensors may experience an increase in impedance after washing, but they offer convenience for patient use [125]. Planar fashionable circuit boards (PFCBs) have also been developed, where the entire circuit is printed on the fabric using a silver paste. These advancements in materials and methods have made wearable electrophysiological sensors an intrinsic part of a person's everyday life [126]. To improve wearable sensors, new sensing modalities must be integrated into a single platform for continuous, simultaneous sensing of multiple parameters. Chemical measurements can offer valuable insights not available from physical or electrophysiological sensors. The Chem-Phys sensor, a wearable device that measures chemical and electrophysiological signals in a single epidermal patch, supports the development of advanced hybrid wearable sensors [127]. Researchers have developed a visual electrocardiogram (ECG) synchronization monitor using perovskite-based multicolor light-emitting diodes (PMCLEDs). These PMCLEDs can emit colorful light based on changes in the ECG signal. By adding polyethylene glycol (PEG) to the charge transport layer (CTL), the color of the PMCLEDs can be switched from deep red to green, allowing for a wide range of colors. The flexible PMCLEDs are stable and resistant to bending, making them suitable for wearable applications on human skin. By converting the ECG signal to a voltage scheme, the PMCLEDs can display real-time ECG information in a cost-effective and easy-to-use manner. This development could lead to the creation of healthcare products with real-time biosignal monitoring capabilities [128]. The importance of self-powered wearable ECG systems is obvious, as they allow for continuous monitoring of the cardiovascular system and detection of any pathological conditions. By using a semi-permanent energy harvester, these systems eliminate inconvenience to the user and ensure stable operation [129]. In this regard, perovskite materials mostly act as a power supply in wearable electrophysiological detection systems, which will fundamentally introduce in another sections.

## 5.5 Mechanical Perovskite-Based Wearable Sensor

Pulse and blood pressure from arteries can be detected on the skin surface due to their elasticity. Direct measurement of aortic pressure requires invasive catheters. Two non-invasive methods use mechanical pressure sensors in arterial tonometry: recording pulse waves over the common carotid artery and detecting pulse waves on the radial artery. Wearable ultrasound devices that detect mechanical signals for long-term monitoring and measure acoustic signals for high-resolution images with minimal tissue penetration depth have been proposed in recent research [118]. On the other hand, gait analysis is used in various fields such as healthcare, animation, and virtual reality. It involves analyzing the way a person walks and can be used for patient monitoring, assessing treatments, identifying movement abnormalities, and evaluating surgical outcomes. Fall risk assessment is important for maintaining health, especially for elderly and disease-impacted individuals. Wearable sensor systems can collect data to analyze fall-related mechanical parameters and detect falls to prevent injuries. For example, A methodology using wearable feet sensors was developed to predict falls using pressure distribution of feet. Previous human gait analysis systems include 3D-acceleration sensors, Vicon 3D motion capture systems, insole-force sensors, Vicon 370 systems, and handheld dynamometers. These methods are effective but expensive and complicated. A potential solution is flexible tactile sensors based on composite materials with piezoresistive properties [14, 130, 131]. Tactile sensors are devices that detect mechanical external stimuli, such as strain, pressure, humidity, sound, and temperature, and convert them into electrical signals. It simulates the process of tactile transmission in biological tissue, where mechanical energy stimulates the skin and triggers changes in the cell membrane, leading to depolarization and the generation of electrical signals. Tactile sensors are designed to provide information about the magnitude, shape, position, and distribution of forces derived from the stimuli. They can be developed into multiarray devices with high resolution and flexibility. They have applications in medical devices, artificial intelligence, and robotics. To be effective in these applications, tactile sensors need to overcome challenges related to sensor arrangement, flexibility, durability, and the ability to detect various physical quantities [132]. Researchers are studying different sensing materials and fabrication techniques to improve the performance of tactile sensors. Self-powered electronics are being explored for wearable devices, using renewable energy sources like solar, kinetic, and bioenergy. This approach offers sustainable power, eliminates complex wiring systems, and reduces battery replacement and maintenance, enhancing functionality, sustainability, and durability [10]. In the past decades, numerous materials, including ceramics (e.g., $BaTiO_3$, PZT, $NaNbO_3$), semiconductors (e.g., ZnO, GaN, CdS), and polymers (e.g., PVDF and its copolymers) have been investigated in detail to fabricate nanogenerators and sensors. Nanogenerators made from halide perovskite and polymer composites can generate electricity from ambient energy sources and can be used as self-powered sensors without relying on a battery. The transduction mechanisms of electromechanical sensors include triboelectricity, piezoelectricity, piezocapacitance, and piezoresistivity, Fig. 5.8 [133].

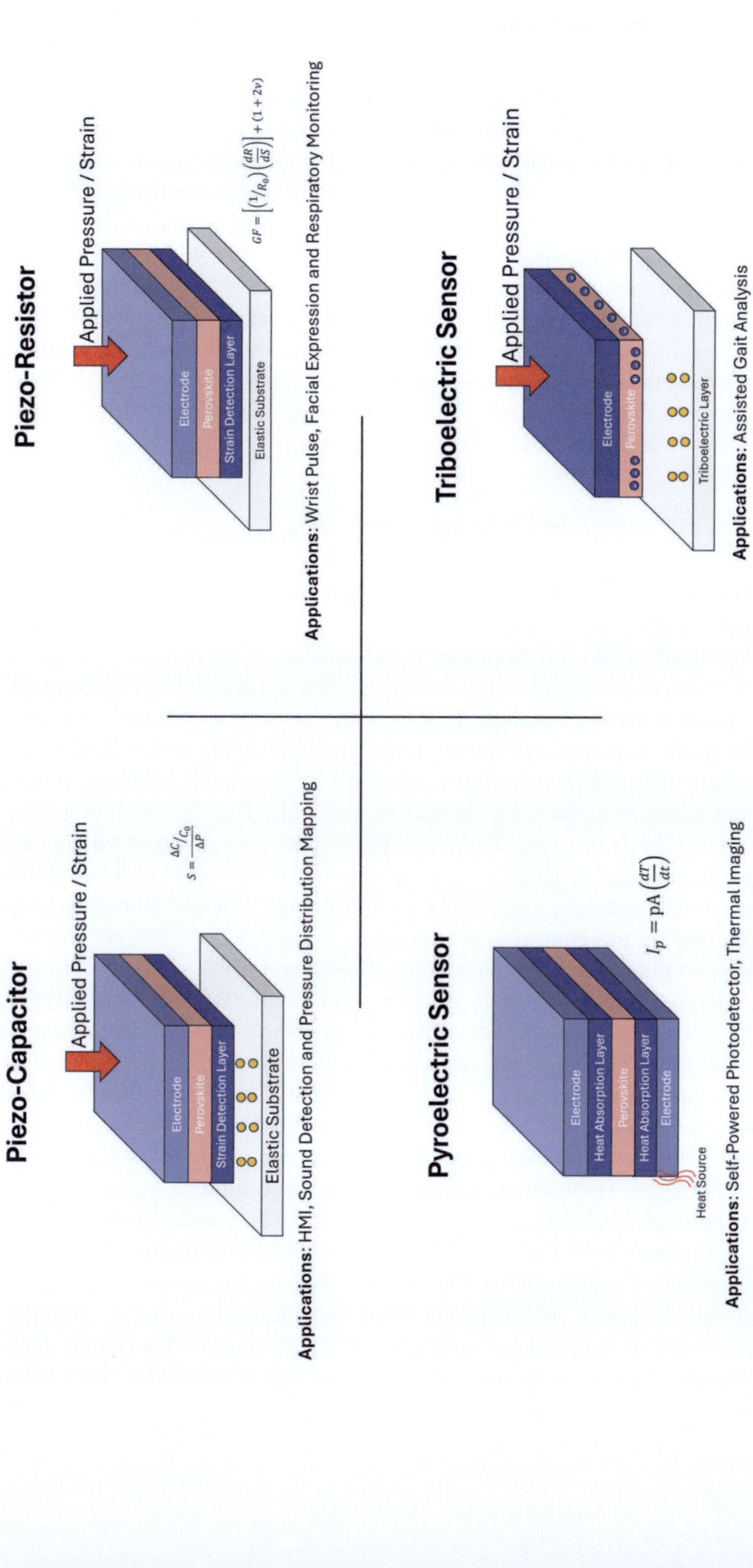


**Fig. 5.8** Overview of perovskite-based electromechanical sensors: showcasing the diverse mechanisms—piezoresistive, piezocapacitive, pyroelectric, and triboelectric—that transform mechanical and thermal stimuli into electrical signals, highlighting their applications in wearable and biomedical technologies

### 5.5.1 Piezoresistive Sensors

Piezoresistive sensors convert mechanical stimuli like pressure or strain into electrical signals, offering high sensitivity, fast response time, and low cost. They are widely used in micro-electromechanical systems (MEMS) devices for health monitoring, physical tracking, and human–computer interaction in wearable electronics. Flexible piezoresistive sensors imitate human skin's properties, offering tactile sensing, toughness, self-healing, and stretchability. Nanocomposite materials, with their combination of a viscoelastic matrix and conductive fillers, are seen as a promising technology for future wearable electronics [115, 134]. Key parameters of piezoresistive sensors include sensitivity, sensing range and linearity, limit of detection, response and recovery time, operating voltage, hysteresis, and stability. The sensitivity of these sensors is expressed through the gauge factor, which is crucial for their effectiveness and is expressed through Eq. (5.1).

$$GF = \frac{1}{R_0}\frac{dR}{ds} + (1 + 2\upsilon) \tag{5.1}$$

where s is the applied strain, R is the resistivity and $R_0$ is the material's zero-strain resistivity, and $\upsilon$ is the Poisson ratio [135]. The resistance change in these sensors is mainly derived from factors such as geometry deformation of the material, changes in resistivity of semiconductor and conductor filters, and microstructures that concentrate stress and increase deformation. Piezoresistive sensors traditionally use rigid and brittle inorganic materials, but for wearable electronic applications, sensors can be made using conductive nanomaterials combined with stretchable polymers, porous conductive materials, or architected conductive materials. The transduction mechanism for these materials involves changes in the contact area of electrical conductors during deformations. Conductive polymeric composites rely on the mobility of conductive fillers within a viscoelastic polymer matrix, while porous conductive materials rely on the contact and separation of conductive elements within a porous matrix. Architected conductive materials have carefully designed microstructures that undergo large contact area changes with small external loadings [134, 136]. Piezoresistive sensors consist of sensing components, signal transfer components, and a substrate. The active materials used in these sensors include conducting polymers [137], metal nanoparticles [138], MXene/Au Dual-Layer [139], carbon nanomaterials/graphene [140], and perovskite materials [141]. Efforts have been made to develop sensors with high gauge factors, and nanocomposites have shown promise in this regard. However, composite sensors have limitations, such as low conductivity and high load/unload hysteresis [142, 143]. To overcome these limitations, some researchers have explored systems where the piezoresistive element is a network of conductive nanoparticles. These networks have higher conductivity and are printable, making them advantageous for sensor fabrication. These networks' piezoresistive behavior is attributed to strain's effect on interparticle charge transport [142]. Besides, these electrodes need to maintain high conductivity under strain and possess excellent stability. The substrate materials used in these sensors include flexible materials like PDMS, block copolymer elastomers, self-healing materials, hydrogels, and UV cross-linkable materials. These substrates provide the necessary flexibility and mechanical properties for stretchable electronics. Piezoresistive sensors use optimized 3D microstructures like hemispheres or pyramids, combined with a high-conductivity sensing layer. These improvements enhance the sensor's sensitivity, linearity, and sensing range for pressure measurement [144–147].

Halide perovskites, specifically $CsPbBr_3$ and $MAPbBr_3$, are promising materials for piezoresistive sensors due to their soft lattice, high defect tolerance, and good carrier transport property. However, halide perovskites are prone to hydrolysis in moisture air, so embedding perovskite nanocrystals into a polymer matrix can enhance their stability. The presence of lead in perovskite structures raises concerns about toxicity and restrictions on their use in consumer goods. Various methods can be used to embed perovskite or other conducting nanocrystals into the elastic substrate, including in-situ preparation, impregnation, spraying, and ion sputtering [148]. Recently, in order to achieve the desired porosity, the development of flexible pressure sensors using $MAPbBr_3$ perovskite nanocrystals embedded in a polymer film was evaluated. The sensor exhibits high sensitivity (0.744 $kPa^{-1}$ in the pressure range of 0–60 kPa), a wide linear range, and a short response/recovery time (102.05 ms/87.75 ms). It also maintains hydrophobicity, allowing it to function in humid environments. A wearable prototype device is created for various applications, such as facial expression monitoring and pulse signal detection [141]. The synthesis of flexible piezoresistive films by dispersing perovskite $MASnBrI_2$ nanocrystals into a thermoplastic polyurethane substrate was reported in another study. The resulting sensors, made by sandwiching the nanocomposite layer between silver electrodes, exhibit piezoresistive behavior due to the uniform distribution of the nanocrystals. The sensors demonstrate high pressure sensitivity (0.796 $kPa^{-1}$ over a wide linear range of 0–0.7 kPa), low detect limitation of 0.2 Pa, fast response and recovery times, and good flexibility (90°). Additionally, the sensors are capable of detecting respiratory airflow, wrist pulse, and finger joint curvature [148]. Moreover, the researchers developed a water-stable luminescent perovskite hydrogel by blending $CsPbBr_3$@PbBr(OH) with stretchable polyacrylamide hydrogels. The resulting hydrogels have excellent optical properties and high mechanical strength. They retain bright green fluorescence even under water, ultraviolet light, and extensive stretching. The researchers used these hydrogels to create a wearable stretchable strain sensor textile that can detect various motions on the human body in real time. The sensor provides electronic output signals from changes in resistance and simultaneously readable optical output signals [149].

### 5.5.2 Piezocapacitive Sensors

Capacitive sensors, which use dielectric layers to encapsulate the electrode materials, have advantages such as reusability, sterilization, and reduced leakage currents.

They have been affixed to the skin to ensure intimate skin contact and avoid motion artifacts. Capacitive sensors have potential for disease identification using machine learning algorithms [150]. Piezocapacitive sensors are versatile, flexible devices that can sense both static pressure and high-frequency vibrations, making them suitable for various applications such as robot haptics, human–machine interfaces, intelligent wearables, and the metaverse. They utilize advanced materials like nanomaterials and conducting polymers in their design. Despite their potential, they have limitations in responding to dynamic stimuli, and improvements are needed for instantaneous and continuous operations [151]. Piezocapacitive sensors typically consist of a parallel plate capacitor with an insulating dielectric layer between two conductive electrodes. The capacitance of the sensor can be calculated using Eq. (5.2).

$$S = \frac{\Delta C/C_0}{\Delta P} \tag{5.2}$$

S represents the sensitivity, ΔC is the relative change of capacitance, $C_0$ represents the initial capacitance when no pressure is applied, and ΔP is the change in applied pressure [151]. The mechanism of a piezocapacitive sensor involves increasing sensitivity by maximizing the relative change of capacitance while keeping the applied pressure range constant. A piezocapacitive sensor uses a flexible substrate, such as elastic polymer films, rubber, sponge, or foam, to provide elasticity to the dielectric layer. However, elastomers have viscoelastic behavior, leading to a long relaxation time, as well as hysteresis, which is the delay in monitoring physiological signals. Porous structures such as sponges or foams, along with microstructures, can be used to reduce viscoelastic behavior and enhance compressibility, thereby improving sensitivity and working range, achieving faster response times, and improving the system's deformation behavior. Signal lag, which negatively affects the sensor's performance, can be reduced by strengthening the interaction between the polymer and nanofiller. Besides, good interfacial contact and interface engineering can also help suppress hysteresis and improve stability [152–154]. To improve response time, researchers have developed a strategy that involves bonding a low-viscosity microstructured dielectric to the electrode, reducing energy dissipation and friction-induced losses. This sensor can quickly respond to stimuli ranging from steady pressures to high-frequency vibrations and exhibits high frequency-resolution and negligible capacitance-pressure hysteresis. The researchers also demonstrate its application in sound detection by designing an artificial ear system [155]. A recent study developed a specific type of piezocapacitive sensors consisting of a structure with a composite dielectric layer sandwiched between two ion-gel thin film electrodes. This sensor system allows for the detection of directional movement of applied pressure and static pressure variation by modulating ion distribution in the ion-gel thin films. They can be used to monitor human physiological signals and finger sliding motions, making them suitable for wearable electronic devices, health care monitoring systems, and artificial skin [152]. A novel wearable self-powered lead-free perovskite piezoelectrodes were utilized in the development of a self-charging supercapacitor for wearable electronics. The supercapacitor had an asymmetric design, with nickel stannate as the positrode and ferrous stannate as the negatrode. The chosen materials were non-toxic, environmentally friendly, and biocompatible. Ferrous stannate exhibited the highest specific capacitance among the perovskite-based electrodes. The self-charging capabilities of the supercapacitor were evaluated, resulting in a maximum of 266 mV. The fabricated supercapacitor was found to be highly flexible, robust, stable, and had significant electrochemical and self-charging potential [156]. In another development, researchers have developed a transparent capacitive proximity and pressure sensor using $Pb(Zr_xTi_{1-x})O_3$ (PZT) nanowires and a polydimethylsiloxane composite dielectric layer, offering a pressure sensitivity of 89 $MPa^{-1}$, ten times higher than a PDMS dielectric layer. The addition of PZT nanowires enhances proximity sensing, potentially benefiting distance monitoring systems and human–machine interfaces [157]. Wearable electronics face challenges in integrating analog, digital, and power electronics. A recent study showcased a multi-scale simulation of a sensor-on-chip for self-powered blood pressure monitoring. The study integrated capacitive pressure sensors, analog interfacing circuits, and lead-free perovskite solar cells. Two analog interfacing circuits were proposed for efficient digital data processing. Lead-free perovskite solar cells were used to power the sensor. The sensor's sensitivity varied based on the pressure range, and the system consumed 50.5 mW in the optimal configuration. These findings highlight the potential of sensor-on-chip designs in wearable health-monitoring technologies [158].

### 5.5.3 Pyroelectric Sensor

Temperature sensors are essential in various fields, including environment monitoring, thermal imaging, biomedical sensing, and microprocessor thermal monitoring. Pyroelectric nanogenerators convert time-varying temperature fluctuations into usable electricity. Two main classes of halide perovskites are inorganic–organic hybrid perovskites and inorganic ferroelectric perovskites. Hybrid perovskites are known for their photovoltaic response, while ferroelectric perovskites are known for piezoelectric, pyroelectric, and thermal energy conversion systems. Perovskites' electrical and ferroelectric properties are crucial for their applications. Ferroelectric perovskites have short diffusion lengths and lifetimes, while hybrid perovskites have better transport properties and long carrier lifetimes [159]. These perovskite generally exhibits temperature-dependent structural/phase transitions, making them suitable for temperature sensor technology. The mechanism behind this effect mostly involves the displacement of the central ion in a non-centrosymmetric perovskite, resulting in a change in output voltage. However, these devices require steady spatial temperature variations, which can significantly affect their performance [133]. The change in polarization ($\Delta P_i$) with temperature change ($\Delta T$) is given as Eq. (5.3):

$$\Delta P_i = p\Delta T \tag{5.3}$$

The pyroelectric coefficient in the direction perpendicular to the electrodes (i.e., in the polarization direction) is denoted as p. More specifically, the induced short-circuit current ($I_P$) for a certain rate of temperature change (dT/dt) and a surface area A is equal to:

$$I_P = AP\frac{dT}{dt} \tag{5.4}$$

One of the first pyroelectric nanogenerator (PyENG) using halide perovskite-polymer composite material, demonstrating temperature-dependent output performance, was fabricated by attaching Ni and Cu-coated fabrics as top and bottom electrodes onto the $MAPbBr_3$-PVDF (($CH_3NH_3PbBr_3$) Methylammonium lead bromide-poly(vinylidene fluoride)) electrospun nanofibers mat, respectively, followed by wiring with copper wires. The addition of $MAPbBr_3$ increases the electroactive phase composition and tensile strength, potentially enhancing output power. The nanogenerator showed fast response time and good reversibility, but its low Curie temperature limits its application in higher temperatures. Stability issues and lack of studies on the pyroelectric properties of halide perovskite materials are still under consideration [160]. A perovskite pyroelectric sensor, known as $(1-x)KNbO_3$-$xBaNi_{1/2}Nb_{1/2}O_{3-\delta}$ (KBNNO), has been developed representing the ability to absorb visible light with a tunable direct bandgap of 1.1–3.8 eV. The sensor exhibits strong piezoelectric and pyroelectric responses, as well as a photovoltaic effect, making it a promising candidate for a multi-source energy harvester or multi-functional sensor. Domain wall movement is hindered by the presence of oxygen vacancies, resulting in reduced dielectric permittivity, loss, resistivity, and piezoelectric properties. This is not the most optimal solution for energy harvesting or sensing. However, the combination of $Ni^{2+}$ ions and oxygen vacancies enhance electronic states in the gap of KN, the parent composition. The optimum KBNNO ceramic feature a direct bandgap of 1.48 eV, a pyroelectric coefficient of 26 $\mu C/m^2K$, piezoelectric coefficients of $d_{33} \sim 23$ pC/N and $g_{33} \sim 4.1 \times 10^{-3}$ Vm/N, and a remanent polarization of 3.4 $\mu C/cm^2$. The KBNNO ceramics exhibit better pyroelectric and piezoelectric coefficients than ZnO and AlN, making them suitable for hybrid energy harvesting and multi-functional sensing devices [161]. Researchers have successfully engineered ZnO-based devices with improved charge-carrier transport by using pyroelectric polarizations, building on the multifunctional potential of perovskite materials. A new self-powered photodetector was made using a $CH_3NH_3PbI_3$ perovskite and ZnO heterostructure. It showed big improvements in response time, detectivity, and responsivity. These improvements show how promising perovskite materials are for making next-generation, high-performance, self-powered optoelectronic and sensing platforms. This is a big step toward making devices that can do many things at once [162].

### 5.5.4 Triboelectric Sensor

Triboelectric technology refers to the use of triboelectric nanogenerators (TENGs) to convert mechanical energy into electrical energy. TENGs can detect different types of mechanical motions and generate electricity through contact electrification and electrostatic induction. By integrating TENGs into wearable devices, mechanical energy from the human body can be harvested to power small electronic devices. TENGs can also be used to detect environmental parameters and degrade various types of pollutants, such as organic water pollutants and particulate matter [163]. Furthermore, these TENGs could be used as potential anti-theft sensors in books, as they can harvest kinetic energy, act as an energy supply, and be a self-powered active sensor [164]. The relative position of the triboelectrically-charged surface changes due to external mechanical motions, resulting in a variation of the induced potential difference between two electrodes. TENGs can detect both static and dynamic processes by analyzing the open-circuit voltage and short-circuit current. They can also be used for self-powered active chemical sensing and monitoring changes in the magnitude of an external magnetic field. TENGs have two main operation modes: vertical contact-separation mode and lateral sliding mode. In the vertical contact-separation mode, a potential difference is generated when two surfaces come into contact and then separate, leading to charge transfer between the surfaces. In the lateral sliding mode, electron flow is induced through the external circuit due to in-plane charge separation when one surface slides outward. There are also two additional operation modes: single electrode mode and free-standing mode. In the single electrode mode, charge transfer occurs between the polymer surface and the electrode, driving electrons to flow from the electrode to the ground. In the free-standing mode, electrons flow between two electrodes to balance the local potential distribution caused by the frictional layer approaching and departing from the electrodes [165–167].

The monitoring of human joint motion is important for diagnosing and treating diseases such as Parkinson's disease and arthritis. Traditional methods of monitoring joint motion have limitations such as slowness, bulkiness, high cost, and the need for a power source. TENGs offer a robust and cost-effective solution for monitoring joint motion by integrating them into insoles of shoes or other part of body. The waist, thigh, and foot were selected as candidate sensor positions [168]. Motion sensors that utilize electrostatic charges on the skin have been found to accurately detect motion in various studies. These sensors are made of a skin-friendly material, allowing them to capture real-time motion data from human joints. The effect of microstructure on triboelectric charges is demonstrated in a study where a micropyramid-patterned PDMS film is utilized as a spacer and contact surface. When this micro-patterned surface comes into contact with human skin, friction occurs, resulting in the generation of triboelectric charges. The PDMS surface is equipped with copper electrodes, enabling the establishment of an electric potential difference and electron flow when there is partial separation between the PDMS and human skin. The maximum accumulation of positive charges on the copper electrodes is observed when the separation area is at its maximum, and this charge decreases as the separation area decreases.

Upon full contact between the skin and PDMS, electrons flow back to the copper electrodes. This mechanism allows for the accurate detection of motion by the motion sensor, as it capture real-time motion data from human joints [169].

Triboroelectric materials have been explored for high-performance outputs, with dielectric materials being preferred due to their long charge-maintaining characteristics. In a nanogenerator, the introduction of perovskite filler material (such as $CsPbBr_3$ and $CsCuCl_3$) into polymers such as polyvinylidene difluoride (PVDF) can enhance output voltage, current density, and power density due to the superior optoelectronic properties of lead-free and holide perovskites. Hybrid perovskites have remarkable dielectric features, making them a promising candidate for high-output triboelectric charges, Fig. 5.9 [170–172]. They were first used as a triboelectric material in a self-powered photodetector, operating in a vertical contact-separation mode [163]. Capacitance optimization using materials with good dielectric constants enhances electrical output. Organic–inorganic hybrid perovskite TENGs with tailored compositions and ion migration have higher output voltage and power density. Lead-free triple perovskite nanoparticles, specifically $Sr_3Co_2WO_9$, have been used to create a triboelectric nanogenerator (TP-TENG) that acts as a sustainable power source and self-powered information generator. The TP-TENG operates in a vertical contact-separation mode and demonstrates stable output with high power density. This research highlights the potential of triple perovskite for triboelectric energy harvesting and its suitability for real-time analysis of different gaits (walk, run, jump, and stomp), potentially extending its use as a healthcare sensor in the future [168]. Research revealed that the involvement of dynamic metal/perovskite Schottky junctions in TENGs resulted in improved output pulse currents because they were able to capture mechanical and light energy at the same time. The interaction between photo-generated carriers in the junction's perovskite layer and mechanically-induced bound back carriers is responsible for this phenomenon [173]. In another study, a photodetector was developed consisting of a copper layer on a PET substrate and a $MAPbI_3$ polycrystalline film with a mesoporous $TiO_2$ scaffold on an FTO substrate. The joint properties of the perovskite material, including the photoelectric effect and the triboelectric effect under vertical contact-separation mode, contribute to its dual sensing mechanism. Upon solar illumination, the $MAPbI_3$-based composite thin film experiences significant alterations in both photoconductivity and surface triboelectric density, resulting in a considerable reduction of the output voltage. When an external force is applied, opposite charges appear on the surfaces of the contacted copper and $MAPbI_3$ films, leading to a drop in electric potential on the friction surfaces. This sensor demonstrates excellent responsivity (7.5 V $W^{-1}$), rapid response time (<80 ms), great repeatability, and a broad detection range that extends from UV to visible regions [174]. Another study revealed that the introduction of $Ba^{2+}$ cation and doping of $Cl^-$ and $I^-$ anions in $CsPbBr_3$ perovskite further enhance the dielectric properties and charge density generation [175].

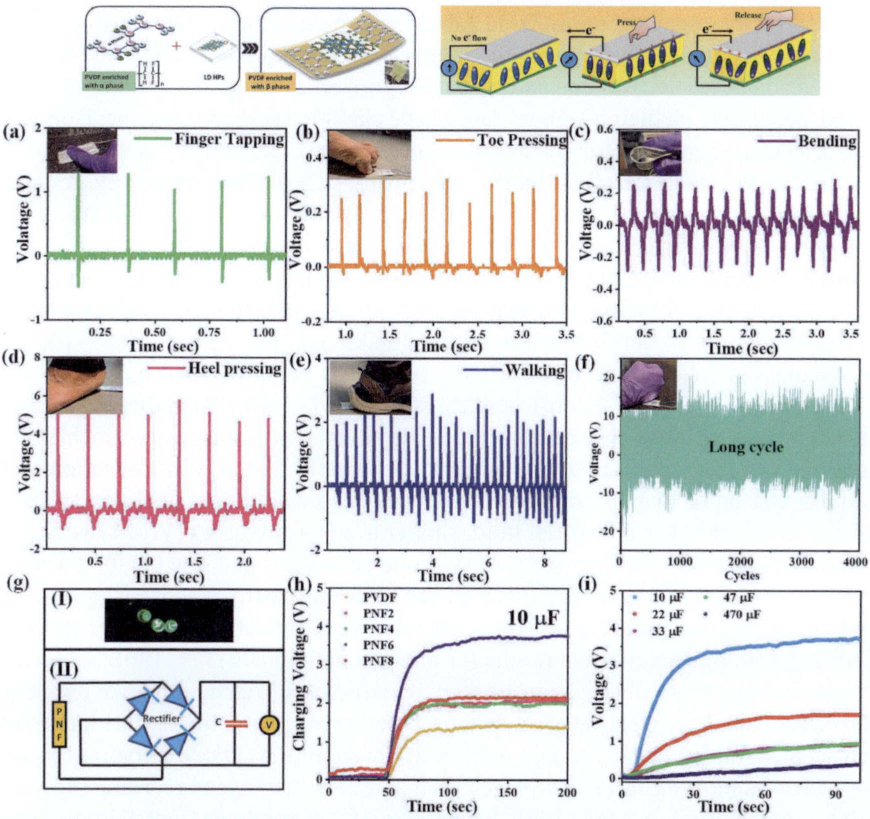


**Fig. 5.9** Schematic illustration of the interfacial interactions in $Cs_4PbBr_6$ high-performance and PVDF matrix, schematic representation of the operational concept of the constructed nanogenerator, and voltage response resulting from **a** finger tapping, **b** toe pressing, **c** film bending, **d** heel pushing, and **e** walking. **f** Voltage response over prolonged cycles using hand hammering action, and **g** (I) powering commercial LEDs with the PNF6 device and (II) schematic illustration of the capacitor charging circuit. **h** Charging a 10 μF capacitor using PVDF and various PNF composites, and **i** charging distinct capacitors with the PNF6 device. Reproduced from Ref. [171] published by Chemical Physics Impact. Licensed under Creative Commons Attribution 4.0

## 5.6 Wearable Electrochemical Perovskite-Based Sensors and Biosensors

The International Union of Pure and Applied Chemistry defines a chemical sensor as: "a device that transforms chemical information, ranging from the concentration of a specific sample component to total composition analysis, into an analytically useful signal" [176]. Chemical sensors are composed of a receptor and a physico-chemical transducer. When the receptor is a biological component, like an enzyme or antibody, the sensor is referred to as a biosensor. The desired analyte is selectively detected by the receptor, minimizing false-positive outcomes. The transducer then converts the signal resulting from the interaction between the receptor and analyte into a readable value. Biosensors can be categorized as either catalytic or affinity-based, depending on their receptors, and can be classified as electrochemical, optical, piezoelectric, or calorimetric sensors based on the type of transducer used [177]. Key features of biosensors include their high sensitivity, excellent selectivity, remarkable stability, consistent reproducibility, low detection limit, rapid response time, and wide range of linearity [34]. The potential for revolutionizing healthcare monitoring and sports analytics lies in wearable biosensors, which can provide real-time multi-analyte physiological information, identify potential threats, and increase the wearer's chances of survival in hazardous conditions. The detection of explosive compounds in a faster and more reliable manner is being pursued by researchers through the utilization of electrochemical sensors [177–179]. Wearable biosensors, also known as lab-on-a-chip devices, are being developed for on-the-go biochemical analysis, offering advantages like automation, parallelization, cost reduction, and cellular and molecular information. These devices detect target analytes in tears, saliva, sweat, and skin interstitial fluid, similar to in vitro sensors [177, 178]. Saliva contains biomarkers that can indicate systemic conditions and oral cavity diseases, making it valuable for detecting diseases like diabetes, gastro-esophageal disorders, and cancer. These sensors can also detect viral antigens or antibodies related to SARS-CoV-2, providing positive test results for infected individuals [179, 180]. Saliva is a convenient and accessible diagnostic tool, by which researchers integrate temperature sensors and pH electrodes on denture bases to overcome limitations like tooth replacement and solution leakage. However, developing wearable intraoral sensors for saliva monitoring faces challenges such as saliva heterogeneity, sampling variations, and biofouling [181–183]. Sweat contains electrolytes, metabolites, trace elements, and small molecules, which can detect and monitor various diseases and apply to fitness tracking. Sweat acidity can be used for the detection of infections and metabolic alkalosis. It also contains glucose for diabetes screening and lactate for tissue viability assessment. Sweat detoxifies heavy metals, allowing stress and immune system-induced muscle damage monitoring [184]. Furthermore, tears are a useful fluid for non-invasive diabetes monitoring due to their correlation with blood glucose levels. However, current instruments have limitations, such as evaporation and eye irritation. Wearable sensors, fabricated using lithographic techniques and low-cost thick film printing technology, can detect analytes directly on the retina. However, they have limitations, such as a lack of electronic integration and potential eye irritation. Soft contact lens-based sensors containing nanomaterials with integrated wireless electronics can address these challenges [177, 185].

The field of wearable biosensors has seen advancements in materials and fabrication strategies, such as the use of polymers and soft lithography for microfluidic methods. These technologies enable the mass production of microchips at low cost. Sampling modalities for wearable biosensors include direct contact, chamber collection, and multistep sampling. Electrochemical sensing is a common modality used in wearable biosensors, allowing for the detection of various analytes in biofluids. Potentiometry, conductometry, and voltammetry/amperometry are electrochemical techniques used for real-time monitoring, Fig. 5.10 [186]. The fabrication technology of electrochemical biosensors involves the use of screen-printed electrodes (SPEs) and stamp transfer electrodes. SPEs have been widely used in various domains due to their low cost, ease of fabrication, and good electroanalytical performance. The electrode geometry is designed, stencils are fabricated, and printing parameters are adjusted. Conductive, active, and insulator layers are then printed onto a textile substrate. The substrate is annealed to stabilize the shape and activate the electrode material. Additional steps such as electrochemical cleaning, deposition of conducting polymers or metal catalysts, and dispensing of reagents or membranes may be performed to enhance detection. Stamp transfer electrodes offer an alternative approach for printing on non-planar substrates, allowing for the patterning of diverse surfaces without compromising the structural integrity of the electrode pattern. This technique has been demonstrated on various substrates, including the skin, and has shown promising electrochemical performance [179].

Materials are being utilized in the creation of biosensors for health monitoring, as they possess distinctive properties, biocompatibility, stability, and modifiability for effective interaction with target biomolecules [187]. Perovskite electrochemical sensors are a promising technology for non-enzymatic detection, offering advantages such as low cost, flexibility, and high catalytic activity due to their unique properties, including tunable redox properties, lattice oxygen vacancies, high oxygen mobility, superoxide oxygen ions, ionic conduction, thermal and chemical stability, and active

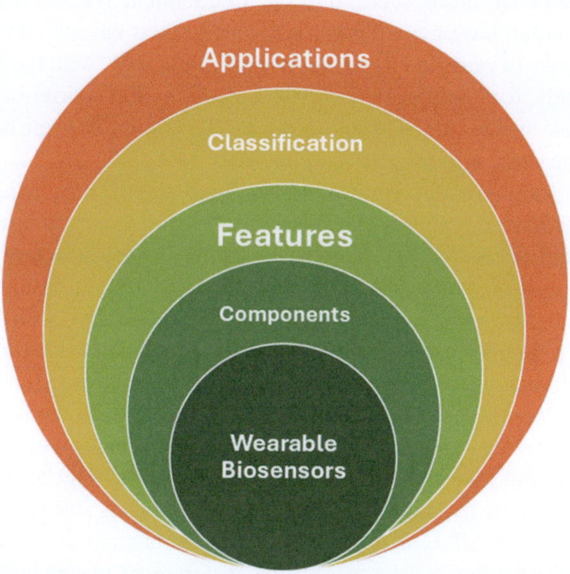


**Fig. 5.10** Overview of wearable biosensors: Components (Receptor (Analyte), Transducer (Readable Values)), Features (High Sensitivity, Excellent Selectivity, Remarkable Stability, Low Detection Limit, Rapid Response Time), Classifications (Catalytic, Affinity-Based, Mechanisms: Electrochemical, Piezoelectric, Calorimetric), and Applications (Sweat, Tears, and Saliva Monitoring)

electrical structure. These features make perovskite an ideal material for efficient detection of important biomarkers. However, perovskite oxides alone have limitations in terms of electron transfer kinetics, stability, and selectivity. To address these limitations, researchers are investigating composites of perovskite with other materials. These composites, including graphitic carbon, metals, metal compounds, conducting organics, and biomolecules, aim to improve the electron transfer kinetics, stability, and selectivity of perovskite electrochemical sensors. The choice of materials for modifying electrode surface greatly impacts the sensor's performance [45, 188]. In the following, the application of perovskite materials in sensing different biomarkers will be introduced.

### 5.6.1 *Sensing of Hydrogen Peroxide ($H_2O_2$)*

Hydrogen peroxide ($H_2O_2$) is a reactive oxygen species (ROS) produced during aerobic metabolism and physiological processes like immune activation, apoptosis, and cell growth. However, the presence of hydrogen peroxide can lead to oxidative damage and various diseases including cancer, cell damage, and inflammatory diseases. Therefore, the detection of hydrogen peroxide has become an important topic in industries such as food, pharmaceutical, cosmetics, and chemical [45]. Different types of sensors have been developed for hydrogen peroxide detection, including colorimetric, fluorescent, and electrochemical sensors. Electrochemical sensors, in particular, have gained popularity due to their ease of commercial development. These sensors can be enzymatic or non-enzymatic, with the latter requiring metal compounds for hydrogen peroxide detection. Perovskite-based materials, specifically those in the $ABO_3$ form, have been commonly used in electrochemical sensors. The most important A cations, such as La, Sr, and Ca, are often from the lanthanide group or alkaline earth metals. They stabilize the oxidation states of the B cations, which are typically transition metals from the fourth group of the periodic table, including Ni, Co, Fe, and Mn. By modifying the structure of perovskites and introducing different cations, their catalytic activity can be altered. One way to enhance the activity is by replacing the A-site metal with a different A' metal cation that has a distinct valence state ($A_{1-X}A'_X BO_3$). This substitution stabilizes the uncommon oxidation state of the B-site cations and leads to the formation of structural defects, such as oxygen vacancies. The redox reaction between the B cations and hydrogen peroxide produces molecular oxygen, water, and electrons, allowing for the detection and measurement of hydrogen peroxide, Fig. 5.11. The development of non-enzymatic hydrogen peroxide sensors involves studying the electrocatalytic performance of different A cations and dopant agents, as well as calcination temperature and its effect on the crystalline structure of the materials [189, 190]. Research indicates a correlation between peroxide decomposition rate and δ-oxygen non-stoichiometry in certain perovskite systems. As δ-oxygen non-stoichiometry decreases, decomposition rate increases in $La_{1-X}Sr_XMnO_{3+\delta}$ and $La_{1-X}K_XMnO_{3+\delta}$ series [191]. Calcium substitution at the A site in $LaMnO_3$ perovskites increases

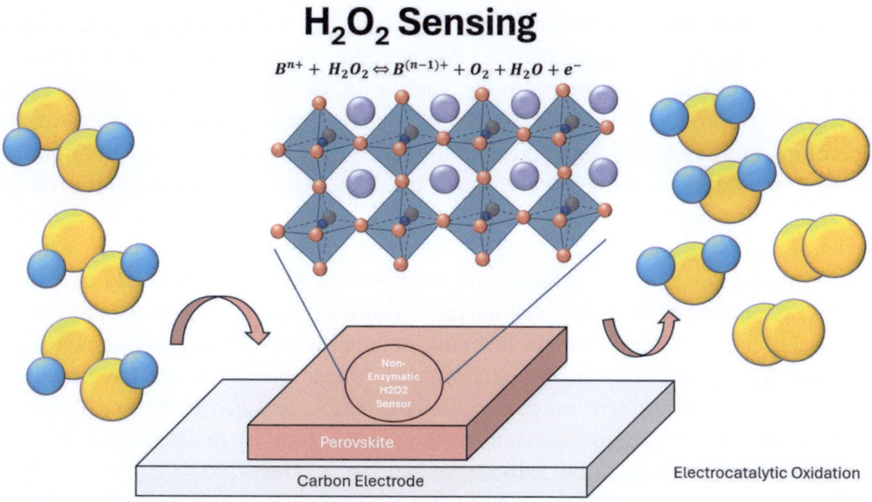


**Fig. 5.11** A non-enzymatic hydrogen peroxide ($H_2O_2$) sensor based on perovskite-related materials

activity with higher calcium content. Factors like δ-oxygen non-stoichiometry and average manganese valence state also influence catalytic performance [190]. Carbon-based electrodes loaded with perovskite-type oxides have been found to be effective in sensing $H_2O_2$. The electromotive force (EMF) of the electrode loaded with $La_{0.6}Ca_{0.4}Ni_{0.7}Fe_{0.3}O_3$ showed a logarithmic relationship with $H_2O_2$ concentration, with a response time of 5 min at 25 °C. The sensor's EMF was not affected by dissolved oxygen concentration. The electrodes also exhibited high activity for electrochemical oxidation of $H_2O_2$, with a limiting current observed between +0.6 and +1.0 V [192]. The morphology of perovskites is crucial in sensor development as it affects the electrode's surface area and active sites for detection and sensing. The synthesis methodology is the most significant variable in the resulting morphology. Perovskite nanostructures have been synthesized using hydrothermal, solvothermal, and electrospinning routes, resulting in a wide variety of morphologies. The final morphology is also influenced by a calcination procedure. For example, electrospinning and sequential calcination of perovskite oxide $LaNiO_3$ nanofibers reduced the fiber's diameter from 400 to 140 nm, resulting in a fast response and long-term stability [45].

### 5.6.2 *Sensing of Glucose*

Glucose is an important energy source and intermediate product of metabolism. It plays a role in physiology and biochemistry by enhancing the body's ability to remember, stimulating calcium absorption, and increasing cellular communication.

However, overexpression of serum glucose can lead to detrimental effects such as diabetes, which can cause chronic damage and dysfunction of various tissues. Traditional glucose detection methods, such as colorimetric methods, have issues with complex production, complicated operation, and susceptibility to interference. Therefore, there is a need for a convenient method to detect glucose in human serum [193]. Perovskite electrochemical sensors, resembling a three-dimensional network microstructure, may support a variety of metallic components inside their lattice structure. They can detect and test blood sugar levels in a human body without any invasion. The sensors measure glucose levels by electrocatalytically oxidizing glucose on the perovskite electrode surface. Various modifications to the perovskite materials, such as doping with noble metals or creating oxygen vacancies, have been explored to improve the sensing performance. These modifications have resulted in sensors with wide linear ranges, high sensitivities, and low detection limits for glucose detection. The perovskite sensors have shown promise in detecting glucose in biological fluids such as blood, saliva, sweat, urine, and tears [45]. $La_{0.6}Sr_{0.4}CoO_{3-\delta}$ (LSC) perovskite oxide shows comparable performance to noble metal nanomaterials in detecting hydrogen peroxide ($H_2O_2$) and glucose without the need for enzymes. The oxidation of $H_2O_2$ and glucose on LSC perovskite is facilitated by the $Co^{3+}/Co^{4+}$ redox couple and the formation of oxygen vacancies. The detection performance is further enhanced by the addition of reduced graphene oxide (rGO). The linear ranges of detection are 0.2–3350 μM and 2–3350 μM for $H_2O_2$ and glucose, respectively [194]. Glucose electro-oxidation in perovskites, known as the lattice-oxygen-mediated oxygen evolution reaction (LOM-OER), is a complex process involving the substitution of cations in the A sites with divalent cations from group IIA of the periodic table, like $Ca^{2+}$, $Sr^{2+}$, and $Ba^{2+}$. This substitution leads to the oxidation of cations in the B sites, formation of $B^{3+}$. Additionally, increasing this substitution reduces the energy required for oxygen vacancy formation, which is consistent with the presence of highly oxidative oxygen species [195].

In an enzymatic biosensor, the hydrolysis reaction of glucose by glucose oxidase produces water, oxygen, and many protons, lowering the pH and harshening the environment of the sensing probe. To eliminate this problem, a variety of modification methods are introduced, including coating, introducing dopant, composite formation, or hybridization. Silicon-coated $CsPbBr_3$ quantum dots ($CsPbBr_3$@$SiO_2$) were used to develop modified perovskite electrochemical sensors. The $SiO_2$ coating protected the quantum dots from water and oxygen, increasing their stability and dispersibility in aqueous solutions. The fluorescence intensity of $CsPbBr_3$@$SiO_2$ remained unchanged for 7 days and only decreased by 11.9% within 2 months. These sensors demonstrated sensitivity to environmental pH, with fluorescence intensity decreasing as pH decreased. This sensitivity was utilized for the detection of glucose, as the pH decreased when glucose was catalyzed by glucose oxidase. $CsPbBr_3$@$SiO_2$ demonstrated excellent selectivity and sensitivity to glucose, with a detection limit of 18.5 μM, while successfully applied to detect glucose in human serum [193]. In another approach, a ratiometric fluorescent nanoprobe was used to successfully determine the amount of glucose in human serums. This nanoprobe was created by combining green emission $CsPbBr_3$ perovskite nanocrystals ($CsPbBr_3$ PNCs) and red emissive copper nanoclusters (Cu NCs). The formation of the $CsPbBr_3$@Cu nanohybrid occurred through the reduction of cupric ions with d-penicillamine in the presence of $CsPbBr_3$ PNCs. The resulting ratiometric fluorescent nanosensor was then developed for the detection of glucose, with $CsPbBr_3$ PNCs serving as a reference. Glucose oxidation by glucose oxidase resulted in the generation of hydrogen peroxide, which quenched the red fluorescence of Cu NCs at 645 nm. However, the green emission of $CsPbBr_3$ PNCs at 517 nm remained unaffected. This nanoprobe showed a good linear response range for glucose from 2.0 to 170.0 μM with a detection limit of 0.8 μM [196]. Additionally, a nonenzymatic perovskite composite oxide-based $LaTiO_3$-$Ag_{0.1}$ electrochemiluminescence glucose sensors have been reported, showing a wide linear range, low detection limit, and high sensitivity. These sensors increase active surface area and enhance electron-transfer rate. They were successfully used to detect glucose in real blood samples, indicating their potential for electrochemical sensing and biosensing applications [197]. Ongoing research into perovskite composites and nanohybrids holds the potential to address current challenges and facilitate the development of intelligent, real-time glucose monitoring technologies applicable in both clinical and consumer health contexts.

### *5.6.3 Gas Sensor*

Global air pollution is a major concern due to industrialization and urbanization, with toxic gases like CO, $SO_2$, $NO_2$, and $H_2S$ posing threats to human health and the environment. Gas sensors are used for diagnostic applications and biomarkers, while wearable sensors are gaining popularity for real-time data transmission, cloud computing, big data analytics, and remote monitoring. Gas detection methods like

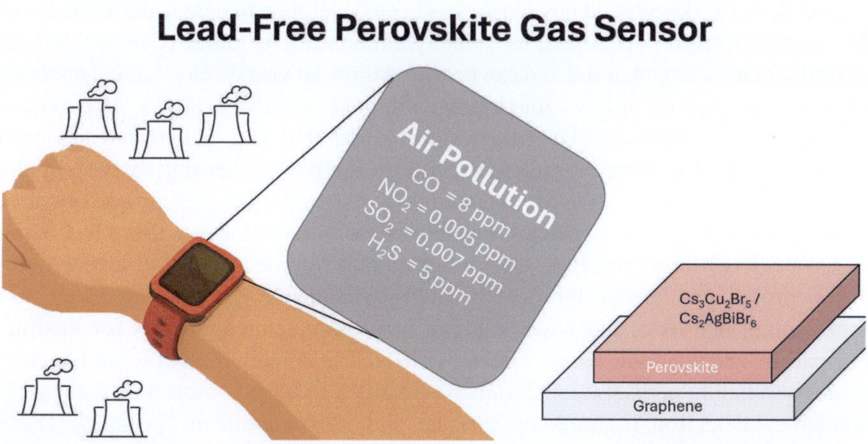


**Fig. 5.12** Schematic representation of a lead-free perovskite gas sensor system for air quality monitoring

gas chromatography, mass spectrometry, and pellistors are reliable but limited due to power consumption, cost, and size. As demand for portable gas sensor devices grows, alternative methods such as capacitive-based, thermometric, optical, field-effect transistor, solid-state electrochemical, and chemiresistive gas sensors have gained attention for their low power consumption and accurate sensing performance. Electrochemical gas sensors are more specific and have ppm- or ppb-level sensitivity, but they suffer from cross-response issues and a limited lifespan. They are susceptible to temperature fluctuations and should be kept constant. The electrochemical interaction at the analyte-electrode interface leads to an electrically detectable signal proportional to the concentration of the analyte [198]. Research is focusing on developing sensitive layers for room-temperature applications using nanomaterials with properties like high surface area-to-volume ratio, low noise, and affordable production costs. Hybrid nanomaterials like lead halide perovskite nanocrystals have been used to improve sensing performance, resulting in low-power and inexpensive gas sensors. The combination of graphene and perovskite nanocrystals enhances sensitivity, while graphene's hydrophobic character protects perovskite nanocrystals against moisture, enabling their use in monitoring applications. However, halide perovskites use lead, a divalent cation, which poses risks to human health and environmental effects, making their implementation in commercial devices limited. For that reason, very recently, some research efforts have redirected toward the use of lead-free halide perovskites for reducing the potential dangerousness and retaining their excellent properties. Two perovskite nanocomposites, $Cs_3Cu_2Br_5$ and $Cs_2AgBiBr_6$, were studied for detecting various gas compounds. The highest sensing performance of lead-free perovskites was obtained, and graphene-supported lead-free perovskites showed even higher gas-sensing performance than graphene-supported lead halide perovskites. The method leverages the high versatility of perovskites and their straightforward synthesis and integration into transducing substrates, making it easily exportable to large-production scale, Fig. 5.12 [199]. A textile-based lead-free halide perovskite-based gas sensor has been developed to detect ammonia gas at ambient temperature. The sensor uses methylammonium tin iodide ($CH_3NH_3SnI_3$ or $MASnI_3$) as an active material and can be traced through visual color change methods and electrical readout. It has a maximum calibrated sensitivity of 85% on exposure to 100 ppm ammonia gas. The sensor can be operated at 2 V bias and has been compared to its lead-based counterpart, confirming the gas-sensing activity by the halide perovskite family [200]. Another study explores the gas-sensing performance of 2D all-inorganic $NbWO_6$ perovskite nanosheets with thicknesses down to 1.5 nm. The nanosheets show fast $H_2S$ sensing speed and high selectivity at low temperatures, promising the development of high-performance gas sensors using low-dimensional, nontoxic, and wide-bandgap perovskite semiconductors for flexible, transparent, and wearable sensing devices [201]. The combination of wearable electronics and halide perovskite materials has resulted in a novel biosensor that excels in chemical detection, is simple to fabricate, and offers significant versatility. These sensors can now detect sweat metabolites, breathing patterns, and physical stresses on surfaces that can bend and move. Future work will be directed toward enhancing material stability in biofluidic environments, exploring scalable patterning strategies for ultrathin device integration, and implementing fully self-powered platforms through energy harvesting. Moreover, the utilization of machine learning algorithms for real-time signal deconvolution and disease prediction represents a significant advancement in the development of adaptive, closed-loop biosensing systems.

## 5.7 Summary

Wearable sensor technology is changing the way we keep an eye on the environment, make sure food is safe, and find diseases. A thorough study of new materials, like perovskites, backs this up even more. Standard analytical methods are accurate, but they are too expensive, heavy, and require trained staff, so they can't be used in real time or on-site. These problems have led to the creation of new sensing technologies that focus on being portable, using as little power as possible, and working with flexible systems. This new technology has made it possible to make electrochemical sensors. They have very low detection thresholds, are very sensitive, and can process very small shapes. There are a lot of different analytes in complicated biofluids, such as saliva, heavy metals, glucose, lactate, and interstitial fluid. This is why they work well for devices that you can wear. Perovskites and polymeric compounds are used to improve the electroactive surfaces of these devices. This makes signals more stable and speeds up the flow of electrons. Recent improvements in microfluidic topologies, screen printing methods, and flexible substrates have made it possible to make skin-fitting electrochemical sensors that are both cheap and easy to throw away.

Wearable technology has progressed in tandem with optical and optoelectronic sensor technology, facilitating the non-invasive assessment of physiological and environmental data. Perovskite photodetectors are very useful in optoelectronics because they have flexible bandgaps, high photoconductivity, and are easy to process in solution. Phototransistors, photodiodes, and photoconductors are different types of these optoelectronics. Each type is better than the others in three ways: sensitivity, speed, and the ability to identify different wavelengths apart. The substrate, electrode, and encapsulating materials used will determine how well the device works in different situations and how long it lasts. Wearable sensors are getting more useful, easier to use, and faster at collecting data as technology improves. In general, these platforms may look for signs of food deterioration, electrolyte balance, and hydration levels to measure the health of people and the environment.

Perovskite composite mechanical sensors have greatly improved non-invasive evaluation of biomechanical activity. Intricate pressure sensors can assess arterial pressure non-invasively. On the other hand, wearable ultrasound equipment can be used to keep an eye on hemodynamic parameters all the time. Piezoresistive and piezoelectric tactile sensors are made from a mix of perovskites and polymers. They are cheap and can be used for many aspects, such as mobility analysis and therapy. They are very beneficial for monitoring an individual's steps, forecasting fall risks, and facilitating collaboration between humans and technology, since they can detect minute variations in motion and pressure. Extensive research is now being conducted on sensor systems that use triboelectric and piezoelectric properties. These sensors can harvest biomechanical energy and convert mechanical impulses into electrical signals. Halide perovskites and conductive polymer matrices are prevalent components. Due to these technological advancements, it is now feasible to create wearable gadgets that possess extended longevity and need minimal maintenance. This significantly impacts smart agriculture, environmental preservation, and personal wellness. The future of wearable sensing looks good because of improvements in signal processing, device engineering, and materials science. Next, we will look into biocompatible perovskite substitutes, real-time analytics powered by machine learning, and scalable manufacturing methods that can be used in healthcare, consumer, and industrial settings.

## References


1. T.H. Bui, J.H. Shin, Perovskite materials for sensing applications: recent advances and challenges. Microchem. J. **191**, 108924 (2023). https://doi.org/10.1016/j.microc.2023.108924
2. J.K. George, V.V. Halali, C.G. Sanjayan, V. Suvina, M. Sakar, R.G. Balakrishna, Perovskite nanomaterials as optical and electrochemical sensors. Inorg. Chem. Front. **7**(14), 2702–2725 (2020). https://doi.org/10.1039/D0QI00306A
3. D. Maity, T. Govindaraju, Highly selective colorimetric chemosensor for Co2+. Inorg. Chem. **50**(22), 11282–11284 (2011). https://doi.org/10.1021/ic2015447
4. A. Hulanicki, S. Glab, F. Ingman, Chemical sensors: definitions and classification. Pure Appl. Chem. **63**(9), 1247–1250 (1991). https://doi.org/10.1351/pac199163091247
5. X. You, J. Wu, Y. Chi, Superhydrophobic silica aerogels encapsulated fluorescent perovskite quantum dots for reversible sensing of SO2 in a 3D-printed gas cell. Anal. Chem. **91**(8), 5058–5066 (2019). https://doi.org/10.1021/acs.analchem.8b05253
6. A. Balilonda et al., Perovskite fiber-shaped optoelectronic devices for wearable applications. J. Mater. Chem. C **10**(18), 6957–6991 (2022). https://doi.org/10.1039/D2TC00532H
7. S. Wang, A comprehensive review of wearable applications and material construction. Open J. Appl. Sci. **10**, 364–408 (2020). https://doi.org/10.4236/ojapps.2020.106027
8. E. Ismar, S. Kurşun Bahadir, F. Kalaoglu, V. Koncar, Futuristic clothes: electronic textiles and wearable technologies. Glob. Challenges **4**(7), 1900092 (2020). https://doi.org/10.1002/gch2.201900092
9. V. Koncar, 1—Introduction to smart textiles and their applications, in *Smart Textiles and their Applications*, ed. by V. Koncar (Woodhead Publishing, Oxford, 2016), pp. 1–8. https://doi.org/10.1016/B978-0-08-100574-3.00001-1
10. A. Pandiyan et al., A comprehensive review on perovskite and its functional composites in smart textiles: progress, challenges, opportunities, and future directions. Prog. Mater. Sci. **140**, 101206 (2023). https://doi.org/10.1016/j.pmatsci.2023.101206
11. J.R. Sempionatto, J.A. Lasalde-Ramírez, K. Mahato, J. Wang, W. Gao, Wearable chemical sensors for biomarker discovery in the omics era. Nat. Rev. Chem. **6**(12), 899–915 (2022). https://doi.org/10.1038/s41570-022-00439-w
12. E. Costa-Rama, M.T. Fernández-Abedul, Chapter 4—Signal detection techniques, in *Wearable Physical, Chemical and Biological Sensors*, ed. by E. Morales-Narvaez, C. Dincer (Elsevier, 2022), pp. 71–122. https://doi.org/10.1016/B978-0-12-821661-3.00008-2

13. X. Marimon et al., Kinematic analysis of human gait in healthy young adults using IMU sensors: exploring relevant machine learning features for clinical applications. Bioengineering **11**(2), 105 (2024). https://www.mdpi.com/2306-5354/11/2/105
14. T.T. Alemayoh, J.H. Lee, S. Okamoto, Leg-joint angle estimation from a single inertial sensor attached to various lower-body links during walking motion. Appl. Sci. **13**(8), 4794 (2023). https://www.mdpi.com/2076-3417/13/8/4794
15. M. Cao, T. Xie, Z. Chen, Wearable sensors and equipment in VR games: a review, in *Transactions on Edutainment XV*, ed. by Z. Pan, A. D. Cheok, W. Müller, M. Zhang, A. El Rhalibi, K. Kifayat (Springer Berlin Heidelberg, Berlin, Heidelberg, 2019), pp. 3–12. https://doi.org/10.1007/978-3-662-59351-6_1
16. D. James, N. Petrone, *Sensors and Wearable Technologies in Sport: Technologies, Trends and Approaches for Implementation* (2016), pp. 1–49
17. B. Zhang, Y. Xie, J. Zhou, K. Wang, Z. Zhang, State-of-the-art robotic grippers, grasping and control strategies, as well as their applications in agricultural robots: a review. Comput. Electron. Agric. **177**, 105694 (2020). https://doi.org/10.1016/j.compag.2020.105694
18. Y. Zhou et al., Halide-exchanged perovskite photodetectors for wearable visible-blind ultraviolet monitoring. Nano Energy **100**, 107516 (2022). https://doi.org/10.1016/j.nanoen.2022.107516
19. P.V. Shinde, A. Patra, C.S. Rout, A review on the sensing mechanisms and recent developments on metal halide-based perovskite gas sensors. J. Mater. Chem. C **10**(28), 10196–10223 (2022). https://doi.org/10.1039/D2TC01980A
20. W. Xuan, Y. Chen, D. Hu, X. Gao, S. Huang, Smart mask based on lead-free perovskite humidity sensor for labor intensity grading by breath monitoring. Sens. Actuators B Chem. **397**, 134622 (2023). https://doi.org/10.1016/j.snb.2023.134622
21. A. Drexelius, Y. Yuan, M. Friedel, M. DeBrosse, J. Heikenfeld, Chapter 9—Wearable biosensors, in *Wearable Physical, Chemical and Biological Sensors*, ed. by E. Morales-Narvaez, C. Dincer (Elsevier, 2022), pp. 235–254. https://doi.org/10.1016/B978-0-12-821661-3.00001-X
22. W.A. Wood et al., Wearable sensor-based performance status assessment in cancer: a pilot multicenter study from the Alliance for Clinical Trials in Oncology (A19_Pilot2). PLOS Digit. Health **2**(1), e0000178 (2023). https://doi.org/10.1371/journal.pdig.0000178
23. M. Shellaiah, K.W. Sun, Review on sensing applications of perovskite nanomaterials. Chemosensors **8**(3), 55 (2020). https://www.mdpi.com/2227-9040/8/3/55
24. L. Li et al., A review of nanomaterials for biosensing applications. J. Mater. Chem. B **12**(5), 1168–1193 (2024). https://doi.org/10.1039/D3TB02648E
25. D.M. Mathkor et al., Multirole of the internet of medical things (IoMT) in biomedical systems for managing smart healthcare systems: an overview of current and future innovative trends. J. Infect. Public Health **17**(4), 559–572 (2024). https://doi.org/10.1016/j.jiph.2024.01.013
26. S. Khan, J. Kim, S. Acharya, W. Kim, Review on the operation of wearable sensors through body heat harvesting based on thermoelectric devices. Appl. Phys. Lett. **118**(20) (2021). https://doi.org/10.1063/5.0049347
27. D.B. Niranjan, J. Jacob, B.R. Vaidehi, M. Peter, J. Medikonda, P.K. Namboothiri, Current status and applications of photovoltaic technology in wearable sensors: a review. Front. Nanotechnol. **5** (2023). https://doi.org/10.3389/fnano.2023.1268931
28. X. Hu, F. Li, Y. Song, Wearable power source: a newfangled feasibility for perovskite photovoltaics. ACS Energy Lett. **4**(5), 1065–1072 (2019). https://doi.org/10.1021/acsenergylett.9b00503
29. A.S. Dahiya et al., Review—energy autonomous wearable sensors for smart healthcare: a review. J. Electrochem. Soc. **167**(3), 37516 (2020). https://doi.org/10.1149/2.0162003JES
30. W. Fan et al., High-performance stretchable thermoelectric generator for self-powered wearable electronics. Adv. Sci. **10**(12), 2206397 (2023). https://doi.org/10.1002/advs.202206397
31. Z. Tabaie, A. Omidvar, Human body heat-driven thermoelectric generators as a sustainable power supply for wearable electronic devices: recent advances, challenges, and future perspectives. Heliyon **9**(4), e14707 (2023). https://doi.org/10.1016/j.heliyon.2023.e14707

32. L. Zhang et al., Advances in the application of perovskite materials. Nano-Micro Lett. **15**(1), 177 (2023). https://doi.org/10.1007/s40820-023-01140-3
33. L. Clark, P. Lightfoot, 13—Magnetic properties of transition metal fluoride perovskites, in *Photonic and Electronic Properties of Fluoride Materials*, ed. by A. Tressaud, K. Poeppelmeier (Elsevier, Boston, 2016), pp. 261–284. https://doi.org/10.1016/B978-0-12-801639-8.00013-1
34. V. Korde, S. Khot, D. Kamble, S. Amalraj, Review: perovskite nanostructures materials versatile platform for advance biosensor applications. Sens. Actuators Rep. **7**, 100201 (2024). https://doi.org/10.1016/j.snr.2024.100201
35. S. Akin, E. Akman, S. Sonmezoglu, FAPbI3-based perovskite solar cells employing hexyl-based ionic liquid with an efficiency over 20% and excellent long-term stability. Adv. Funct. Mater. **30**(28), 2002964 (2020). https://doi.org/10.1002/adfm.202002964
36. X. Li et al., Dimensional diversity (0D, 1D, 2D, and 3D) in perovskite solar cells: exploring the potential of mixed-dimensional integrations. J. Mater. Chem. A **12**(8), 4421–4440 (2024). https://doi.org/10.1039/D3TA06953B
37. R. Zhao, Z. Gu, P. Li, Y. Zhang, Y. Song, Flexible and wearable optoelectronic devices based on perovskites. Adv. Mater. Technol. **7**(3), 2101124 (2022). https://doi.org/10.1002/admt.202101124
38. D. Li, P. Zhuang, C. Sun, Unlocking the potential of perovskite-based nanomaterials for revolutionary smartphone-based sensor applications. J. Mater. Chem. C **12**(13), 4544–4561 (2024). https://doi.org/10.1039/D3TC04505F
39. Q. Chen et al., Under the spotlight: the organic–inorganic hybrid halide perovskite for optoelectronic applications. Nano Today **10**(3), 355–396 (2015). https://doi.org/10.1016/j.nantod.2015.04.009
40. J. Zhang et al., Printed 1d perovskite photodetector for indoor/outdoor non-contact and real-time sports training monitoring. Adv. Sens. Res. **3**(7), 2300158 (2024). https://doi.org/10.1002/adsr.202300158
41. Y. Hu et al., Perovskite-based photodetector for real-time and quantitative monitoring of sports motion. iScience **26**(11) (2023). https://doi.org/10.1016/j.isci.2023.108298
42. G. Li, S. Li, J. Ahmed, W. Tian, L. Li, Flexible perovskite photodetector with room-temperature self-healing capability without external trigger. InfoMat e12594. https://doi.org/10.1002/inf2.12594
43. Y. Tang et al., Enabling low-drift flexible perovskite photodetectors by electrical modulation for wearable health monitoring and weak light imaging. Nat. Commun. **14**(1), 4961 (2023). https://doi.org/10.1038/s41467-023-40711-1
44. X. Liu et al., Biocompatible metal-free perovskite membranes for wearable X-ray detectors. ACS Appl. Mater. Interfaces **16**(13), 16300–16308 (2024). https://doi.org/10.1021/acsami.4c01069
45. C.I. Piñón-Balderrama et al., Electrochemical perovskite-based sensors for the detection of relevant biomarkers for human kidney health. Chemosensors **11**(9), 507 (2023). https://www.mdpi.com/2227-9040/11/9/507
46. N.J. Jeon, J.H. Noh, Y.C. Kim, W.S. Yang, S. Ryu, S. Il Seok, Solvent engineering for high-performance inorganic–organic hybrid perovskite solar cells. Nat. Mater. **13**(9), 897–903 (2014). https://doi.org/10.1038/nmat4014
47. W. Pisula, Inorganic semiconductors in electronic applications. Electron. Mater. **4**(3), 136–138 (2023). https://www.mdpi.com/2673-3978/4/3/11
48. C. Perumal Veeramalai, S. Feng, X. Zhang, S.V.N. Pammi, V. Pecunia, C. Li, Lead-halide perovskites for next-generation self-powered photodetectors: a comprehensive review. Photon. Res. **9**(6), 968–991 (2021). https://doi.org/10.1364/PRJ.418450
49. Y. Zhao et al., Multifunctional perovskite photodetectors: from molecular-scale crystal structure design to micro/nano-scale morphology manipulation. Nano-Micro Lett. **15**(1), 187 (2023). https://doi.org/10.1007/s40820-023-01161-y
50. X. Wu et al., Trap states in lead iodide perovskites. J. Am. Chem. Soc. **137**(5), 2089–2096 (2015). https://doi.org/10.1021/ja512833n

51. X. Hu et al., High-performance flexible broadband photodetector based on organolead halide perovskite. Adv. Funct. Mater. **24**(46), 7373–7380 (2014). https://doi.org/10.1002/adfm.201402020
52. F. Mei et al., Recent progress in perovskite-based photodetectors: the design of materials and structures. Adv. Phys. X **4**(1), 1592709 (2019). https://doi.org/10.1080/23746149.2019.1592709
53. X. Zhao et al., Interface engineering for gain perovskite photodetectors with extremely high external quantum efficiency. RSC Adv. **10**(54), 32976–32983 (2020). https://doi.org/10.1039/D0RA06618D
54. H. Tsai et al., Quantum efficiency gain in 2D perovskite photo and X-Ray detectors. Adv. Opt. Mater. **11**(23), 2300847 (2023). https://doi.org/10.1002/adom.202300847
55. M. Ahmadi, T. Wu, B. Hu, A review on organic-inorganic halide perovskite photodetectors: device engineering and fundamental physics. Adv. Mater. **29**(41), 1605242 (2017). https://doi.org/10.1002/adma.201605242
56. L. Dou et al., Solution-processed hybrid perovskite photodetectors with high detectivity. Nat. Commun. **5**(1), 5404 (2014). https://doi.org/10.1038/ncomms6404
57. F. Wang, T. Zhang, R. Xie, Z. Wang, W. Hu, How to characterize figures of merit of two-dimensional photodetectors. Nat. Commun. **14**(1), 2224 (2023). https://doi.org/10.1038/s41467-023-37635-1
58. G. Li, Y. Wang, L. Huang, W. Sun, Research progress of high-sensitivity perovskite photodetectors: a review of photodetectors: noise, structure, and materials. ACS Appl. Electron. Mater. **4**(4), 1485–1505 (2022). https://doi.org/10.1021/acsaelm.1c01349
59. Y. Wang, D. Li, L. Chao, T. Niu, Y. Chen, W. Huang, Perovskite photodetectors for flexible electronics: recent advances and perspectives. Appl. Mater. Today **28**, 101509 (2022). https://doi.org/10.1016/j.apmt.2022.101509
60. H. Chen et al., Flexible optoelectronic devices based on metal halide perovskites. Nano Res. **13**(8), 1997–2018 (2020). https://doi.org/10.1007/s12274-020-2805-x
61. D. Wu et al., Recent progress of narrowband perovskite photodetectors: fundamental physics and strategies. Adv. Dev. Instrum. **4**, 6 (2023). https://doi.org/10.34133/adi.0006
62. Z. Yang, J. Dou, M. Wang, J. Li, J. Huang, J. Shao, Flexible all-inorganic photoconductor detectors based on perovskite/hole-conducting layer heterostructures. J. Mater. Chem. C **6**(25), 6739–6746 (2018). https://doi.org/10.1039/C8TC02093K
63. C. Li, Y. Ma, Y. Xiao, L. Shen, L. Ding, Advances in perovskite photodetectors. InfoMat **2**(6), 1247–1256 (2020). https://doi.org/10.1002/inf2.12141
64. H. Wang et al., A review of perovskite-based photodetectors and their applications. Nanomaterials **12**(24), 4390 (2022). https://www.mdpi.com/2079-4991/12/24/4390
65. W.-H. Gao, C. Chen, Perovskites and their constructed near-infrared photodetectors. Nano Energy **128**, 109904 (2024). https://doi.org/10.1016/j.nanoen.2024.109904
66. J. Miao, F. Zhang, Recent progress on highly sensitive perovskite photodetectors. J. Mater. Chem. C **7**(7), 1741–1791 (2019). https://doi.org/10.1039/C8TC06089D
67. S. Cai, X. Xu, W. Yang, J. Chen, X. Fang, Materials and designs for wearable photodetectors. Adv. Mater. **31**(18), 1808138 (2019). https://doi.org/10.1002/adma.201808138
68. M. Tabassum, Q. Zia, Y. Zhou, Y. Wang, M.J. Reece, L. Su, A review of recent developments in smart textiles based on perovskite materials. Textiles **2**(3), 447–463 (2022). https://www.mdpi.com/2673-7248/2/3/25
69. X. Zhang et al., Review on flexible perovskite photodetector: processing and applications. Front. Mech. Eng. **18**(2), 33 (2023). https://doi.org/10.1007/s11465-023-0749-z
70. S. Chen, J. Qi, S. Fan, Z. Qiao, J.C. Yeo, C.T. Lim, Flexible wearable sensors for cardiovascular health monitoring. Adv. Healthc. Mater. **10**(17), 2100116 (2021). https://doi.org/10.1002/adhm.202100116
71. M. Brosch et al., An optically transparent multi-electrode array for combined electrophysiology and optophysiology at the mesoscopic scale. J. Neural Eng. **17**(4), 46014 (2020). https://doi.org/10.1088/1741-2552/aba1a4

72. C.F. Guo, Z. Ren, Flexible transparent conductors based on metal nanowire networks. Mater. Today **18**(3), 143–154 (2015). https://doi.org/10.1016/j.mattod.2014.08.018
73. Y. Qin, X. Ouyang, Y. Lv, W. Liu, Q. Liu, S. Wang, A review of carbon-based conductive inks and their printing technologies for integrated circuits. Coatings **13**(10), 1769 (2023). https://www.mdpi.com/2079-6412/13/10/1769
74. W. Deng et al., All-sprayed-processable, large-area, and flexible perovskite/MXene-based photodetector arrays for photocommunication. Adv. Opt. Mater. **7**(6), 1801521 (2019). https://doi.org/10.1002/adom.201801521
75. G. Yang, J. Li, M. Wu, X. Yu, J. Yu, Recent advances in materials, structures, and applications of flexible photodetectors. Adv. Electron. Mater. **9**(10), 2300340 (2023). https://doi.org/10.1002/aelm.202300340
76. C. Bao, X. Peng, L. Ying, Y. Mei, B. Zhang, H. Long, Two-step spin coating $CsPbBr_3$ thin films and photodetectors in the atmosphere. Opt. Lett. **48**(9), 2472–2475 (2023). https://doi.org/10.1364/OL.485858
77. V.R.F. Schröder et al., Using combinatorial inkjet printing for synthesis and deposition of metal halide perovskites in wavelength-selective photodetectors. Adv. Eng. Mater. **24**(4), 2101111 (2022). https://doi.org/10.1002/adem.202101111
78. Y. Liu et al., Inkjet-printed photodetector arrays based on hybrid perovskite CH3NH3PbI3 microwires. ACS Appl. Mater. Interfaces **9**(13), 11662–11668 (2017). https://doi.org/10.1021/acsami.7b01379
79. J.T. Matondo, H. Hu, Y. Ding, M. Mateen, G. Cheng, J. Ding, Slot-die coating for scalable fabrication of perovskite solar cells and modules. Adv. Mater. Technol. **9**(10), 2302082 (2024). https://doi.org/10.1002/admt.202302082
80. X. Feng et al., Spray-coated perovskite hemispherical photodetector featuring narrow-band and wide-angle imaging. Nat. Commun. **13**(1), 6106 (2022). https://doi.org/10.1038/s41467-022-33934-1
81. T.I. Alanazi, Current spray-coating approaches to manufacture perovskite solar cells. Results Phys. **44**, 106144 (2023). https://doi.org/10.1016/j.rinp.2022.106144
82. F. Fang et al., Blade-coating of a highly crystallized lead-free silver-bismuth halide double perovskite thin film with improved stability for high-performance photodetection. J. Mater. Chem. C **11**(21), 7048–7058 (2023). https://doi.org/10.1039/D3TC01031G
83. Y.-R. Shi, C.-H. Chen, Y.-H. Lou, Z.-K. Wang, Strategies of perovskite mechanical stability for flexible photovoltaics. Mater. Chem. Front. **5**(20), 7467–7478 (2021). https://doi.org/10.1039/D1QM00616A
84. J. Ye et al., Crack-free perovskite layers for high performance and reproducible devices via improved control of ambient conditions during fabrication. Appl. Surf. Sci. **407**, 427–433 (2017). https://doi.org/10.1016/j.apsusc.2017.02.135
85. H. Chen, H. Liu, Z. Zhang, K. Hu, X. Fang, Nanostructured photodetectors: from ultraviolet to terahertz. Adv. Mater. **28**(3), 403–433 (2016). https://doi.org/10.1002/adma.201503534
86. H. Wang, D.H. Kim, Perovskite-based photodetectors: materials and devices. Chem. Soc. Rev. **46**(17), 5204–5236 (2017). https://doi.org/10.1039/C6CS00896H
87. K.O.M. Salih, T.A. Rashid, D. Radovanovic, N. Bacanin, A comprehensive survey on the internet of things with the industrial marketplace. Sensors **22**(3), 730 (2022). https://www.mdpi.com/1424-8220/22/3/730
88. Y. Cao, Y. Zheng, X. Wang, Y. Liu, Y. Liu, Research and prospect on key technologies of indoor positioning based on visible light communication. J. Phys. Conf. Ser. **2160**(1), 12074 (2022). https://doi.org/10.1088/1742-6596/2160/1/012074
89. L. Min et al., Frequency-selective perovskite photodetector for anti-interference optical communications. Nat. Commun. **15**(1), 2066 (2024). https://doi.org/10.1038/s41467-024-46468-5
90. T. Wang et al., A dual strategy to enhance the photoelectric performance of perovskite-based photodetectors for potential applications in optical communications. Chem. Eng. J. **488**, 151068 (2024). https://doi.org/10.1016/j.cej.2024.151068

91. L. Li, S. Ye, J. Qu, F. Zhou, J. Song, G. Shen, Recent advances in perovskite photodetectors for image sensing. Small **17**(18), 2005606 (2021). https://doi.org/10.1002/smll.202005606
92. Y. Liu et al., Perovskite-based color camera inspired by human visual cells. Light Sci. Appl. **12**(1), 43 (2023). https://doi.org/10.1038/s41377-023-01072-y
93. L. Gu et al., A biomimetic eye with a hemispherical perovskite nanowire array retina. Nature **581**(7808), 278–282 (2020). https://doi.org/10.1038/s41586-020-2285-x
94. F. Liu et al., Highly efficient and stable self-powered mixed tin-lead perovskite photodetector used in remote wearable health monitoring technology. Adv. Sci. **10**(5), 2205879 (2023). https://doi.org/10.1002/advs.202205879
95. K.J. Riisnaes et al., 2D hybrid perovskite sensors for environmental and healthcare monitoring. ACS Appl. Mater. Interfaces **16**(24), 31399–31406 (2024). https://doi.org/10.1021/acsami.4c02966
96. S. Kim, X. Xiao, J. Chen, Advances in photoplethysmography for personalized cardiovascular monitoring. Biosensors **12**(10) (2022). https://doi.org/10.3390/bios12100863
97. Y. Lv, G. Cen, W. Li, C. Zhao, W. Mai, Highly sensitive fast-response near-infrared photodetectors based on triple cation Sn-Pb perovskite for pulse oximetry system. Sci. China Mater. **66**(12), 4704–4710 (2023). https://doi.org/10.1007/s40843-023-2613-5
98. T. Zheng et al., Multifunctional and multicolor perovskite-CdSe quantum dots diodes for pulse oximetry. ACS Appl. Mater. Interfaces **15**(17), 20753–20760 (2023). https://doi.org/10.1021/acsami.3c01891
99. J.H. Heo, J.K. Park, Y. Yang, D.S. Lee, S.H. Im, Self-powered flexible all-perovskite X-ray detectors with high sensitivity and fast response. iScience **24**(8) (2021). https://doi.org/10.1016/j.isci.2021.102927
100. Z. Li, F. Zhou, H. Yao, Z. Ci, Z. Yang, Z. Jin, Halide perovskites for high-performance X-ray detector. Mater. Today **48**, 155–175 (2021). https://doi.org/10.1016/j.mattod.2021.01.028
101. C.-F. Lin, K.-W. Huang, Y.-T. Chen, S.-L. Hsueh, M.-H. Li, P. Chen, Perovskite-based X-ray detectors. Nanomaterials **13**(13), 2024 (2023). https://www.mdpi.com/2079-4991/13/13/2024
102. H. Li et al., A solar-blind perovskite scintillator realizing portable X-ray imaging. ACS Energy Lett. **7**(9), 2876–2883 (2022). https://doi.org/10.1021/acsenergylett.2c01440
103. L. Xiang et al., X-ray sensitive hybrid organic photodetectors with embedded $CsPbBr_3$ perovskite quantum dots. Org. Electron. **98**, 106306 (2021). https://doi.org/10.1016/j.orgel.2021.106306
104. X. Zhang et al., Tailoring performance of perovskite-based tunneling photodetector for portable monitoring of ultraviolet radiation risk. Nano Energy **122**, 109282 (2024). https://doi.org/10.1016/j.nanoen.2024.109282
105. W. Tian, Y. Wang, L. Chen, L. Li, Self-powered nanoscale photodetectors. Small **13**(45), 1701848 (2017). https://doi.org/10.1002/smll.201701848
106. S.-F. Leung et al., A self-powered and flexible organometallic halide perovskite photodetector with very high detectivity. Adv. Mater. **30**(8), 1704611 (2018). https://doi.org/10.1002/adma.201704611
107. D. Liu et al., Self-powered and broadband lead-free inorganic perovskite photodetector with high stability. ACS Appl. Mater. Interfaces **12**(27), 30530–30537 (2020). https://doi.org/10.1021/acsami.0c05636
108. H. Sun, W. Tian, F. Cao, J. Xiong, L. Li, Ultrahigh-performance self-powered flexible double-twisted fibrous broadband perovskite photodetector. Adv. Mater. **30**(21), 1706986 (2018). https://doi.org/10.1002/adma.201706986
109. T.M.H. Nguyen, S.K. Lee, S. Kim, C.W. Bark, Practical demonstration of deep-ultraviolet detection with wearable and self-powered halide perovskite-based photodetector. ACS Appl. Mater. Interfaces **13**(48), 57609–57618 (2021). https://doi.org/10.1021/acsami.1c18099
110. P. Helen, Sensor developments for electrophysiological monitoring in healthcare, in *Applied Biomedical Engineering*, ed. by D.G. Gaetano, M. Alistair (IntechOpen, Rijeka, 2011), p. Ch. 12. https://doi.org/10.5772/22958

111. J. Heikenfeld et al., Wearable sensors: modalities, challenges, and prospects. Lab Chip **18**(2), 217–248 (2018). https://doi.org/10.1039/C7LC00914C
112. E.S. Dahiya, A.M. Kalra, A. Lowe, G. Anand, Wearable technology for monitoring electrocardiograms (ECGs) in adults: a scoping review. Sensors **24**(4), 1318 (2024). https://www.mdpi.com/1424-8220/24/4/1318
113. Y.-G. Park, S. Lee, J.-U. Park, Recent progress in wireless sensors for wearable electronics. Sensors **19**(20), 4353 (2019). https://www.mdpi.com/1424-8220/19/20/4353
114. R. Lin et al., Applications of flexible electronics related to cardiocerebral vascular system. Mater. Today Bio **23**, 100787 (2023). https://doi.org/10.1016/j.mtbio.2023.100787
115. H. Ullah et al., Recent advances in stretchable and wearable capacitive electrophysiological sensors for long-term health monitoring. Biosensors **12**(8), 630 (2022). https://www.mdpi.com/2079-6374/12/8/630
116. S. Ramasamy, A. Balan, Wearable sensors for ECG measurement: a review. Sens. Rev. **38**(4), 412–419 (2018). https://doi.org/10.1108/SR-06-2017-0110
117. T.W. Kang, J. Lee, Y. Kwon, Y.J. Lee, W.-H. Yeo, Recent progress in the development of flexible wearable electrodes for electrocardiogram monitoring during exercise. Adv. NanoBiomed. Res. **4**(8), 2300169 (2024). https://doi.org/10.1002/anbr.202300169
118. P. Mao, H. Li, Z. Yu, A review of skin-wearable sensors for non-invasive health monitoring applications. Sensors **23**(7), 3673 (2023). https://www.mdpi.com/1424-8220/23/7/3673
119. A. Albulbul, Evaluating major electrode types for idle biological signal measurements for modern medical technology. Bioengineering **3**(3) (2016). https://doi.org/10.3390/bioengineering3030020
120. C.R. Merritt, H.T. Nagle, E. Grant, Textile-based capacitive sensors for respiration monitoring. IEEE Sens. J. **9**(1), 71–78 (2009). https://doi.org/10.1109/JSEN.2008.2010356
121. H. Lee, J. Jang, J. Lee, M. Shin, J.S. Lee, D. Son, Stretchable gold nanomembrane electrode with ionic hydrogel skin-adhesive properties. Polymers **15**(18) (2023). https://doi.org/10.3390/polym15183852
122. M. Chi, J. Zhao, Y. Dong, X. Wang, Flexible carbon nanotube-based polymer electrode for long-term electrocardiographic recording. Materials **12**(6) (2019). https://doi.org/10.3390/ma12060971
123. A.A. Chlaihawi, B.B. Narakathu, S. Emamian, B.J. Bazuin, M.Z. Atashbar, Development of printed and flexible dry ECG electrodes. Sens. Bio-Sensing Res. **20**, 9–15 (2018). https://doi.org/10.1016/j.sbsr.2018.05.001
124. Y. Yang, Y. Wang, Y. Qiao, Y. Song, Flexible transparent electrodes based on metallic micro–nano architectures for perovskite solar cells. J. Mater. Chem. C **10**(7), 2349–2363 (2022). https://doi.org/10.1039/D1TC04101K
125. A. Lymberis, R. Paradiso, Smart fabrics and interactive textile enabling wearable personal applications: R&D state of the art and future challenges, in *Annu Int Conf IEEE Eng Med Biol Soc*, vol. 2008 (2008), pp. 5270–5273. https://doi.org/10.1109/iembs.2008.4650403
126. J. Yoo, L. Yan, S. Lee, H. Kim, H.J. Yoo, A wearable ECG acquisition system with compact planar-fashionable circuit board-based shirt. IEEE Trans. Inf. Technol. Biomed. **13**(6), 897–902 (2009). https://doi.org/10.1109/titb.2009.2033053
127. S. Imani et al., A wearable chemical–electrophysiological hybrid biosensing system for real-time health and fitness monitoring. Nat. Commun. **7**(1), 11650 (2016). https://doi.org/10.1038/ncomms11650
128. X. Luo et al., Visual electrocardiogram synchronization monitor using perovskite-based multicolor light-emitting diodes. ACS Photon. **8**(11), 3337–3345 (2021). https://doi.org/10.1021/acsphotonics.1c01296
129. C.S. Kim et al., Self-powered wearable electrocardiography using a wearable thermoelectric power generator. ACS Energy Lett. **3**(3), 501–507 (2018). https://doi.org/10.1021/acsenergylett.7b01237
130. S. Subramaniam, A.I. Faisal, M.J. Deen, Wearable sensor systems for fall risk assessment: a review. Front Digit. Heal. **4**, 921506 (2022). https://doi.org/10.3389/fdgth.2022.921506

131. V. Bucinskas et al., Wearable feet pressure sensor for human gait and falling diagnosis. Sensors (Basel) **21**(15) (2021). https://doi.org/10.3390/s21155240
132. T.-D. Nguyen, J.S. Lee, Recent development of flexible tactile sensors and their applications. Sensors **22**(1), 50 (2022). https://www.mdpi.com/1424-8220/22/1/50
133. S. Ippili, V. Jella, A.M. Thomas, S.-G. Yoon, The recent progress on halide perovskite-based self-powered sensors enabled by piezoelectric and triboelectric effects. Nanoenergy Adv. **1**(1), 3–31 (2021). https://www.mdpi.com/2673-706X/1/1/2
134. J. Li, L. Fang, B. Sun, X. Li, S.H. Kang, Review—recent progress in flexible and stretchable piezoresistive sensors and their applications. J. Electrochem. Soc. **167**(3), 37561 (2020). https://doi.org/10.1149/1945-7111/ab6828
135. W. Yan et al., Giant gauge factor of Van der Waals material based strain sensors. Nat. Commun. **12**(1), 2018 (2021). https://doi.org/10.1038/s41467-021-22316-8
136. Y. Ding, T. Xu, O. Onyilagha, H. Fong, Z. Zhu, Recent advances in flexible and wearable pressure sensors based on piezoresistive 3D monolithic conductive sponges. ACS Appl. Mater. Interfaces **11**(7), 6685–6704 (2019). https://doi.org/10.1021/acsami.8b20929
137. A. Mukherjee et al., Electrically conductive and highly stretchable piezoresistive polymer nanocomposites via oxidative chemical vapor deposition. ACS Appl. Mater. Interfaces **15**(26), 31899–31916 (2023). https://doi.org/10.1021/acsami.3c06015
138. D. Lee, J. Kim, H. Kim, H. Heo, K. Park, Y. Lee, High-performance transparent pressure sensors based on sea-urchin shaped metal nanoparticles and polyurethane microdome arrays for real-time monitoring. Nanoscale **10**(39), 18812–18820 (2018). https://doi.org/10.1039/C8NR05843A
139. N. Yang et al., High-performance wearable piezoresistive sensor with a wide temperature range via a Ti3C2Tx MXene/Au dual-layer conductive network and microspike structure. ACS Appl. Nano Mater. **7**(14), 16964–16974 (2024). https://doi.org/10.1021/acsanm.4c03060
140. S. Chen et al., Fabrication and piezoresistive/piezoelectric sensing characteristics of carbon nanotube/PVA/Nano-ZnO flexible composite. Sci. Rep. **10**(1), 8895 (2020). https://doi.org/10.1038/s41598-020-65771-x
141. Y. Zhuang, R. Li, M. Jiao, X. He, X. Gu, S. Huang, A flexible piezoresistive pressure sensor based on a perovskite $MAPbBr_3$ nanocrystal-embedded polymer composite. J. Mater. Chem. C **12**(28), 10494–10503 (2024). https://doi.org/10.1039/D4TC01439A
142. E. Caffrey, J.R. Garcia, D. O'Suilleabhain, C. Gabbett, T. Carey, J.N. Coleman, Quantifying the piezoresistive mechanism in high-performance printed graphene strain sensors. ACS Appl. Mater. Interfaces **14**(5), 7141–7151 (2022). https://doi.org/10.1021/acsami.1c21623
143. W. Chiappim, M.A. Fraga, H. Furlan, D.C. Ardiles, R.S. Pessoa, The status and perspectives of nanostructured materials and fabrication processes for wearable piezoresistive sensors. Microsyst. Technol. **28**(7), 1561–1580 (2022). https://doi.org/10.1007/s00542-022-05269-w
144. J. He et al., Recent advances of wearable and flexible piezoresistivity pressure sensor devices and its future prospects. J. Mater. **6**(1), 86–101 (2020). https://doi.org/10.1016/j.jmat.2020.01.009
145. K.-Y. Chen, Y.-T. Xu, Y. Zhao, J.-K. Li, X.-P. Wang, L.-T. Qu, Recent progress in graphene-based wearable piezoresistive sensors: from 1D to 3D device geometries. Nano Mater. Sci. **5**(3), 247–264 (2023). https://doi.org/10.1016/j.nanoms.2021.11.003
146. G. Hu, F. Huang, C. Tang, J. Gu, Z. Yu, Y. Zhao, High-performance flexible piezoresistive pressure sensor printed with 3D microstructures. Nanomaterials **12**(19), 3417 (2022). https://www.mdpi.com/2079-4991/12/19/3417
147. A.M. Kamat, Y. Pei, B. Jayawardhana, A.G.P. Kottapalli, Biomimetic soft polymer microstructures and piezoresistive graphene MEMS sensors using sacrificial metal 3D printing. ACS Appl. Mater. Interfaces **13**(1), 1094–1104 (2021). https://doi.org/10.1021/acsami.0c21295
148. Y. Zhuang et al., Fabrication of lead-free perovskite MASnBrI2 nanocrystals-embedded polymer composites for flexible strain sensors. Ceram. Int. **50**(3, Part B), 5766–5774 (2024). https://doi.org/10.1016/j.ceramint.2023.11.375
149. X. Li et al., Stretchable luminescent perovskite-polymer hydrogels for visual-digital wearable strain sensor textiles. Adv. Fiber Mater. **5**(5), 1671–1684 (2023). https://doi.org/10.1007/s42765-023-00302-7

150. S. Ying et al., A flexible piezocapacitive pressure sensor with microsphere-array electrodes. Nanomaterials **13**(11), 1702 (2023). https://www.mdpi.com/2079-4991/13/11/1702
151. Z. Ma, Y. Zhang, K. Zhang, H. Deng, Q. Fu, Recent progress in flexible capacitive sensors: structures and properties. Nano Mater. Sci. **5**(3), 265–277 (2023). https://doi.org/10.1016/j.nanoms.2021.11.002
152. S.G. Yoon, B.J. Park, S.T. Chang, Highly sensitive piezocapacitive sensor for detecting static and dynamic pressure using ion-gel thin films and conductive elastomeric composites. ACS Appl. Mater. Interfaces **9**(41), 36206–36219 (2017). https://doi.org/10.1021/acsami.7b11700
153. M.-J. Yin, Z. Yin, Y. Zhang, Q. Zheng, A.P. Zhang, Micropatterned elastic ionic polyacrylamide hydrogel for low-voltage capacitive and organic thin-film transistor pressure sensors. Nano Energy **58**, 96–104 (2019). https://doi.org/10.1016/j.nanoen.2019.01.032
154. S.C.B. Mannsfeld et al., Highly sensitive flexible pressure sensors with microstructured rubber dielectric layers. Nat. Mater. **9**(10), 859–864 (2010). https://doi.org/10.1038/nmat2834
155. Y. Zhang et al., Ultrafast piezocapacitive soft pressure sensors with over 10 kHz bandwidth via bonded microstructured interfaces. Nat. Commun. **15**(1), 3048 (2024). https://doi.org/10.1038/s41467-024-47408-z
156. B. Padha, S. Verma, S. Arya, Fabric-based wearable self-powered asymmetric supercapacitor comprising lead-free perovskite piezoelectrodes. Adv. Mater. Technol. **7**(10), 2200079 (2022). https://doi.org/10.1002/admt.202200079
157. L. Pang et al., Enhanced pressure & proximity sensitivities of a flexible transparent capacitive sensor with PZT nanowires, in *IOP Conf. Ser. Mater. Sci. Eng.*, vol. 479, no. 1 (2019), p. 12035. https://doi.org/10.1088/1757-899X/479/1/012035
158. S.O. Abdellatif, A. Moustafa, A. Khalid, R. Ghannam, Integration of capacitive pressure sensor-on-chip with lead-free perovskite solar cells for continuous health monitoring. Micromachines **14**(9), 1676 (2023). https://www.mdpi.com/2072-666X/14/9/1676
159. R. Pandey et al., Mutual insight on ferroelectrics and hybrid halide perovskites: a platform for future multifunctional energy conversion. Adv. Mater. **31**(43), 1807376 (2019). https://doi.org/10.1002/adma.201807376
160. A. Sultana et al., Organo-lead halide perovskite regulated green light emitting poly(vinylidene fluoride) electrospun nanofiber mat and its potential utility for ambient mechanical energy harvesting application. Nano Energy **49**, 380–392 (2018). https://doi.org/10.1016/j.nanoen.2018.04.057
161. Y. Bai, T. Siponkoski, J. Peräntie, H. Jantunen, J. Juuti, Ferroelectric, pyroelectric, and piezoelectric properties of a photovoltaic perovskite oxide. Appl. Phys. Lett. **110**(6) (2017). https://doi.org/10.1063/1.4974735
162. Z. Wang et al., Light-induced pyroelectric effect as an effective approach for ultrafast ultraviolet nanosensing. Nat. Commun. **6**(1), 8401 (2015). https://doi.org/10.1038/ncomms9401
163. R. Ding, M.-C. Wong, J. Hao, Recent advances in hybrid perovskite nanogenerators. EcoMat **2**(4), e12057 (2020). https://doi.org/10.1002/eom2.12057
164. L. Zhang, F. Xue, W. Du, C. Han, C. Zhang, Z. Wang, Transparent paper-based triboelectric nanogenerator as a page mark and anti-theft sensor. Nano Res. **7**(8), 1215–1223 (2014). https://doi.org/10.1007/s12274-014-0484-1
165. S. Wang, L. Lin, Z.L. Wang, Triboelectric nanogenerators as self-powered active sensors. Nano Energy **11**, 436–462 (2015). https://doi.org/10.1016/j.nanoen.2014.10.034
166. J.P. Lee, J.W. Lee, J.M. Baik, The progress of PVDF as a functional material for triboelectric nanogenerators and self-powered sensors. Micromachines (Basel) **9**(10) (2018). https://doi.org/10.3390/mi9100532
167. P. Munirathinam et al., A comprehensive review on triboelectric nanogenerators based on real-time applications in energy harvesting and self-powered sensing. Mater. Sci. Eng. B **297**, 116762 (2023). https://doi.org/10.1016/j.mseb.2023.116762
168. M. Sahu, S. Hajra, J. Bijelic, D. Oh, I. Djerdj, H.J. Kim, Triple perovskite-based triboelectric nanogenerator: a facile method of energy harvesting and self-powered information generator. Mater. Today Energy **20**, 100639 (2021). https://doi.org/10.1016/j.mtener.2021.100639

169. W. Yang et al., Triboelectrification based motion sensor for human-machine interfacing. ACS Appl. Mater. Interfaces **6**(10), 7479–7484 (2014). https://doi.org/10.1021/am500864t
170. S. Mondal, T. Paul, S. Maiti, B.K. Das, K.K. Chattopadhyay, Human motion interactive mechanical energy harvester based on all inorganic perovskite-PVDF. Nano Energy **74**, 104870 (2020). https://doi.org/10.1016/j.nanoen.2020.104870
171. N.H. Makani, P. Kumar, T. Paul, A. Sahoo, S. Maiti, R. Banerjee, Low-dimensional halide perovskite/PVDF nanocomposite with enhanced piezoelectricity as flexible biomechanical energy harvester. Chem. Phys. Impact **8**, 100564 (2024). https://doi.org/10.1016/j.chphi.2024.100564
172. S. Mondal, S. Poddar, S. Bhattacharjee, S. Maiti, A. Banerjee, K.K. Chattopadhyay, Lead free halide perovskite embedded PVDF based efficient mechanical energy harvester: self-driven respiratory sensor. Nano Energy **115**, 108689 (2023). https://doi.org/10.1016/j.nanoen.2023.108689
173. Z. Hao et al., Co-harvesting light and mechanical energy based on dynamic metal/perovskite Schottky junction. Matter **1**(3), 639–649 (2019). https://doi.org/10.1016/j.matt.2019.05.003
174. L. Su et al., High-performance organolead halide perovskite-based self-powered triboelectric photodetector. ACS Nano **9**(11), 11310–11316 (2015). https://doi.org/10.1021/acsnano.5b04995
175. J. Du, X. Yang, J. Duan, Y. Wang, Q. Tang, Tailoring all-inorganic cesium lead halide perovskites for robust triboelectric nanogenerators. Nano Energy **70**, 104514 (2020). https://doi.org/10.1016/j.nanoen.2020.104514
176. C.I.L. Justino, A.C. Freitas, R. Pereira, A.C. Duarte, T.A.P. Rocha Santos, Recent developments in recognition elements for chemical sensors and biosensors. TrAC Trends Anal. Chem. **68**, 2–17 (2015). https://doi.org/10.1016/j.trac.2015.03.006
177. A.J. Bandodkar, J. Wang, Non-invasive wearable electrochemical sensors: a review. Trends Biotechnol. **32**(7), 363–371 (2014). https://doi.org/10.1016/j.tibtech.2014.04.005
178. S. Ye, S. Feng, L. Huang, S. Bian, Recent progress in wearable biosensors: from healthcare monitoring to sports analytics. Biosensor **10**(12) (2020). https://doi.org/10.3390/bios10120205
179. J.R. Windmiller, J. Wang, Wearable electrochemical sensors and biosensors: a review. Electroanalysis **25**(1), 29–46 (2013). https://doi.org/10.1002/elan.201200349
180. N. Pisanic et al., COVID-19 serology at population scale: SARS-CoV-2-specific antibody responses in saliva. J. Clin. Microbiol. **59**(1) (2020). https://doi.org/10.1128/jcm.02204-20
181. J. Timpel et al., Sensors for in situ monitoring of oral and dental health parameters in saliva. Clin. Oral Investig. **27**(10), 5719–5736 (2023). https://doi.org/10.1007/s00784-023-05206-9
182. F.C. Thompson, F. Brudevold, A micro-antimony electrode designed for intraoral pH measurements in man and small experimental animals. J. Dent. Res. **33**(6), 849–853 (1954). https://doi.org/10.1177/00220345540330061301
183. M. Farella et al., Simultaneous wireless assessment of intra-oral pH and temperature. J. Dent. **51**, 49–55 (2016). https://doi.org/10.1016/j.jdent.2016.05.012
184. F. Gao et al., Wearable and flexible electrochemical sensors for sweat analysis: a review. Microsyst. Nanoeng. **9**(1), 1 (2023). https://doi.org/10.1038/s41378-022-00443-6
185. Q. Yan, B. Peng, G. Su, B.E. Cohan, T.C. Major, M.E. Meyerhoff, Measurement of tear glucose levels with amperometric glucose biosensor/capillary tube configuration. Anal. Chem. **83**(21), 8341–8346 (2011). https://doi.org/10.1021/ac201700c
186. W. Clarke, P. D'Orazio, Chapter 9—Electrochemistry, in *Contemporary Practice in Clinical Chemistry*, 4th edn, ed. by W. Clarke, M.A. Marzinke (Academic Press, 2020), pp. 159–170. https://doi.org/10.1016/B978-0-12-815499-1.00009-0
187. H. Ehtesabi, Carbon nanomaterials for salivary-based biosensors: a review. Mater. Today Chem. **17**, 100342 (2020). https://doi.org/10.1016/j.mtchem.2020.100342
188. J. He, X. Xu, M. Li, S. Zhou, W. Zhou, Recent advances in perovskite oxides for non-enzymatic electrochemical sensors: a review. Anal. Chim. Acta **1251**, 341007 (2023). https://doi.org/10.1016/j.aca.2023.341007

189. A. Navrotsky, Thermochemistry of perovskite-related oxides with high oxidation states: superconductors, sensors, fuel cell materials. Pure Appl. Chem. **66**(9), 1759–1764 (1994). https://doi.org/10.1351/pac199466091759
190. M. Soleymani, A. Moheb, D. Babakhani, Hydrogen peroxide decomposition over nanosized La1–XCaXMnO3 ($0 \leq X \leq 0.6$) perovskite oxides. Chem. Eng. Technol. **34**(1), 49–55 (2011). https://doi.org/10.1002/ceat.201000293
191. Y.N. Lee, R.M. Lago, J.L.G. Fierro, J. González, Hydrogen peroxide decomposition over Ln1−xAxMnO3 (Ln = La or Nd and A = K or Sr) perovskites. Appl. Catal. A Gen. **215**(1), 245–256 (2001). https://doi.org/10.1016/S0926-860X(01)00536-1
192. Y. Shimizu, H. Komatsu, S. Michishita, N. Miura, N. Yamazo, Sensing characteristics of hydrogen peroxide sensor using carbon-based electrode loaded with perovskite-type oxide. Sens. Actuators B Chem. **34**(1), 493–498 (1996). https://doi.org/10.1016/S0925-4005(97)80021-4
193. M. Gao et al., Glucose and pH responsive fluorescence detection system based on simple synthesis of silicon-coated perovskite quantum dots. Spectrochim. Acta Part A Mol. Biomol. Spectrosc. **289**, 122212 (2023). https://doi.org/10.1016/j.saa.2022.122212
194. J. He et al., High-performance non-enzymatic perovskite sensor for hydrogen peroxide and glucose electrochemical detection. Sens. Actuators B Chem. **244**, 482–491 (2017). https://doi.org/10.1016/j.snb.2017.01.012
195. I. Boubezari, A. Zazoua, A. Errachid, N. Jaffrezic-Renault, Sensitive electrochemical detection of bioactive molecules (hydrogen peroxide, glucose, dopamine) with perovskites-based sensors. Chemosensors **9**(10), 289 (2021). https://www.mdpi.com/2227-9040/9/10/289
196. X. Niu, H. Gao, J. Du, $CsPbBr_3$ perovskite nanocrystals decorated with Cu nanoclusters for ratiometric detection of glucose. ACS Appl. Nano Mater. **5**(2), 2350–2357 (2022). https://doi.org/10.1021/acsanm.1c04037
197. F.-F. Jia et al., A novel nonenzymatic ECL glucose sensor based on perovskite LaTiO3-Ag0.1 nanomaterials. Sens. Actuators B Chem. **212**, 174–182 (2015). https://doi.org/10.1016/j.snb.2015.02.011
198. S. Hooshmand et al., Wearable nano-based gas sensors for environmental monitoring and encountered challenges in optimization. Sensors **23**(20), 8648 (2023). https://www.mdpi.com/1424-8220/23/20/8648
199. J. Casanova-Chafer, R. Garcia-Aboal, P. Atienzar, E. Llobet, Unraveling the gas-sensing mechanisms of lead-free perovskites supported on graphene. ACS Sens. **7**(12), 3753–3763 (2022). https://doi.org/10.1021/acssensors.2c01581
200. A. Maity, S. Mitra, B. Ghosh, Textile based lead-free halide perovskite CH3NH3SnI3 ammonia gas sensor working at room temperature. ACS Appl. Electron. Mater. **6**(4), 2677–2682 (2024). https://doi.org/10.1021/acsaelm.4c00230
201. J. Wang et al., Ultrathin 2D NbWO6 perovskite semiconductor based gas sensors with ultrahigh selectivity under low working temperature. Adv. Mater. **34**(2), 2104958 (2022). https://doi.org/10.1002/adma.202104958

# Chapter 6
# Perovskite-Based Energy Devices for Energy Harvesting and Storage Systems

This chapter presents a detailed exploration of perovskite-based energy devices, emphasizing their critical role in advancing energy harvesting and storage systems for next-generation wearable technologies. Key developments in perovskite thermoelectrics, photovoltaics, and piezoelectrics are discussed, with a focus on their outstanding material properties, fabrication techniques, and energy conversion efficiencies. Furthermore, the chapter investigates the incorporation of perovskites into batteries and supercapacitors, showcasing their potential in enabling flexible, lightweight, and high-performance energy storage solutions. By linking the distinctive attributes of perovskite materials to emerging applications, this chapter highlights their transformative impact on the development of sustainable and self-powered wearable devices.

## 6.1 Thermoelectrics

### *6.1.1 Evolution of Thermoelectric Effects*

The study of thermoelectric (TE) effects has a rich history dating back over two centuries, with foundational discoveries that have profoundly shaped the field of thermal energy management and conversion. The genesis of this field can be traced back to the early nineteenth century, specifically in the year 1821. It was during this period that the German physicist Thomas Johann Seebeck first observed a fundamental phenomenon in thermoelectricity: the generation of an electrical voltage when two dissimilar metals, such as copper and bismuth, are joined at two junctions maintained at different temperatures. This observation, now termed the Seebeck effect, serves as the cornerstone of thermoelectric technology (refer to Fig. 6.1) [1].

When two conductive materials form a thermocouple, they create a closed circuit. The application of heat to one of the junctions induces excitation of electrons,



M. Darab et al., *Perovskites for Next-Generation Wearable Technology*, Engineering Materials, https://doi.org/10.1007/978-3-032-03931-6_6

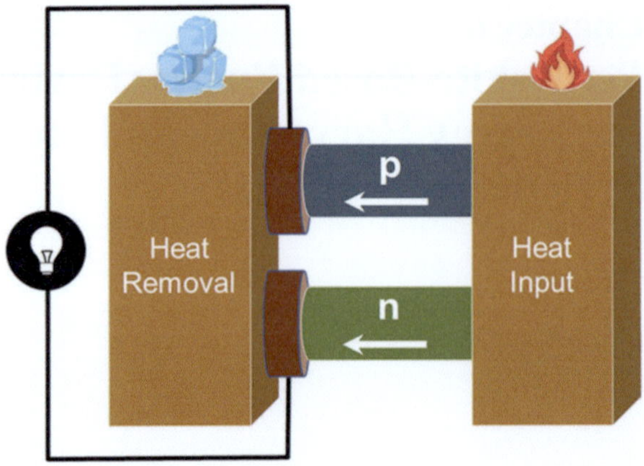


**Fig. 6.1** Schematic illustration of the Seebeck effect

primarily those in the valence band. These thermally excited electrons migrate from the hotter junction to the cooler one, driven by the combined forces of thermal energy and their inherent kinetic energy. This migration creates an electron density imbalance, resulting in a flow of direct current (DC) when connected in a closed circuit.

The voltage generated as a function of the temperature difference across the thermocouple junctions can be mathematically described by the Eq. (6.1) [2]:

$$\Delta V = S \Delta T \tag{6.1}$$

where ($\Delta V$) represents the voltage produced, ($\Delta T$) is the temperature gradient, and (S) is the Seebeck coefficient, a material-dependent property that varies with temperature.

Although the voltage generated by the Seebeck effect is typically diminutive—measured in microvolts per kelvin (μV/K)—larger temperature differentials can produce millivolt-level outputs. To augment both the efficiency and voltage output of thermoelectric generators (TEGs), numerous thermocouples can be connected in series (to elevate voltage) or in parallel (to increase current). This adaptability allows large configurations of Seebeck-effect devices to provide practical electrical power, contingent upon sustained temperature gradients. The efficiency of thermoelectric devices is significantly influenced by the Seebeck coefficient of the materials used. Higher Seebeck coefficients are desirable, driving ongoing research into nanostructured materials and optimizing TE properties through material engineering. Concurrently, the interplay of electrical and thermal conductivity within a material also plays a crucial role in energy conversion efficiency, as minimizing thermal losses is essential for effective performance. Consequently, advancements in materials science and the synthesis of novel thermoelectric materials underscore a vibrant area of research focused on elevating the parameters that dictate thermoelectric performance [3].

The exploration of thermoelectricity deepened in 1834 when French physicist Jean Charles Athanase Peltier elucidated the reverse thermoelectric phenomenon known

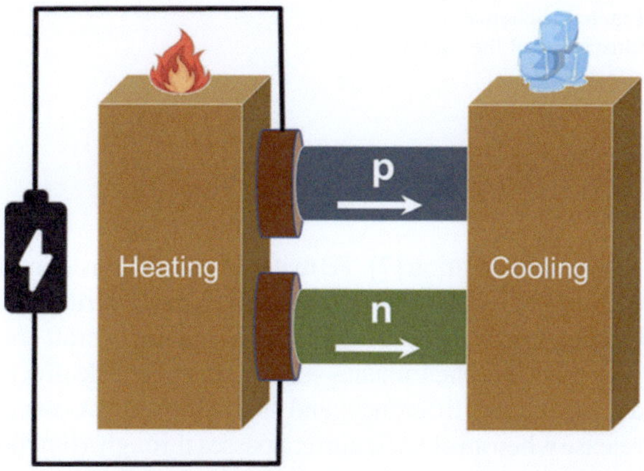


**Fig. 6.2** Schematic illustration of the Peltier effect

as the Peltier effect. This effect describes the heat absorption or release occurring at the junction between two dissimilar conductive materials when an electric current flows through them (see Fig. 6.2). The Peltier effect's utility lies in its capacity for active thermal management, thus enabling the development of thermoelectric refrigeration and heating systems. A landmark demonstration of this effect was conducted by Friedrich Emil Lenz in 1838, exemplifying the principles of TE refrigeration. By applying a current through a bismuth-antimony junction, Lenz was able to freeze a drop of water; reversing the current subsequently caused the frozen water to melt. The relationship governing the heat exchange in the Peltier effect can be delineated by the Eq. (6.2) [4]:

$$\Delta Q = \Pi \cdot I \cdot \Delta T \tag{6.2}$$

where ($\Delta Q$) is the heat transferred, ($I$) is the electric current, and ($\Pi$) represents the Peltier coefficient, indicative of the heat transfer efficiency between the two materials. The Peltier coefficient also varies based on the materials and operational conditions.

Further complicating the thermoelectric landscape was British scientist William Thomson's discovery of the Thomson effect in 1851, which captures the interaction of electrical current through materials with existing temperature gradients. The Thomson effect describes how charge carriers respond to temperature differentials in conductors and semiconductors, either absorbing or releasing heat as they conduct current (refer to Fig. 6.3). This can be mathematically expressed via the equation:

$$\dot{Q} = -K \cdot J \cdot \nabla T \tag{6.3}$$

where ($\dot{Q}$) signifies the heat generated per unit volume, ($J$) is the current density, ($\nabla T$) represents the temperature gradient, and ($K$) is the Thomson coefficient, unique to each material [5].

The significance of the Thomson effect extends to various applications in thermoelectric devices, where it plays a pivotal role in systems designed for power generation

**Fig. 6.3** Schematic illustration of the Thomson effect

[6] and refrigeration [7]. TE materials are of considerable interest in the field of energy conversion technologies, especially for recovering waste heat and converting it into usable electrical energy, thereby enhancing overall energy efficiency. In conjunction with other TE phenomena—such as the Seebeck effect, which generates voltage from a temperature difference, and the Peltier effect, which involves heat absorption or release when an electric current passes through a junction of two different materials—the Thomson effect contributes to a comprehensive understanding of thermoelectric behaviors and applications. Each of these effects presents unique characteristics and implications based on the material properties involved and the specific operating conditions.

The evolution of thermoelectric science has significantly progressed since the foundational discoveries of the Seebeck, Peltier, and Thomson effects. Between the late 19th and early twentieth centuries, a deeper understanding of these TE phenomena emerged, largely due to advances in materials science and thermodynamic principles. Researchers began systematically investigating various materials and the conditions necessary to optimize thermoelectric effects, paving the way for innovative applications. During World War II, there was a notable surge in the exploration of thermoelectric devices, driven by military requirements for advanced technologies. These devices were considered essential for powering instruments and for ensuring efficient cooling in military applications, highlighting their potential beyond academic inquiry [8].

The 1950s marked a turning point in thermoelectric research with the advent of semiconductor materials, which catalyzed improvements in device efficiency. Among these developments, bismuth telluride emerged as a premier thermoelectric material, recognized for its favorable thermoelectric properties. This period set a significant milestone in the commercial viability of thermoelectric devices [9].

Interest in thermoelectric materials experienced a renaissance in the 1990s and 2000s, primarily fueled by the increasing global demand for energy efficiency and the need for effective waste heat recovery technologies. Researchers focused on developing strategies to enhance the thermoelectric performance of materials by minimizing thermal conductivity ($K$) while maximizing electrical conductivity ($\sigma$), Seebeck coefficient ($S$), and temperature ($T$), epitomized in the pursuit of the "figure of merit" (FOM). This dimensionless number quantitatively describes the efficiency of TE materials and has become a focal point in research and development, as follows [10]:

$$FOM = \frac{S^2\sigma}{K}T \tag{6.4}$$

In recent years, the thermoelectric field has experienced significant advancements, particularly with the discovery and investigation of novel materials such as skutterudites, half-Heusler alloys, nanostructured materials, and perovskites [11–13]. These materials have exhibited exceptional thermoelectric performance, leading to substantial enhancements in figure of merit (FOM) values. The landscape of thermoelectric research is dynamic and continuously evolving. Current research efforts are concentrated on the synthesis of innovative materials, optimization of manufacturing processes, and the scaling of thermoelectric technologies for practical and commercial applications. Interdisciplinary collaboration among physicists, materials scientists, and engineers is crucial for addressing the challenges inherent in this field and for realizing the full potential of thermoelectric devices within everyday technologies. Innovations in design methodologies, advanced manufacturing techniques, and integration with complementary energy systems represent promising paths for the future of thermoelectric applications. As the global emphasis on sustainable energy solutions grows, thermoelectric materials are likely to assume a critical role in enhancing energy efficiency and enabling waste heat recovery. Figure 6.4 illustrates the temperature-dependent FOMs for cutting-edge bulk thermoelectric materials that have been developed in recent years [14]. In the subsequent section, we will review the application of perovskite materials within the realm of thermoelectric technology.

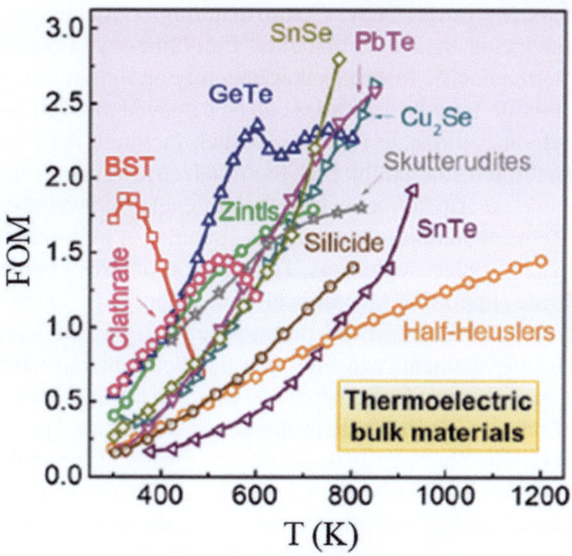


**Fig. 6.4** Temperature-dependent FOMs for cutting-edge bulk thermoelectric materials. Reproduced with permission from [14] © 2020, Elsevier. The underlying datasets used to generate this figure are credited in the original source

## *6.1.2 Perovskites-Based Wearable Thermoelectrics*

### 6.1.2.1 Perovskite Oxides

The application of perovskite materials in thermoelectric devices has attracted substantial scholarly attention in recent decades due to their exceptional thermoelectric characteristics. Investigations into the thermoelectric properties of various materials—including simple oxides and carbides—began in the latter part of the twentieth century. Research revealed that the unique crystal structure inherent to perovskites could facilitate improved thermoelectric performance, attributable to their distinctive electronic and phononic attributes. Interest in oxide-based perovskites reached a zenith in the 1990s, as materials such as $SrTiO_3$ [15] and $LaCoO_3$ [16] began to be scrutinized for their thermoelectric potential. These perovskite oxides demonstrate a compelling combination of performance, stability, and environmental sustainability, positioning them as promising candidates for future thermoelectric applications, particularly in energy conversion and waste heat recovery.

Figure 6.5 summarizes the numerous advantages that oxide-based perovskites offer within this domain. The general $ABO_3$ structure of these oxides accommodates a diverse array of chemical compositions and substitutions, allowing for the tuning of electronic and thermal transport properties to optimize thermoelectric effectiveness. Moreover, many of these materials consist of abundant and non-toxic elements, enhancing their environmental friendliness and cost-effectiveness relative to other thermoelectric materials that may rely on rare or hazardous elements. The synthesis of oxide-based perovskites can be achieved through various methods, including sol–gel and solid-state reactions, which facilitate their fabrication into devices. These materials often exhibit high thermoelectric figures of merit (FOM) at elevated temperatures, a critical factor for efficient energy conversion. Their excellent thermal and chemical stability renders them suitable for high-temperature applications prevalent in thermoelectric systems. The lattice structure of oxides can hinder phonon transport while supporting favorable electrical conductivity, resulting in low thermal conductivity and an improved thermoelectric figure of merit. Additionally, doping with specific elements can fine-tune the electronic structure, enabling a transition from metallic to insulating behavior, which is advantageous for enhancing thermoelectric efficiency. The intrinsic properties of oxide-based perovskites allow for their integration with other energy technologies, thus broadening their applicability within hybrid systems. Furthermore, engineering these materials at the nanoscale through techniques such as nanostructuring and composite formation can significantly enhance their thermoelectric performance [17, 18].

The perovskite families comprising titanium ($ATiO_3$), manganese ($AMnO_3$), niobium ($ANbO_3$), and iron ($AFeO_3$) are among the most promising oxide-based perovskites can be utilized in wearable technologies. Each of these families exhibits distinct structural, electronic, and thermal characteristics that significantly affect their efficacy in thermoelectric applications. An overview of the properties of these families in wearable thermoelectric applications is presented in Table 6.1.

**Fig. 6.5** Advantages of oxide-based perovskites for thermoelectric applications

### *Ti-Based Perovskite Oxides ($ATiO_3$)*

The $ATiO_3$ perovskite family has emerged as a focal point of research in thermoelectric applications, attributed to a variety of inherent characteristics that facilitate efficient thermoelectric conversion—the direct transformation of temperature differentials into electrical voltage and vice versa. Members of the $ATiO_3$ family can achieve a commendable FOM owing to their favorable electrical conductivity and moderate thermal conductivity, intrinsic properties that are pivotal for thermoelectric efficiency. The ionic nature of the perovskite framework permits the tuning of electronic properties through methods such as doping or variation of the A-site cation's composition, thereby enhancing electrical conductivity essential for optimizing the FOM. Enhanced thermoelectric efficiency is associated with lower thermal conductivity, which minimizes heat dissipation. The perovskite structure

**Table 6.1** An overview of the properties of various perovskite families in wearable thermoelectric applications

| Oxide perovskite family | A-site elements | Structure | Electronic properties | Seebeck coefficient | Thermal conductivity | Thermoelectric performance |
|---|---|---|---|---|---|---|
| $ATiO_3$ | Sr [19]<br>Ca [20]<br>Ba [21]<br>Fe [22] | Cubic, orthorhombic, tetragonal | Wide band gaps and limited electrical conductivity at room temperature | Moderate | Moderate | Moderate |
| $AMnO_3$ | La [23]<br>Pr [24]<br>Tb [25]<br>Ca [26]<br>Bi [27] | Orthorhombic, rhombohedral | Potential half-metallic behavior and magnetism | High | Low | High |
| $ANbO_3$ | Na [28]<br>K [29]<br>Rb [30]<br>La [31] | Cubic, tetragonal, orthorhombic, rhombohedral | Semiconducting behavior | High | Low | High |
| $AFeO_3$ | La [32]<br>Sr [33]<br>Ba [34]<br>Bi [35] | Cubic, rhombohedra, orthorhombic | Ferroelectric and magnetic behavior | High | Low | High |

can facilitate phonon scattering, particularly at lattice defects and grain boundaries, contributing to lower thermal conductivity. Notably, the structural stability of perovskites is vital for maintaining performance across diverse temperatures and environmental conditions, thereby supporting applications in domains such as waste heat recovery. Specific $ATiO_3$ materials retain effective thermoelectric properties over extensive temperature ranges, enhancing their versatility for high-temperature applications. Certain A-site cations, including calcium and strontium, are not only abundant and non-toxic but also render the $ATiO_3$ perovskites much more environmentally friendly compared to alternative thermoelectric materials, which may contain rare or harmful constituents [20–22].

Among the titanium-based perovskite oxides, strontium titanate ($SrTiO_3$) is well recognized for its favorable thermoelectric characteristics, particularly when doped. Its compatibility with flexible substrates positions it as an ideal candidate for wearable technology. $SrTiO_3$ showcases a high thermoelectric figure of merit at elevated temperatures, indicative of its efficiency in converting temperature gradients into electrical voltage. Furthermore, its relatively low thermal conductivity aids in sustaining a temperature gradient, thereby enhancing thermoelectric performance. With robust thermal and chemical stability, $SrTiO_3$ is particularly suited for long-term use in varying environmental conditions, which is crucial in wearable applications that may be subject to moisture and perspiration. The mechanical properties of $SrTiO_3$ can be engineered for compatibility with flexible substrates, a critical feature for user comfort in wearable technologies. The raw materials for $SrTiO_3$

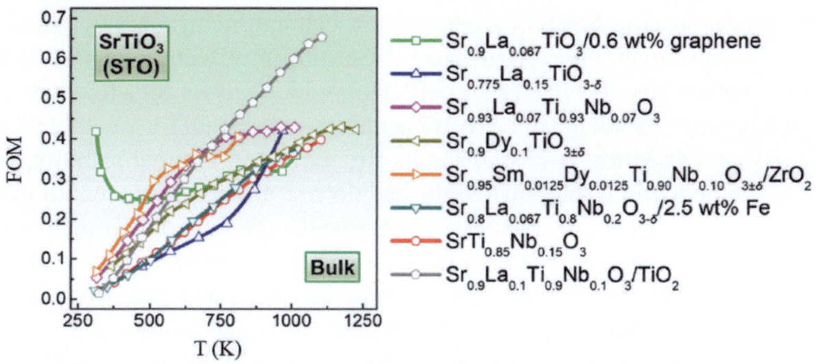


**Fig. 6.6** Temperature-dependent FOMs of various doped $SrTiO_3$ thermoelectrics [14]. Copyright 2020, Elsevier

synthesis are relatively inexpensive and environmentally benign, further supporting its appeal for large-scale production. The thermoelectric performance of $SrTiO_3$ can be fine-tuned through doping and the formation of heterostructures, facilitating optimization for specific applications. The ability to integrate $SrTiO_3$ with other materials for doped configurations enhances its thermoelectric performance and broadens functional applications in wearable devices, as shown in Fig. 6.6 [14, 19, 36, 37].

### *Mn-Based Perovskite Oxides ($AMnO_3$)*

Similarly, the $AMnO_3$ perovskite family has attracted considerable interest for thermoelectric applications, showcasing a combination of high thermoelectric performance, stability, tunability, and sustainable characteristics. This family of materials often demonstrates an advantageous balance of these properties, leading to higher FOM values. The partially filled d-orbitals of manganese ions contribute to a complex electronic band structure that can enhance the Seebeck coefficient, which is favorable for thermoelectric applications. The perovskite crystal structure usually exhibits low thermal conductivity due to phonon scattering, essential for maintaining the temperature gradient needed for efficient thermoelectric conversion. $AMnO_3$ materials also display commendable thermal and chemical stability, preserving their properties under operational conditions—a critical aspect for long-term thermoelectric applications. The perovskite framework's ability to accommodate various A-site cations enables researchers to optimize electronic and thermal characteristics for specific performance needs. Moreover, manganese oxides may exhibit intriguing magnetic properties, potentially enhancing thermoelectric performance through magnetic ordering effects. Many $AMnO_3$ materials can be synthesized from non-toxic and abundant elements, making them environmentally friendly alternatives to traditional thermoelectric materials composed of rare or hazardous components [23–25].

Focusing on specific Mn-based perovskites, calcium manganate ($CaMnO_3$) has exhibited promising thermoelectric properties with an effective balance between electrical conductivity and Seebeck coefficient, which allows for effective conversion of temperature gradients into electrical energy—essential for wearable thermoelectric devices. $CaMnO_3$ is lauded for its structural and thermal stability, crucial for wearable applications that must withstand diverse environmental conditions and physical strain. Compared to conventional thermoelectric materials containing toxic elements, $CaMnO_3$ is constructed from non-toxic components, rendering it a safer choice for direct skin applications. The raw materials required for $CaMnO_3$ synthesis are relatively inexpensive and abundantly available, paving the way for large-scale production. Research efforts have also been directed towards improving the flexibility of $CaMnO_3$-based thermoelectric devices, a vital trait for seamless integration with clothing or skin. Additionally, $CaMnO_3$ possesses noteworthy catalytic attributes, beneficial for energetic efficiency in thermoelectric generators or converters. The thermoelectric properties of $CaMnO_3$ can be enhanced through doping practices, enabling customization for specific application requirements. In the realm of environmentally sustainable materials, $CaMnO_3$ offers a green alternative for thermoelectric applications as opposed to conventional options [26, 38].

Another promising Mn-based perovskite for wearable thermoelectric applications is lanthanum strontium manganate ($La_{1-x}Sr_xMnO_3$), or LSM. This notable perovskite demonstrates a high Seebeck coefficient, crucial for thermoelectric voltage generation from temperature differences. The material's mixed-valence characteristic promotes high electrical conductivity, which is vital for effective thermoelectric conversion, with proper doping with strontium enhancing overall performance. The perovskite structure can be synthesized into thin films or nanostructures, aligning with the flexibility requirements typical of wearable applications. LSM materials generally exhibit good thermal stability across variable temperatures, suitable for execution in wearable technology environments. Versatile synthesis methods, such as sol–gel processing and sputtering, allow for adaptation to thin-film deposition techniques, facilitating integration onto flexible substrates utilized in wearable devices. Through systematic doping and compositional adjustments, the properties of LMS can be tailored for enhanced thermoelectric performance. The relative abundance and low cost of lanthanum and strontium compared to other thermoelectric materials make LSM a promising choice for large-scale applications, particularly in wearable technologies where energy can be harvested from body heat or ambient temperature variations [39, 40].

***Nb-Based Perovskite Oxides ($ANbO_3$)***

The $ANbO_3$ perovskite family is also recognized for its promising prospects in thermoelectric applications, shaped by several fundamental properties. The crystal structure of these perovskites fosters electronic, optical, and thermal characteristics conducive to thermoelectric performance. Certain members of this family exhibit high FOM values attributable to optimized carrier concentration, low thermal conductivity, and a high Seebeck coefficient. The low thermal conductivity observed in some $ANbO_3$ compounds arises from their lattice structure, which efficiently scatters phonons—thus minimizing thermal transport. The electronic properties can often be tuned by varying the A-site cation or through doping approaches, which allows for the enhancement of electrical conductivity and Seebeck coefficient—both critical factors in bolstering thermoelectric performance. $ANbO_3$ materials also manifest strong thermal and chemical stability, suitable for use in high-temperature scenarios where other materials may fail. Unlike certain conventional thermoelectric materials containing toxic elements (e.g., lead or bismuth), $ANbO_3$ often comprises non-toxic and environmentally-friendly elements, underscoring its sustainability in materials research. These compounds lend themselves to integration into composite structures with additional materials, potentially enhancing thermoelectric performance through synergies in electrical and thermal traits. The operational mechanism of $ANbO_3$ thermoelectric materials is characterized by their unique electronic properties, which can markedly influence their thermoelectric performance. Typically displaying either a direct or indirect band gap, depending on the influence of cation selection and concentration, $ANbO_3$ materials are often associated with enhanced electrical conductivity through careful doping or compositional adjustments. Effective thermoelectric materials are distinguished by their low thermal conductivity, crucial for preserving the necessary temperature gradient for thermoelectric conversion. The crystal lattice and phonon scattering behavior in $ANbO_3$ can yield relatively low thermal conductivity, translating into favorable FOM values. The Seebeck coefficient—indicative of a material's capability to generate voltage from a thermal difference—varies with the choice of A-site cation and doping level, influenced further by the band's overall structure and carrier concentration. Presence of defects such as vacancies or substitutions introduces localized states within the band gap, which can enhance thermoelectric performance through altered scattering or changes in carrier mobility. The thermoelectric behavior of $ANbO_3$ materials can also vary with temperature, necessitating characterization across different conditions to understand their suitability for thermoelectric applications [28–31].

Among Nb-based perovskites, $(Na_{0.5}K_{0.5})NbO_3$ has recently gained significant attention for its potential in thermoelectric applications, particularly in the realm of wearable technology. This compound is characterized by an advantageous Seebeck coefficient and low thermal conductivity, both of which are critical for optimizing thermoelectric efficiency. Figure 6.7 illustrates the variation of the Seebeck coefficient, electrical conductivity, and thermal conductivity of this perovskite oxide as a function of temperature. With a high thermoelectric figure of merit, $(Na_{0.5}K_{0.5})NbO_3$ excels in converting thermal gradients into electrical energy. Furthermore, its robust chemical and thermal stability renders it well-suited for a variety of environmental conditions, a particularly desirable trait for wearable devices that must endure fluctuations in temperature, humidity, and mechanical stress. The material's inherent flexibility contributes to its comfort and usability when integrated into devices

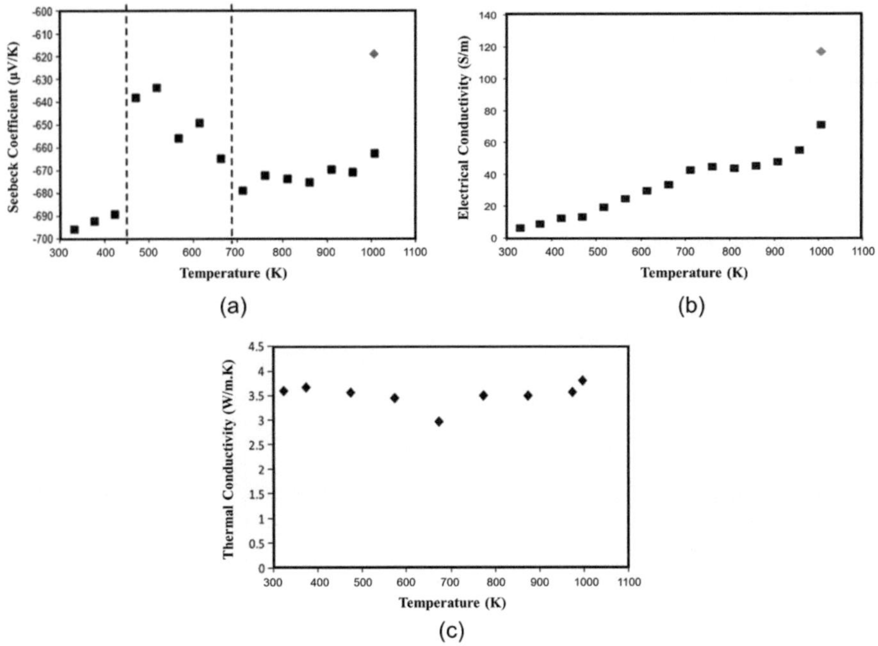


**Fig. 6.7** Variation of the Seebeck coefficient, electrical conductivity, and thermal conductivity of $(Na_{0.5}K_{0.5})NbO_3$ perovskite oxide as a function of temperature. Reproduced with permission [41]. Copyright 2016, Elsevier

intended for human contact. In addition to its favorable performance characteristics, $(Na_{0.5}K_{0.5})NbO_3$ is composed of relatively abundant elements—sodium, potassium, niobium, and oxygen—making it a sustainable and cost-effective alternative compared to other thermoelectric materials that rely on scarce resources. Its nontoxic nature further enhances its appeal for applications in direct contact with the skin, thereby mitigating potential health risks associated with wearable technologies. The thermoelectric properties of $(Na_{0.5}K_{0.5})NbO_3$ can be fine-tuned through doping and compositional modifications, providing opportunities for optimization tailored to specific applications. Additionally, the material's compatibility with conventional ceramic processing techniques facilitates its integration into existing manufacturing processes for thermoelectric devices, thus streamlining the production of wearable technology. Collectively, the advantageous characteristics of high thermoelectric efficiency, stability, mechanical flexibility, low toxicity, and tunable properties position $(Na_{0.5}K_{0.5})NbO_3$ as a promising candidate for thermoelectric applications aimed at efficient energy harvesting from body heat and temperature differentials [41, 42].

### *Fe-Based Perovskite Oxides ($AFeO_3$)*

Another oxide-based perovskite family that demonstrates significant potential for thermoelectric applications is $AFeO_3$. These materials typically exhibit favorable Seebeck coefficients, electrical conductivities, and low thermal conductivities, critical parameters for achieving high thermoelectric performance. The synergy among these properties results in a high dimensionless figure of merit, indicative of their effective heat-to-electricity conversion capabilities. The perovskite structure's inherent capacity to accommodate various cations at the A-site allows for customization of the material's properties via substitution and alloying. Such adaptability enables the optimization of electrical and thermal transport properties for specific thermoelectric applications. Certain $AFeO_3$ compounds showcase relatively low thermal conductivity, a characteristic that helps to sustain the necessary temperature gradients for effective thermoelectric conversion, thereby enhancing the efficiency of the thermoelectric process. Moreover, specific members of the $AFeO_3$ family exhibit noteworthy magnetic properties due to the presence of iron ions, introducing additional degrees of freedom for property tuning and facilitating novel thermoelectric behaviors that may not occur in non-magnetic materials. These perovskites generally demonstrate good chemical stability, particularly in high-temperature and reactive environments, which is essential for ensuring device durability and longevity. Due to their composition of non-toxic and abundant elements, $AFeO_3$ materials represent a more environmentally friendly option compared to conventional thermoelectric materials that may contain rare or hazardous elements. Furthermore, the potential for processing these materials into nano-scale forms enhances their thermoelectric performance by further reducing thermal conductivity while simultaneously improving electrical conductivity through enhanced phonon scattering and charge carrier mobility [32–34].

In the context of the iron-based perovskite family, bismuth ferrite ($BiFeO_3$) emerges as a particularly promising material for wearable thermoelectric applications. $BiFeO_3$ has demonstrated reasonable thermoelectric figures of merit under specific doping and processing conditions, positioning it as a viable candidate for effectively converting temperature gradients into electrical energy. Figure 6.8 shows the variation of the Seebeck coefficient and electrical conductivity of $BiFeO_3$ perovskites doped with different weight ratios of cobalt as a function of temperature.

Composed of non-toxic and readily available elements, $BiFeO_3$ offers an environmentally friendly alternative compared to other thermoelectric materials, particularly those containing lead or tellurium, making it suitable for consumer products and wearable technologies alike. The structural stability and mechanical strength of $BiFeO_3$ further enhance its suitability for flexible applications prevalent in wearable devices, allowing it to endure the mechanical stresses associated with bending and stretching. Additionally, the presence of both ferroelectric and magnetic properties in $BiFeO_3$ lends to the possibility of multifunctional device applications, thereby enhancing energy conversion efficiencies through innovative design approaches. Operating effectively across moderate temperature ranges, $BiFeO_3$ is ideal for applications involving conditions akin to body temperature. Its versatility in synthesis,

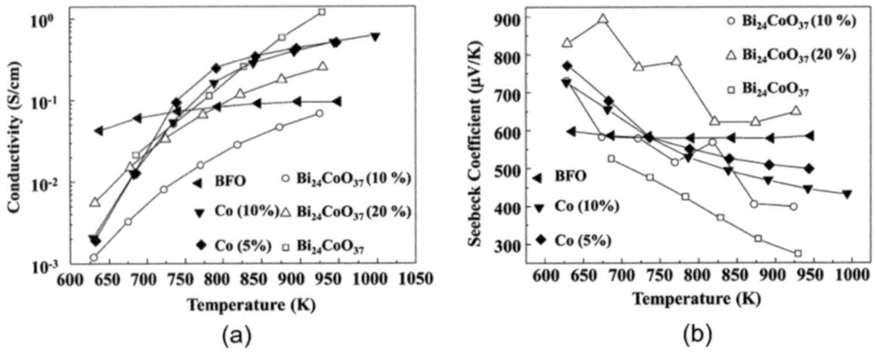


**Fig. 6.8** **a** Electrical conductivity and **b** Seebeck coefficient of the Co-doped (0, 5, 10%) $BiFeO_3$, $Bi_{24}CoO_{37}$ (10, 20%) perovskites. Reproduced with permission [35]. Copyright 2013, Ceramic Society of Japan

which includes methods such as sol–gel and solid-state reactions, facilitates scalability and integration into flexible substrates, thereby ensuring ease of fabrication for commercial uses. The capacity to integrate $BiFeO_3$ thin films onto flexible substrates promotes the development of lightweight and bendable thermoelectric devices, enhancing user comfort. Ongoing research focused on enhancing its thermoelectric properties through doping and nanostructuring continues to broaden its applicability within the domain of wearable technologies [35, 43].

## 6.2 Photovoltaics

### *6.2.1 Evolution of Photovoltaic Effect*

The evolution of the photovoltaic (or PV) effect represents a significant chapter in the history of science and technology, tracing its origins to the early nineteenth century and culminating in the advanced applications of modern photovoltaics. The phenomenon was first demonstrated in 1839 by French physicist Edmond Becquerel, who observed the generation of an electric current when two platinum or gold plates, immersed in an acidic, neutral, or alkaline solution, were exposed unevenly to solar radiation. Becquerel documented his findings in Comptes rendus de l'Académie des sciences, providing the first scientific account of the photovoltaic effect and establishing a foundation for subsequent research into the conversion of light into electrical energy [44].

The next major milestone in the development of photovoltaic technology occurred in 1884, when American inventor Charles Fritts constructed the first solar cell. Fritts' device utilized a layer of selenium coated with a thin film of gold, marking the first attempt to create a solid-state photovoltaic device. Although this early solar cell demonstrated the feasibility of converting light into electricity, its efficiency was exceedingly low, rendering it impractical for widespread use. Nevertheless, Fritts' work represented a critical step forward, as it introduced the concept of using solid materials for photovoltaic energy conversion and inspired further experimentation in the field [45].

The modern understanding and application of the photovoltaic effect are largely based on solid-state devices, particularly photodiodes. In these devices, incident light with sufficient energy excites electrons from the valence band to the conduction band, where they become free to move. These excited electrons diffuse through the material, and some reach a rectifying junction, typically a p–n junction in a semiconductor diode. At this junction, the built-in potential, or Galvani potential, accelerates the electrons into the n-type semiconductor material, generating an electromotive force and an electric current. This process effectively converts light energy into electrical energy, forming the basis of contemporary photovoltaic systems. In addition to the conventional single-photon absorption mechanism, the photovoltaic effect can also occur through more complex processes, such as the two-photon photovoltaic effect. In this phenomenon, two photons are absorbed simultaneously, enabling energy conversion under specific conditions and expanding the theoretical and practical possibilities for photovoltaic technology [46].

From its initial discovery by Becquerel to the pioneering work of Fritts and the development of modern solid-state photodiodes, the evolution of the photovoltaic effect reflects a continuous interplay between scientific inquiry and technological innovation. Today, photovoltaic technology stands as a cornerstone of renewable energy, playing a pivotal role in addressing global energy challenges and advancing sustainable development. This historical progression underscores the transformative potential of scientific discovery and its capacity to shape the future of energy systems [47].

### 6.2.2 *Photovoltaic Cells*

Photovoltaic cells, or solar cells, convert sunlight directly into electricity. Over the years, different types of solar cells have emerged, each with its own unique materials, efficiencies, and applications. The National Renewable Energy Laboratory (NREL) chart highlights various technology platforms of solar cells, their efficiency trends, and their performance over time, as shown in Fig. 6.9 [48].

Below is a detailed introduction to these different technology platforms of solar cells and an explanation of how they work [48].

#### 6.2.2.1 III–V Multijunction Cells

Multijunction cells are composed of multiple layers of semiconductor materials, each designed to absorb different portions of the solar spectrum. The most common

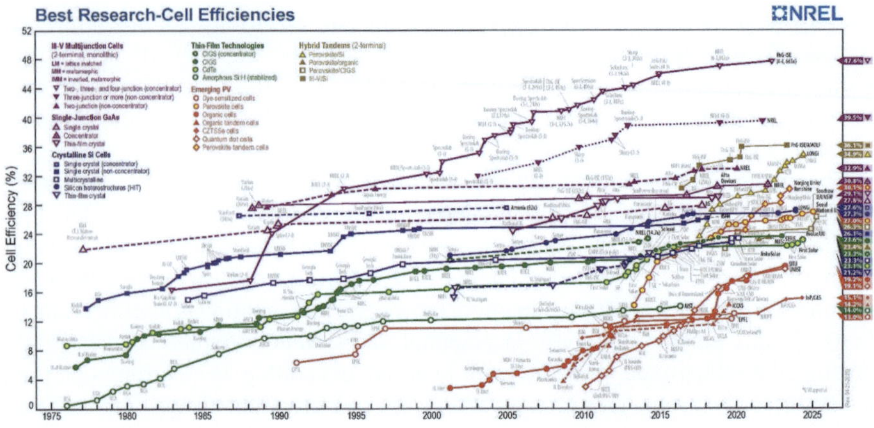


**Fig. 6.9** NREL's best research-cell efficiency chart (updated April 2025) [48]

semiconductors used in these cells are Gallium Arsenide (GaAs) and its alloys with other III-V materials. These cells have multiple junctions, or layers, each optimized for a specific range of wavelengths of light. Each junction in a multijunction solar cell absorbs light of a different wavelength, which increases the overall efficiency of the device. The sunlight is split and absorbed by different materials in the cell, enabling it to harness more energy compared to single-junction cells. The layered approach allows multijunction cells to convert a broader spectrum of light into electricity. III–V multijunction cells are among the highest-efficiency solar cells available. As shown in the NREL chart, these cells have reached efficiencies up to 47.6%, setting world records in solar cell technology. These high efficiencies are largely due to their ability to capture a broader range of sunlight through the multiple junctions [48, 49].

#### 6.2.2.2 Single-Junction Gallium Arsenide (GaAs) Cells

Gallium Arsenide (GaAs) cells use a single layer of GaAs to absorb light and convert it into electricity. GaAs is a compound semiconductor that offers better performance than silicon in certain applications, especially in terms of energy efficiency and bandgap matching for solar energy. In single-junction GaAs cells, sunlight excites electrons in the semiconductor material, which are then collected as electrical current. While these cells are limited to the efficiency of a single junction, they are well-suited for high-performance applications, including space solar panels and concentrated solar power systems. As of April 2025, single-junction GaAs cells have reached efficiencies of approximately 39.5%. Though lower than multijunction cells, they still represent a highly efficient technology, particularly for specialized applications where cost is less of an issue [48, 50].

#### 6.2.2.3 Crystalline Silicon Cells

Crystalline silicon (Si) cells are the most widely used type of solar cells today. They are made from high-purity silicon, which is either single-crystal or multicrystalline. Single-crystal silicon cells are made from a single continuous crystal, while multicrystalline cells are made from silicon crystals that are melted and then solidified. Silicon-based solar cells rely on the photovoltaic effect, where light strikes the semiconductor material and excites electrons, creating electron–hole pairs that generate electricity. The cells are typically arranged in modules, and the efficiency is influenced by factors such as the quality of the silicon, the type of cell, and the presence of antireflection coatings. Crystalline silicon cells are commonly found in residential and commercial solar installations due to their balance of cost, efficiency, and longevity. The efficiency of crystalline silicon cells, according to the NREL chart, ranges from approximately 21.2–27.6% as of April 2025. Notable companies such as SunPower, Panasonic, and LG have contributed to the development of high-efficiency silicon-based cells [48, 51].

#### 6.2.2.4 Thin-Film Technologies

Thin-film solar cells are made from layers of photovoltaic materials that are just a few micrometers thick. These materials can include cadmium telluride (CdTe), copper indium gallium selenide (CIGS), and amorphous silicon (a-Si). These cells are less efficient than crystalline silicon cells but are cheaper to produce, making them attractive for large-scale installations. Thin-film solar cells work similarly to crystalline silicon cells, with the photovoltaic material absorbing light and generating electricity. However, because they are much thinner than silicon cells, they require less material to produce, which reduces costs. As shown in the NREL chart, the efficiency of thin-film technologies has increased over time, reaching up to 23.6% by April 2025, with companies like First Solar leading the way in improving the performance of CdTe and CIGS-based cells [48, 52].

#### 6.2.2.5 Emerging Photovoltaics

Emerging PV technologies are at the forefront of solar research and development. These include several novel materials and designs that have the potential to surpass the performance of conventional silicon-based cells. Some of the most promising emerging PV technologies are:

- Dye-Sensitized Solar Cells (DSSC): These cells use organic dyes to capture sunlight and convert it into electrical current. They are typically less efficient than traditional cells but offer a low-cost, flexible alternative [53].
- Perovskite Solar Cells: These cells use a class of materials known as perovskites, which have shown impressive efficiency improvements in recent years. They are

cheap to manufacture and flexible but face challenges regarding stability and scalability [54].
- Organic Solar Cells: Organic solar cells use conductive polymers to absorb light. They are flexible, lightweight, and cost-effective but have lower efficiencies compared to silicon-based cells [55].
- Quantum Dot Solar Cells: Quantum dots are tiny semiconductor particles that exhibit unique electronic properties. They have the potential to capture a wide range of the solar spectrum, which could lead to higher efficiencies [56].
- CZTSSe Cells: Copper zinc tin sulfide selenide (CZTSSe) cells are a promising alternative to silicon, with lower material costs and relatively high efficiencies [57].

Emerging PV technologies currently have lower efficiencies compared to traditional options but are showing rapid improvements. For example, perovskite cells have reached around 27% efficiency by April 2025, while quantum dot and dye-sensitized cells are still in the research phase but show significant promise [58].

#### 6.2.2.6 Hybrid Tandem Cells

Hybrid tandem cells combine multiple materials or technologies into a single device. For instance, a perovskite/Si tandem cell combines the high efficiency of silicon with the tunable bandgap of perovskite materials. The aim of these cells is to overcome the efficiency limits of single-junction cells by utilizing different materials to capture more of the solar spectrum. In tandem cells, the top layer absorbs high-energy photons, while the bottom layer absorbs lower-energy photons, maximizing the range of sunlight that can be converted into electricity. Hybrid tandem cells combine the benefits of silicon with newer materials, such as perovskites or organic materials, to improve efficiency. Hybrid tandem cells, like perovskite/Si and perovskite/organic tandems, have shown significant efficiency improvements, reaching up to 36.1% by April 2025, as demonstrated by research institutions such as FhG-ISE/AMOLF [48, 58].

Table 6.2 provides an overview of the peak power conversion efficiencies achieved by various technology platforms as of April 2025 [48].

### *6.2.3 Perovskite Solar Cells*

#### 6.2.3.1 Mechanism

Perovskite solar cells operate through a complex multilayered architecture that comprises several essential components: a transparent conducting oxide (TCO) layer, a blocking layer, an electron transport layer (ETL), the perovskite layer itself, a hole transport layer (HTL), and a metal electrode. This structure is fundamental to the

**Table 6.2** NREL's overview of the peak power conversion efficiencies achieved by various technology platform of solar cells (updated April 2025) [48]

| Technology platform | Solar cell | Maximum efficiency (%) |
|---|---|---|
| III–V multijunction cells | Two-, three-, and four-junction (concentrator) | 47.6 |
| | Three-junction or more (non-concentrator) | 39.5 |
| | Two-junction (non-concentrator) | 32.9 |
| Single-junction GaAs | Single crystal | 27.8 |
| | Concentrator | 30.8 |
| | Thin-film crystal | 29.1 |
| Crystalline Si cells | Single crystal (concentrator) | 27.6 |
| | Single crystal (non-concentrator) | 26.1 |
| | Multicrystalline | 23.3 |
| | Silicon heterostructures | 27.3 |
| | Thin-film crystal | 21.2 |
| Thin-film technologies | CIGS (concentrator) | 23.3 |
| | CIGS | 23.6 |
| | CdTe | 23.1 |
| | Amorphous Si:H (stabilized) | 14.0 |
| Emerging PV | Dye-sensitized cells | 13.0 |
| | Perovskite cells | 27.0 |
| | Organic cells | 19.2 |
| | Organic tandem cells | 14.2 |
| | CZTSSe cells | 15.1 |
| | Quantum dot cells | 19.1 |
| | Perovskite tandem cells | 30.1 |
| Hybrid tandems (2-terminal) | Perovskite/Si | 34.6 |
| | Perovskite/organic | 23.4 |
| | Perovskite/CIGS | 24.6 |
| | III–V/Si | 36.1 |

cell's functionality and overall efficiency, as schematically illustrated in Fig. 6.10 [59].

The TCO Layer serves as the topmost layer and facilitates the transmission of sunlight into the solar cell. It also functions as an electron collector, playing a pivotal role in both light transmission and charge collection, which are essential for efficient electricity generation. The ETL, usually composed of mesoporous materials such as $TiO_2$, ZnO, or $SnO_2$, is positioned between the perovskite and TCO layers [60]. The primary role of the ETL is to capture and transport electrons generated by the perovskite layer to the metal electrode. Efficient electron transport is critical to the

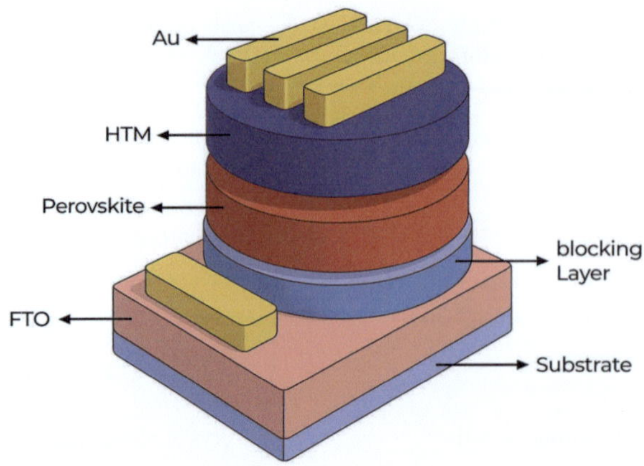


**Fig. 6.10** Schematic illustration of perovskite solar cells' structure

device's overall efficiency, as it ensures that the electrons can flow freely towards the electrode, generating an electrical current. The blocking layer prevents charge recombination by blocking contact between TCO and ETL. It ensures efficient electron extraction, proper energy alignment, and reduces shunt pathways, enhancing cell performance, stability, and efficiency. The perovskite layer, which constitutes the core of the cell, is responsible for absorbing photons from sunlight, leading to the formation of excitons (bound electron–hole pairs). The unique electronic properties of the perovskite material facilitate the separation of these excitons at the interfaces within the device, a process essential for efficient charge generation. The HTL is located between the perovskite layer and the metal electrode. It collects the holes generated in the perovskite layer and transports them to the electrode. This function complements that of the ETL, contributing to the charge separation and overall performance of the solar cell. Finally, the metal electrode collects the free charges—both electrons and holes—produced in the perovskite layer. This movement of charges across the device generates a potential difference, allowing electrical current to flow through an external circuit and producing usable electricity. When sunlight strikes the perovskite layer, it generates charge carriers that are transported through the ETL and HTL, culminating in electricity production. The efficiency of this transport process is a key determinant of the PSC's performance. The efficiency of carrier movement in the ETL is particularly influential, as it directly affects how quickly and effectively the electrons are transferred to the electrode [61].

#### 6.2.3.2 Optoelectronic Properties of Perovskites

The optoelectronic properties of perovskites were first reviewed by Mitzi et al. [62] in the early 1990s, highlighting their strong excitation characteristics for applications such as LEDs, transistors, and solar cells. The photovoltaic effect in perovskite materials was later observed by Kojima et al. [63] in 2009, marking the beginning of their use in solar cells. Initially, perovskites were utilized as liquid sensitizers in dye-sensitized solar cells (DSSC) architecture. Since their first application in solar cells, the efficiency of perovskite solar cells has increased from 3.81% in 2009 to 27.0% by 2025. This dramatic efficiency improvement can be attributed to advancements in material engineering, device architecture, and fabrication methods [48].

Halide organic–inorganic perovskites represent a transformative class of materials which play a pivotal role in the development of PSCs. In these perovskites, the "A" site can be occupied by cations such as methylammonium (MA), formamidinium (FA), or cesium (Cs), while the "B" site can contain elements like lead (Pb) or tin (Sn). The "X" site typically holds halide ions such as iodide ($I^-$), bromide ($Br^-$), or chloride ($Cl^-$). Their outstanding optoelectronic properties, coupled with advancements in stability and sustainability, position them as leading candidates for next-generation solar technologies. Nevertheless, ongoing research is necessary to address remaining challenges related to environmental stability and lead toxicity, which are critical for ensuring their feasibility in large-scale and environmentally friendly energy production. These materials have garnered significant attention for their ability to outperform traditional photovoltaic materials such as silicon, particularly owing to the following characteristics [64], as shown in Fig. 6.11.

Halide organic–inorganic perovskites are recognized for their exceptional optical and electronic properties, including a high absorption coefficient that allows thin films, merely a few hundred nanometers thick, to absorb sunlight effectively. Moreover, these materials exhibit a broad absorption spectrum, covering wavelengths from the visible to the near-infrared regions of the solar spectrum, thereby making them highly efficient at solar energy harvesting. Their direct bandgap, typically ranging from 1.5 to 1.6 eV, makes them particularly suitable for photovoltaic applications by facilitating efficient photon absorption and charge carrier generation [65].

The fabrication of halide organic–inorganic perovskites can be achieved using solution-based [66], low-temperature processes such as spin coating [67], spray deposition [68], and inkjet printing [69]. These methods enable the production of thin-film solar cells at significantly lower costs compared to silicon-based photovoltaics, which rely on energy-intensive manufacturing processes. Furthermore, these solution-based techniques are compatible with large-scale production, highlighting their potential for industrial-scale applications. Halide perovskites also exhibit excellent charge transport properties, including balanced electron and hole mobilities, which minimize charge carrier recombination losses. Their long charge carrier diffusion lengths—extending up to several micrometers even in polycrystalline films—enable efficient charge collection at the electrodes. Additionally, the low density of electronic defects in these materials further reduces non-radiative recombination losses, enhancing overall performance [70].

A defining characteristic of halide perovskites lies in their chemical tunability. By adjusting the composition of organic cations (e.g., methylammonium [$MA^+$], formamidinium [$FA^+$], or cesium [$Cs^+$]) and halides (e.g., iodide [$I^-$], bromide [$Br^-$], or chloride [$Cl^-$]), the bandgap can be precisely tuned to meet specific application requirements. This tunability enables the optimization of material properties for advanced photovoltaic configurations, such as tandem solar cells. Tandem devices,

**Fig. 6.11** Superior characteristics of halide organic–inorganic perovskites for PSCs

which pair perovskites with other photovoltaic technologies (e.g., perovskite/silicon or perovskite/perovskite combinations), demonstrate the ability to exceed the efficiency limits of single-junction solar cells, making them particularly promising for next-generation solar energy systems [71].

In contrast to conventional silicon-based solar panels, halide perovskites offer the advantages of being lightweight and compatible with flexible substrates. This feature facilitates the creation of innovative and versatile solar energy applications, such as building-integrated photovoltaics (BIPVs), wearable solar devices, and portable energy systems, thereby expanding the range of applications beyond stationary installations [72].

Despite their many benefits, early perovskite materials encountered significant challenges related to their instability under exposure to moisture, oxygen, heat, and light. However, substantial progress has been made in improving their stability. For instance, the incorporation of mixed cations (e.g., $MA^+$, $FA^+$, and $Cs^+$) and mixed halides (e.g., $I^-$ and $Br^-$) has considerably enhanced both the thermal and structural stability of these materials. Furthermore, advancements in encapsulation techniques and additive engineering have addressed several degradation mechanisms, thereby extending the operational lifespan of perovskite solar cells (PSCs) [73].

Nevertheless, the widespread adoption of halide perovskites also raises concerns regarding the toxicity of lead, a common component in their formulation. To address this issue, ongoing research is focused on the development of lead-free alternatives, such as tin-based perovskites, which aim to reduce environmental and health risks while retaining desirable photovoltaic performance. Achieving these goals is crucial for enabling the large-scale, environmentally sustainable deployment of PSCs [74].

Beyond their application in solar cells, halide perovskites have shown considerable promise in other optoelectronic devices, including light-emitting diodes (LEDs), photodetectors, and lasers. This multifunctionality underscores their importance in contemporary materials science and highlights their versatility for a broad range of energy and communication technologies [75].

### *6.2.4 Perovskites-Based Wearable Photovoltaics*

Perovskite solar cells have demonstrated significant potential in revolutionizing the field of wearable technologies due to their exceptional combination of lightweight structure, flexibility, and PCE. The development of flexible perovskite solar cells (FPSCs) represents a natural progression from broader advancements in perovskite-based photovoltaic technologies, which have gained attention only in recent decades. Notably, around 2013, researchers began to explore the potential of flexible perovskite solar cells. This investigation was driven by the unique properties of perovskite materials, including their solution-processable nature, which facilitates deposition onto flexible substrates such as plastics or metal foils, and their compatibility with low-temperature fabrication processes (below 150 °C). These characteristics have been particularly advantageous for integrating perovskite materials with temperature-sensitive substrates [76].

Today, FPSCs are regarded as promising candidates for various commercial applications, including portable energy solutions, wearable electronic devices, and architecturally innovative designs, such as curved surfaces. Despite significant improvements in their efficiency, the widespread adoption of FPSCs faces several barriers, including long-term stability, mechanical durability, and challenges associated with scaling up production processes. Researchers are actively addressing these limitations by optimizing manufacturing techniques and developing new materials, with the expectation that such advancements will facilitate the integration of FPSCs into mainstream solar energy markets. For wearable applications specifically, FPSCs must meet stringent design and functional requirements to ensure reliability and practicality. This necessitates features such as mechanical robustness, self-healing properties, and enhanced durability under repetitive stress, as illustrated in Fig. 6.12.

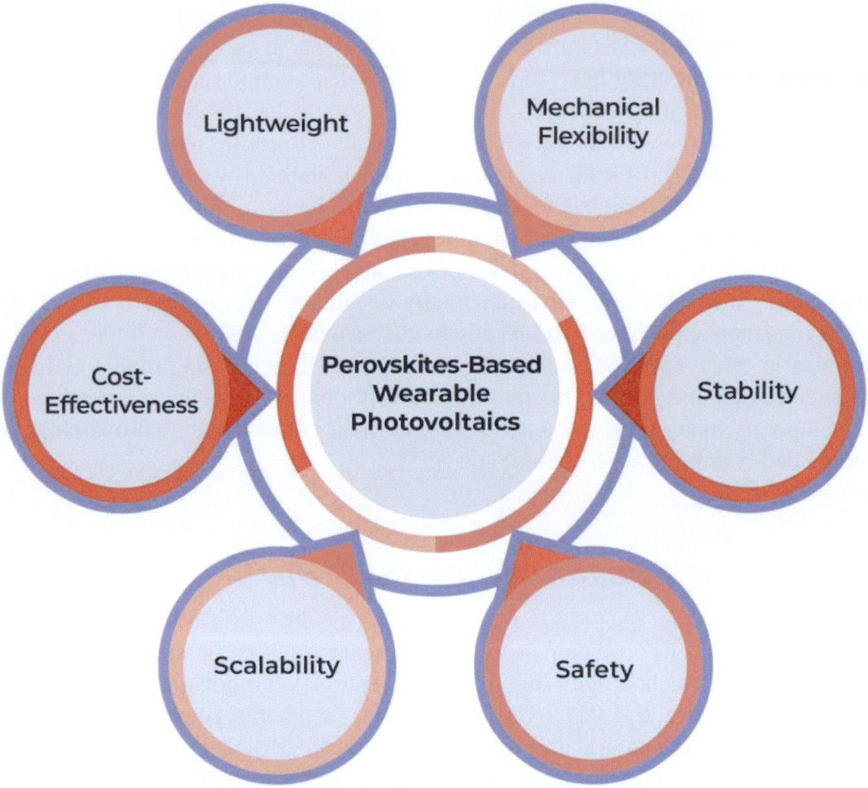


**Fig. 6.12** Required features for perovskites-based wearable photovoltaics

Successfully addressing these challenges will be central to realizing the full potential of FPSCs as a key enabler of next-generation wearable and portable energy technologies [77].

#### 6.2.4.1 Mechanical Flexibility

The operational viability of FPSCs hinges on their capacity to conform to non-planar surfaces, withstand repetitive mechanical deformation, and maintain performance stability under stress. To meet these requirements, fabrication techniques such as roll-to-roll (R2R) slot-die coating and sequential deposition have been adopted. Further improvements in film processing, such as shallow-angle blowing, minimize structural defects, thereby enhancing mechanical robustness and ensuring long-term reliability. A primary cause of mechanical failure in FPSCs is fracture within the perovskite thin film or at interfaces [78]. Crack formation typically initiates at grain boundaries or pre-existing micro-cracks [79], which ultimately leads to performance degradation under mechanical stress [80]. Furthermore, crack formation and stress-induced lattice distortions can induce unfavorable phase transitions and accelerate the degradation of the perovskite material [81]. These cracks also serve as pathways for moisture ingress and the release of decomposition products, further compromising device stability [82].

Docampo et al. [83] in 2013, demonstrated efficient organometal trihalide perovskite planar-heterojunction solar cells on flexible polymer substrates, representing a promising avenue for low-cost, easily manufactured solar technology that is compatible with large-scale, low-temperature solution processing. Within a year, PCEs had surpassed 15%, with further improvements anticipated. Their research highlighted the compatibility of this technology with electron acceptor and donor materials commonly used in organic photovoltaics. They demonstrated that a single thin film of the solution-processed organometal trihalide perovskite absorber $CH_3NH_3PbI_{3-x}Cl_x$, when integrated between organic contacts, could achieve PCEs of up to 10% on glass substrates and over 6% on flexible polymer substrates. This work marked a significant advancement, facilitating the adoption of perovskite technology by the organic photovoltaic community and enabling the leveraging of existing knowledge of hybrid interfaces for further device enhancements and flexible processing methodologies.

Li et al. [84] in 2016, reported high-efficiency, mechanically robust perovskite solar cells on ultrathin flexible substrates. The increasing demand for portable power sources in personal consumer electronics has driven the need for lightweight and mechanically flexible devices. Perovskite solar cells offer a compelling combination of low-cost production and high device performance. They demonstrated high-performance planar heterojunction PSCs fabricated on highly flexible and ultrathin silver-mesh/conducting polymer substrates. These devices exhibited performance comparable to those on rigid glass/indium tin oxide substrates, achieving a PCE of 14.0% and a specific power of 1.96 kW $kg^{-1}$, owing to the use of a 57 μm-thick polyethylene terephthalate substrate. The flexible devices also demonstrated excellent mechanical stability, retaining over 95% of their initial efficiency after 5,000 full bending cycles. These results confirmed the compatibility of perovskite thin films with flexible substrates, suggesting their potential for applications in flexible and bendable solar cells.

Li et al. [85] in 2018, presented an efficient flexible perovskite solar module exceeding 8% efficiency, achieved through an ultrafast $PbI_2$ deposition rate. Large-area, pinhole-free $CH_3NH_3PbI_3$ perovskite thin films were successfully fabricated on 5 cm × 5 cm flexible indium tin oxide coated polyethylene naphthalate (ITO-PEN) substrates using a sequential evaporation/spin-coating deposition method. The study investigated the influence of the rate-controlled evaporation of $PbI_2$ films on the quality of the perovskite layer and the performance of planar-structured perovskite solar cells. An ultrafast evaporation rate of 20 Å $s^{-1}$ was found to be optimal for the conversion of $PbI_2$ to $CH_3NH_3PbI_3$ perovskite. Based on this high-quality $CH_3NH_3PbI_3$ film, a flexible perovskite solar sub-module (active area of 16 $cm^2$) achieved a PCE of over 8%, and a 1.2 $cm^2$ flexible perovskite solar cell achieved a PCE of 12.7%.

Pisoni et al. [86] in 2018, detailed tailored lead iodide growth for efficient flexible perovskite solar cells and thin-film tandem devices. Among the various deposition methods for the perovskite absorber, the two-step hybrid vacuum-solution approach allows precise control over the thickness and morphology of $PbI_2$. However, efficient conversion to perovskite is limited by the diffusion of organic cations into the compact lead halide layer. To address this, a multistage absorber deposition method was developed, involving thermal evaporation of $PbI_2$ and spin coating of $CH_3NH_3I$ (MAI). The process leverages the differential growth of vacuum-deposited $PbI_2$ on amorphous and crystalline surfaces. This approach effectively increases the absorber thickness while addressing the limited MAI diffusion in the compact $PbI_2$ film through a two-step deposition process. The efficiency of FPSCs was improved from 14.2 to 15.8% using multistage deposition. Furthermore, the use of an amorphous transparent conductive oxide (TCO), InZnO, enhanced the mechanical resistance against bending compared to conventional crystalline TCO-based flexible devices. Near-infrared transparent FPSCs were developed with an efficiency of 14.0% and an average transmittance of approximately 74% between 800 and 1000 nm. Flexible perovskite/CIGS thin-film tandem devices were demonstrated with an efficiency of 19.6% measured in the four-terminal configuration.

Zhao et al. [87] in 2018, reported efficient two-terminal all-perovskite tandem solar cells achieved through high-quality low-bandgap absorber layers. Multi-junction all-perovskite tandem solar cells are a promising option for next-generation solar cells, offering high efficiency and low fabrication costs. However, the lack of high-quality low-bandgap perovskite absorber layers has significantly hindered the development of efficient and stable two-terminal monolithic all-perovskite tandem solar cells. They reported a bulk-passivation strategy involving the incorporation of chlorine to enlarge grains and reduce electronic disorder in mixed tin–lead low-bandgap (~1.25 eV) perovskite absorber layers. This enabled the fabrication of efficient low-bandgap perovskite solar cells using thick absorber layers (~750 nm), which is essential for efficient tandem solar cells. This improvement facilitated the fabrication of two-terminal all-perovskite tandem solar cells with a champion PCE of 21% and a steady-state efficiency of 20.7%. The efficiency was retained at 85% of its initial performance after 80 h of operation under continuous illumination.

Gao et al. [88] in 2019, introduced flexible, transparent nanocellulose paper-based perovskite solar cells. Current FPSCs utilize non-biodegradable petroleum-based polymer substrates, the disposal of which contributes to environmental pollution. Therefore, the development of green, biodegradable, and low-cost flexible substrates presents a viable alternative. They developed transparent nanocellulose paper (NCP) coated with acrylic resin as substrates for FPSCs, offering biodegradability and ease of disposal. The PCE of these NCP-based PSCs reached 4.25%, while the power per weight was as high as 0.56 W $g^{-1}$. The FPSCs also exhibited good stability, retaining over 80% of their original efficiency after 50 bending cycles. The NCP-based substrates can also be applied to other electronic systems, potentially fostering the development of next-generation green flexible electronics.

Driven by recent improvements in the efficiency and stability of PSCs, upscaling of PSCs has emerged as the next critical step. Specifically, a high-throughput, low-cost

R2R process would be a significant breakthrough for the commercialization of PSCs, requiring uniform formation of the precursor wet film and complete conversion to the perovskite phase via R2R-compatible processes. Kin et al. [89] in 2020, detailed roll-to-roll gravure-printed FPSCs using eco-friendly antisolvent bathing with a wide processing window. They demonstrated the pilot-scale, fully R2R manufacturing of all layers except for the electrodes in PSCs. Tert-butyl alcohol (tBuOH) was introduced as an eco-friendly antisolvent with a wide processing window. Highly crystalline, uniform formamidinium (FA)-based perovskite formation via tBuOH:EA bathing was confirmed by achieving high PCEs of 23.5% for glass-based spin-coated PSCs and 19.1% for gravure-printed FPSCs. Further work involved R2R gravure-printing and tBuOH:EA bathing, resulting in the highest PCE reported for R2R-processed PSCs: 16.7% for PSCs with R2R-processed $SnO_2$/FA-perovskite and 13.8% for fully R2R-produced PSCs.

Li et al. [90] in 2022, reported flexible all-perovskite tandem solar cells approaching 25% efficiency with a molecule-bridged hole-selective contact. Lightweight FPSCs hold promise for building-integrated photovoltaics, wearable electronics, portable energy systems, and aerospace applications. However, their highest certified efficiency of 19.9% lags behind their rigid counterparts, mainly due to defective interfaces at charge-selective contacts with perovskites on top. They used a mixture of two hole-selective molecules based on carbazole cores and phosphonic acid anchoring groups to form a self-assembled monolayer and bridge perovskite with a low-temperature-processed NiO nanocrystal film. The hole-selective contact mitigated interfacial recombination and facilitated hole extraction. They demonstrated flexible all-perovskite tandem solar cells with an efficiency of 24.7% (certified 24.4%), outperforming all types of flexible thin-film solar cells. They also reported a 23.5% efficiency for larger device areas of 1.05 $cm^2$. The molecule-bridged interfaces enabled significant bending durability, with the flexible all-perovskite tandem solar cells retaining their initial performance after 10,000 bending cycles at a radius of 15 mm.

Xu et al. [91] in 2024, reported that multifunctional entinostat enhances the mechanical robustness and efficiency of flexible perovskite solar cells and mini-modules. Flexible perovskite solar cells have attracted considerable attention due to their potential for soft and high power–weight compatibility. However, the inferior quality of the buried perovskite–substrate interface due to low interfacial adhesion and large deformation of flexible substrates have greatly limited the performance of FPSCs. They added the organic molecule entinostat into the hole extraction material poly(bis(4-phenyl)(2,4,6-trimethylphenyl)amine) to enhance adhesion at the perovskite–substrate interface, utilizing the interaction of entinostat with perovskites, poly(bis(4-phenyl)(2,4,6-trimethylphenyl)amine), and indium tin oxide through its multiple functional groups. In addition, entinostat reduced the density of voids at the bottom of the perovskite film owing to its capability to tune the crystallization of perovskites. They demonstrated inverted small-area flexible perovskite solar cells with a power conversion efficiency of 23.4%. Flexible perovskite minimodules with an area of 9 $cm^2$ achieved a certified aperture efficiency of approximately 19.0%. The optimized unencapsulated flexible minimodule retained 84% of its initial efficiency after 5,000 bending cycles and 90% of its initial power conversion efficiency after light soaking for >750 h.

Li et al. [92] in 2025, reported boosting mechanical durability under high humidity by a bioinspired multisite polymer for high-efficiency flexible perovskite solar cells. FPSCs with high stability in moist air are required for their practical applications. However, the poor mechanical stability under high humidity air remains a critical challenge for flexible perovskite devices. Inspired by the exceptional wet adhesion of mussels via dopamine groups, they proposed a multidentate-cross-linking strategy. This strategy combines a multibranched structure and adequate dopamine anchor sites in a three-dimensional hyperbranched polymer to directly chelate perovskite materials in multiple directions, constructing a vertical scaffold across the bulk of perovskite films from the bottom to the top interfaces. This intimately binds to the perovskite grains and substrates with a strong adhesion ability, and enhances mechanical durability under high humidity. Consequently, the modified rigid PSCs achieved a superior PCE of up to 25.92%, while FPSCs exhibited a PCE of 24.43% and maintained 94.1% of their initial PCE after 10,000 bending cycles with a bending radius of 3 mm under exposure to 65% humidity.

Hybrid organic–inorganic perovskite solar cells (HPSCs) have emerged as a highly promising next-generation PV technology, achieving remarkable power conversion efficiencies of up to 30.1% in laboratory settings [5]. Despite these impressive achievements, the transition from lab-scale production to industrial-scale manufacturing has encountered significant challenges. Traditional methods, such as sheet-to-sheet processing, vacuum-based evaporation, and subtractive laser patterning, while effective for producing individual cells, introduce complexities and costs that hinder scalable and cost-effective manufacturing of large-area modules [89].

Flexible HPSCs present a transformative alternative by enabling high-throughput R2R manufacturing techniques, which are compatible with lightweight, flexible plastic substrates. This approach facilitates continuous production, eliminates the need for costly and time-intensive vacuum processes, and offers the potential to produce PV devices with competitive specific power (power-to-weight ratios). Such attributes make flexible HPSCs particularly appealing for emerging applications, including space exploration, vehicle-integrated PV systems, and building-integrated PV solutions. However, R2R fabrication is not without its challenges, particularly in terms of time and temperature constraints associated with continuous film deposition on moving substrates. Recent advancements in R2R fabrication have addressed many of the limitations associated with flexible HPSC production. A significant breakthrough was the development of entirely R2R-fabricated HPSCs under ambient room conditions [93].

Weerasinghe et al. [94] in 2024, demonstrated small-area cells achieving a record-high certified PCE of 15.5%, alongside larger interconnected modules (~50 $cm^2$ active area) exhibiting PCEs of up to 11%. A schematic illustration of a roll-to-roll fabrication of perovskite solar cell modules under ambient room conditions is illustrated in Fig. 6.13. A key innovation enabling this achievement was the replacement of high-cost vacuum-processed gold electrodes with printable carbon electrodes. These perovskite-compatible carbon inks were engineered to provide reliable electrical conductivity, ensure compatibility with delicate perovskite layers, reduce material and processing costs, and enhance mechanical durability. The integration of automated R2R systems has significantly increased experimental throughput, enabled the fabrication and tested of thousands of cells daily. Notable advancements include the incorporation of high-throughput experimental platforms into R2R manufacturing workflows, which have optimized deposition and drying parameters to produce high-quality films and improve device efficiencies. The printing-friendly sequential deposition (PFSD) technique has been instrumental in this optimization, leveraging controlled crystallization dynamics to produce amorphous precursor films that rapidly convert to perovskite under ambient conditions without compromising film quality or production speed. Additional refinements, such as shallow-angle blowing techniques during film drying, have reduced crystal defects and improved humidity tolerance. The replacement of vacuum-dependent electrodes and the adoption of scalable R2R deposition techniques have significantly reduced production costs. Techno-economic analyses predict manufacturing costs as low as \$0.7 USD per watt for an annual production volume of approximately 1 million $m^2$ in Australia. By further eliminating high-cost components, projected costs could drop below \$0.5 USD per watt, positioning flexible HPSCs as competitive alternatives to silicon-based solar cells. While silicon modules dominate utility-scale PV markets with costs below \$0.30 USD per watt, HPSCs are expected to excel in niche applications requiring lightweight and flexible solutions, such as portable devices and wearable electronics.

In June 2024, the first flexible perovskite/silicon tandem solar cell utilizing ultrathin silicon, with a thickness of approximately 30 $\mu$m, was successfully developed by researchers at the Ningbo Institute of Materials Technology and Engineering, Chinese Academy of Sciences [95]. While significant progress has been achieved in rigid perovskite/silicon tandem solar cells, with efficiencies reported as high as 33.9%, the development of their flexible counterparts has been far more challenging. This difficulty stems from the need to balance enhanced light absorption in ultrathin silicon bottom cells with the mechanical flexibility necessary for such designs. In their study, the research team addressed these challenges by reducing the thickness of silicon wafers and optimizing the dimensions of light-trapping textures, thereby improving the substrate's flexibility without compromising the efficiency of light absorption. Additionally, the incorporation of capped perovskite top cells enhanced the mechanical durability of the device, mitigating issues related to silicon surface

**Fig. 6.13** Schematic illustration of a roll-to-roll fabrication of perovskite solar cell modules under ambient room conditions [94]. Copyright 2016, Elsevier

fractures. As a result, the team successfully fabricated a flexible perovskite/silicon tandem solar cell that achieved a certified power conversion efficiency of 22.8%, setting a performance benchmark for flexible solar cells. The device also demonstrated an outstanding power-to-weight ratio of 3.12 W $g^{-1}$, making it highly efficient for lightweight and portable applications. Furthermore, it exhibited remarkable bending durability, retaining 98.2% of its initial efficiency even after 3,000 bending cycles at a radius as small as 1 cm. This achievement represents a significant step forward in the development of flexible tandem solar cells and their potential for use in advanced, portable energy systems.

#### 6.2.4.2 Stability

The integration of PSCs into wearable technologies necessitates resilience against dynamic environmental stressors, including humidity fluctuations, thermal cycling, and mechanical deformation. To address these challenges, recent advancements in materials engineering have prioritized the development of humidity-tolerant crystallization protocols, solvent-resistant carbon-based conductive inks, and advanced hermetic encapsulation architectures. These innovations collectively enhance the operational stability of PSCs under real-world conditions, enabling sustained photovoltaic performance in high-flexibility applications exposed to atmospheric variability. A critical materials science focus lies in optimizing organic–inorganic hybrid perovskites, which exhibit exceptional optoelectronic properties such as tunable bandgaps, high absorption coefficients, and ambipolar charge transport [96]. While polycrystalline perovskite films dominate current research due to scalable processing advantages [97], single-crystal variants demonstrate superior charge carrier mobilities (exceeding 100 $cm^2\ V^{-1}\ s^{-1}$) and enhanced environmental stability, attributed to their anisotropic charge transport characteristics [98] and reduced grain boundary defect densities [99]. However, the epitaxial growth of single-crystal perovskites on flexible substrates presents substantial crystallographic challenges, particularly in achieving thickness uniformity and compositional gradation [100]. Lei et al. [101] pioneered a lithography-assisted epitaxial-growth-and-transfer (LAEGT) methodology enabling the fabrication of compositionally graded single-crystal perovskites (e.g., $MAPbI_3$ to $MAPb_{0.5}Sn_{0.5}I_3$ gradients) with precise thickness control (600 nm–100 μm) over 5.5 × 5.5 $cm^2$ flexible substrates. This technique yielded mechanically compliant crystals (<3 mm bending radius) with graded bandgaps (1.3–1.6 eV), achieving 18.77% mean photovoltaic efficiency while suppressing carrier recombination through bandgap engineering.

Electrode optimization represents another critical frontier, as conventional noble metal contacts (Au/Ag) suffer from ion migration-induced degradation and high vacuum deposition costs. Babu et al. [102] demonstrated a p–i–n architecture employing solution-processed carbon electrodes modified with a 5 nm chromium interfacial layer, which simultaneously functioned as a halide diffusion barrier and mechanical stress buffer. The Cr/C composite electrode enabled flexible polyethylene terephthalate (PET)-based devices (1 $cm^2$ active area) to achieve 15.18%

power conversion efficiency (PCE) with <10% performance decay after 1,000 h of maximum power point tracking under AM1.5G illumination. Notably, the chromium interlayer reduced series resistance ($R_s$) by 68% compared to unmodified carbon contacts through enhanced interfacial adhesion energy ($\gamma = 2.1$ J m$^{-2}$ vs. 0.9 J m$^{-2}$ for bare carbon). Thermal aging tests (85 °C, $N_2$ atmosphere) revealed <5% PCE loss over 1,000 h, contrasting sharply with Cr/Ag reference devices that degraded by 42% under identical conditions.

Interfacial reinforcement strategies have emerged to address the mechanical vulnerability of charge transport layer/perovskite interfaces in flexible PSCs (FPSCs). Li et al. [103] engineered a polyamide-amine hyperbranched polymer (HBP) interfacial adhesive featuring multidentate hydrogen-bonding networks and $Pb^{2+}$-chelating macrocyclic cavities. The HBP modification doubled the interfacial fracture energy ($G_c = 2.13$ J m$^{-2}$ vs. 1.08 J m$^{-2}$ control) while reducing lead leakage by 93% under mechanical failure conditions. This dual functionality enabled record PCEs of 25.05% (rigid ITO) and 23.86% (flexible PET-ITO), maintaining >95% initial efficiency after 10,000 bending cycles ($r = 5$ mm). The HBP's dynamic bond reconfiguration capability dissipated 78% of interfacial stress during flexural loading, as quantified by finite element analysis.

Recent advances in grain boundary (GB) engineering address strain-induced microcracking in polycrystalline perovskite films under cyclic bending. Jin et al. [104] developed a bifacial capping strategy using 4-methoxybenzylamine hydrobromide (MeOBABr) additives that spontaneously segregate to both top and bottom GBs during crystallization. The MeOBABr capping layers reduced GB groove depth by 82% (AFM-measured) and residual tensile strain by 64% (XRD analysis), enabling FPSCs to retain 92% of initial PCE (23.7% baseline) after 20,000 bending cycles ($r = 3$ mm). The polar methoxy groups induced a 0.3 eV upward band bending at the perovskite/NiOx interface, enhancing hole extraction efficiency ($\tau_h = 12$ ns vs. 23 ns control). This bifunctional modification yielded devices with 42.46% indoor efficiency under 1000 lx LED illumination, demonstrating broad-spectrum applicability.

#### 6.2.4.3 Lightweight

FPSCs have garnered significant attention as a promising energy source for wearable devices and portable electronics, primarily due to their inherent advantages of being lightweight and possessing a high specific power output. The capacity R2R manufacturing of PSCs on flexible plastic substrates further augments their suitability, enabling the creation of ultra-lightweight modules characterized by a high power-to-weight ratio, thereby positioning them as ideal candidates for portable and wearable applications. In 2015, Kaltenbrunner et al. [105] demonstrated the potential of FPSCs by fabricating devices with chromium oxide–metal contacts to enhance stability in ambient conditions, achieving a high power-per-weight performance. The research emphasized the need for photovoltaic technologies to employ light-absorbing materials that exhibit high efficiency, low weight, cost-effectiveness, and operational stability. Organolead halide perovskites emerged as a promising material class, albeit with limitations in ambient stability, which were addressed through the utilization of a chromium oxide–chromium interlayer to protect the metal top contacts from reacting with the perovskite. The employment of a transparent polymer electrode treated with dimethylsulphoxide as the bottom layer facilitated the deposition of pinhole-free perovskite films on thin plastic foils at low temperatures, showcasing a stabilized 12% efficiency and a power-per-weight ratio of 23 W $g^{-1}$. These ultra-lightweight solar cells were successfully employed to power aviation models, suggesting potential applications in unmanned aerial vehicles (UAVs) for environmental and industrial monitoring, rescue operations, emergency response, and tactical security applications.

Kang et al. [106] further advanced the field in 2019 by reporting lightweight and flexible perovskite solar cells with enhanced power-per-weight performance. Their study highlighted the increasing interest in lightweight and flexible photovoltaic devices for specialized applications, such as miniaturized drones, blimps, and aerospace electronics. The researchers developed ultralight and flexible PSCs utilizing orthogonal silver nanowire (AgNW) transparent electrodes on 1.3 μm-thick polyethylene naphthalate foils. The smooth surface morphologies of the orthogonal AgNW transparent electrodes mitigated the formation of nonconducting silver halide, which arises from the chemical reaction between the AgNWs and iodine in the active layer. The resulting PSCs exhibited a significantly improved PCE of 15.18%, compared to devices with random AgNW network electrodes (10.43% PCE). Furthermore, these ultralight and flexible PSCs achieved a power-per-weight ratio of 29.4 W $g^{-1}$, representing the highest reported value for a lightweight solar cell device, with potential applications in wearable optoelectronic devices.

The pursuit of higher efficiency in FPSCs continued with the work of Li et al. [90] in 2022, who reported flexible all-perovskite tandem solar cells approaching 25% efficiency through the implementation of a molecule-bridged hole-selective contact. The study underscored the promise of lightweight flexible perovskite solar cells for various applications, including building-integrated photovoltaics, wearable electronics, portable energy systems, and aerospace applications. Addressing the efficiency gap between flexible and rigid PSCs, the researchers introduced a mixture of two hole-selective molecules based on carbazole cores and phosphonic acid anchoring groups to form a self-assembled monolayer, bridging the perovskite with a low-temperature-processed NiO nanocrystal film. This hole-selective contact effectively mitigated interfacial recombination and facilitated hole extraction, leading to flexible all-perovskite tandem solar cells with a certified efficiency of 24.4%. The molecule-bridged interfaces also conferred significant bending durability, enabling the flexible devices to retain their initial performance after 10,000 cycles of bending at a radius of 15 mm.

More recently, Grabowska et al. [107] addressed the limitations of conventional Poly(ethylene terephthalate) (PET) substrates by exploring the fabrication of inverted PSCs on colorless polyimide (CPI) substrates capable of withstanding harsh environmental conditions. A sequential sputtering technique was employed to deposit indium tin oxide (bottom electrode) and nickel oxide (hole transport layer) as a base stack for the perovskite. The integration of MeO-2PACz into the hole transport bilayer further enhanced the NiOx interface, resulting in a power conversion efficiency (PCE) of 15.52% and a remarkable power-to-weight ratio (PWR) of 4.39 W/g, which is five times higher than that of devices on PET (0.87 W/g). The versatility of the active stack developed in this study suggests its potential application on a variety of transparent substrates.

These collective advancements demonstrate the continuous progress in the development of FPSCs, with a focus on enhancing efficiency, stability, and power-to-weight ratio. The diverse range of applications, from powering aviation models and UAVs to wearable electronics and building-integrated photovoltaics, underscores the transformative potential of FPSCs in addressing the growing demand for lightweight and high-performance energy solutions.

#### 6.2.4.4 Scalability

The transition from laboratory-based experimentation to industrial-scale manufacturing of PSCs is contingent upon overcoming several significant challenges. The scalability of PSC production is intrinsically linked to a complex interplay of factors, encompassing materials selection, fabrication methodologies, and long-term operational stability. Among the array of industrial fabrication techniques available, printing technologies are of paramount importance, offering a viable pathway towards the mass production of high-efficiency, cost-effective, and versatile photovoltaic devices. Achieving efficient scalability in PSC technology necessitates targeted solutions addressing challenges spanning materials science, manufacturing processes, and stability considerations. Material selection constitutes a primary determinant of PSC scalability. The quintessential architecture of PSCs, typically comprising lead-based perovskites and organic halides, endows these solar cells with exceptional attributes, including elevated light absorption coefficients and extended charge carrier diffusion lengths, which are fundamentally crucial for their high-power conversion efficiencies. However, the environmental ramifications associated with these constituent materials, coupled with concerns regarding their cost and large-scale availability, present substantial obstacles to industrial scalability. In response to these challenges, the research community has directed considerable effort towards the exploration of alternative material compositions. This research trajectory aims to mitigate reliance on toxic elements while concurrently enhancing the durability and operational performance of PSCs under realistic operating conditions [108].

Manufacturing methodologies employed in PSC production are equally pivotal in facilitating scalability. R2R printing, alongside other solution-processed techniques such as inkjet printing and slot-die coating, present compelling advantages for large-scale fabrication. These techniques enable the efficient deposition of thin films across expansive areas, thereby enhancing production throughput while simultaneously minimizing material wastage and reducing overall production costs. R2R processing is particularly promising for the industrial scale-up of PSC fabrication, offering the flexibility to manufacture lightweight and flexible photovoltaic panels suitable for a diverse range of commercial applications. Furthermore, the inherent modular design of PSCs demonstrates intrinsic compatibility with such scalable manufacturing techniques, facilitating seamless integration into existing solar energy production infrastructure [109].

The long-term operational stability of PSCs remains a critical bottleneck hindering their widespread industrial adoption. While power conversion efficiencies have dramatically increased from 3.8% to over 27% within a decade, maintaining sustained performance under real-world operational conditions, including exposure to humidity, thermal stress, and ultraviolet radiation, is indispensable for successful commercialization. Advanced material formulations, sophisticated encapsulation strategies, and interface optimization are crucial parameters for enhancing the stability and extending the operational lifespan of PSC devices. Ongoing research endeavors focused on elucidating and mitigating degradation mechanisms, such as ion migration within the perovskite structure, are of paramount importance in bridging the translational gap between laboratory-scale experimentation and practical field applications [110].

Printing technologies, encompassing inkjet printing, R2R processing, and screen printing methodologies, constitute integral components of industrial-scale PSC production. These techniques are instrumental in enabling scalability, improving cost-effectiveness, and fostering design innovation within the field. A significant advantage inherent to printing methods is their intrinsic compatibility with thin-film technologies. Thin-film approaches necessitate substantially reduced quantities of semiconductor material in comparison to conventional silicon-based solar cells. For instance, screen printing techniques facilitate the uniform deposition of active layers across large areas, thereby optimizing material utilization, minimizing waste generation, and concurrently lowering production expenses [111].

The advent of three-dimensional (3D) printing and other advanced additive manufacturing methodologies is expanding the functional capabilities of PSCs. Through techniques such as direct ink writing (DIW) and electrohydrodynamic printing (EHDP), researchers are capable of fabricating intricate light-trapping structures directly onto the surface of solar cells. These micro and nano-structured features enhance light absorption and consequently improve energy conversion efficiency. Furthermore, these advanced fabrication techniques facilitate rapid prototyping and the expedited development of novel device architectures, thereby broadening the scope of potential applications for PSCs in domains such as BIPV and other specialized sectors. The capacity for rapid design iteration and experimental validation through these methods supports continuous innovation within PSC production [112].

#### 6.2.4.5 Cost-Effectiveness

The cost of FPSCs is determined by a variety of interconnected factors, including the cost of materials, manufacturing techniques, system-related expenses, and performance metrics such as efficiency. These elements collectively shape the economic feasibility and market competitiveness of FPSCs when compared to conventional silicon-based photovoltaic technologies. Among the most important determinants of cost are the durability and efficiency of FPSCs. For these solar cells to achieve widespread adoption in commercial applications, their lifespan must be comparable to that of silicon solar panels, which can exceed 25 years with minimal loss in efficiency. Current studies indicate that FPSCs must exhibit a minimum operational life of 21 years and consistently maintain an efficiency level of 25% to be financially viable. Meeting these performance criteria presents significant technological hurdles, especially considering that silicon-based modules are projected to reach efficiency levels of approximately 24% by 2030, coupled with a levelized cost of electricity (LCOE) of only 5 US cents per kilowatt-hour. Consequently, advancing the durability and efficiency of FPSCs is key to lowering production costs and achieving market competitiveness in the renewable energy sector [113].

Another major cost determinant is the expense of manufacturing, which accounts for only about 12% of the total system installation costs associated with perovskite modules. Flexible solar cells leverage innovative R2R manufacturing technologies that promote scalability and cost optimization. However, the manufacturing costs of perovskite modules vary widely, ranging from €10 to €100 per square meter depending on the specific substrate used and whether the design is flexible or rigid. This cost disparity underscores the need for further refinements in manufacturing methods to reduce costs consistently. Tandem cell designs, which integrate perovskite layers with silicon substrates, offer potential benefits for efficiency improvements but also add complexity and additional production expenses. R2R manufacturing, on the other hand, excels in its ability to handle large-scale module production while enabling customization, making it ideal for mass production to meet diverse market demands. In the wearable solar technology market, R2R production methods play a pivotal role in catering to the demand for high-volume output and rapid prototyping. Automated R2R manufacturing allows the fabrication and testing of thousands of solar cells daily, accelerating development cycles and facilitating faster commercialization. Wearable electronics require cost-efficient solutions, making affordability crucial. Studies have shown that R2R manufacturing can significantly bring down production costs—for example, as low as $0.7 USD per watt peak (Wp)—by incorporating scalable solution-based techniques and inexpensive materials like flexible plastic substrates and carbon inks. Additionally, replacing high-cost components like transparent conductive electrodes (TCEs) and silver grids with more affordable alternatives could further enhance cost-effectiveness for wearable solar devices. R2R processes have successfully replaced vacuum-deposited electrodes, such as gold or silver, with printable carbon electrodes, eliminating expensive fabrication steps while aligning with the economic constraints of the wearable technology sector [113, 114].

Material costs, another critical component, significantly influence the overall cost structure of FPSCs. The lightweight and flexible nature of perovskite materials provides an economically attractive alternative to traditional crystalline silicon. Nonetheless, ensuring long-term stability and operational reliability requires continuous improvement in material processing approaches. These advancements, while essential for achieving competitive performance, may lead to higher upfront material expenses. Furthermore, tandem designs, which strategically pair perovskite and silicon layers, can enhance energy conversion efficiency but introduce additional costs due to their technical complexity [114].

System-related expenses, including installation, operational maintenance, and auxiliary costs, also play a substantial role in shaping the total cost of FPSC deployment. These costs often constitute a large proportion of the overall system budget and require a comprehensive cost-reduction strategy across the supply chain. Operational costs (OPEX), including labor and energy inputs, combined with capital investment (CAPEX) for manufacturing facilities and equipment, also present challenges to scalability and cost management for FPSCs [113].

Looking ahead, advances in manufacturing innovation and production efficiency promise significant reductions in FPSC costs over time. Long-term techno-economic predictions suggest that FPSCs could achieve LCOE levels comparable to those of existing solar technologies by 2050. However, achieving this requires breakthroughs in material processing, optimization of manufacturing scalability, and improvements in efficiency and durability. These innovations will be vital to ensuring that FPSCs meet or exceed the cost-effectiveness and adaptability of rigid silicon-based modules, positioning them as a competitive option in the global renewable energy sector [114].

#### 6.2.4.6 Safety

The widespread commercialization of solar cells, particularly FPSCs, is contingent upon addressing significant environmental and human health concerns associated with their production, use, and disposal. A major challenge lies in the presence of hazardous materials, particularly lead, in perovskite formulations. Lead's toxicity and its potential release into the environment throughout the solar cell lifecycle pose serious risks to ecosystems and human health. Eliminating harmful substances like lead is crucial for sustainable production and broader adoption, especially as industries increasingly prioritize eco-friendly technologies that align with global environmental standards [115].

Various strategies have been explored to mitigate or eliminate lead in perovskite solar cells. One prominent approach involves substituting lead with less toxic elements such as tin (Sn), bismuth (Bi), or antimony (Sb). Tin-based perovskites have shown promise due to their non-toxic nature, although challenges related to stability and efficiency necessitate further research and optimization. Another significant development is the use of advanced encapsulation techniques to seal perovskite solar cells, preventing lead leakage during degradation or disposal. These protective barriers not only reduce environmental risks but also enhance the durability and operational lifespan of FPSCs. Additionally, circular economy initiatives, including recycling systems for end-of-life solar cells, are being implemented to recover hazardous materials like lead and other valuable components, thereby minimizing the environmental impact of disposal [116–118].

The mass production of solar cells depends on balancing three key factors: sustainability, efficiency, and economic feasibility. Solar cells must incorporate environmentally friendly materials that minimize risks to human health and ecosystems. Furthermore, non-toxic alternatives must match or exceed the efficiency and durability of lead-based perovskites to remain competitive in the photovoltaic market. Solar technologies with power conversion efficiencies exceeding 20% and long-term operational stability are more likely to achieve commercialization. Finally, cost-effective manufacturing processes, scalable material sourcing, and affordability for end-users are essential for widespread adoption across industries and regions globally. While challenges related to the environmental and health risks of materials like lead persist, ongoing efforts to develop lead-free perovskite compositions, enhance protective measures, and implement recycling frameworks demonstrate a clear pathway toward sustainable mass production. Combining these advancements with regulatory support and cost optimization will enable solar technologies to transition into a new era of environmentally responsible and competitive renewable energy solutions [119].

Li et al. [120] investigated the environmental and biological impacts of lead-containing halide perovskites, particularly in the context of perovskite solar cells (PSCs). Halide perovskites are widely regarded as revolutionary materials for next-generation solar cells and optoelectronic devices due to their exceptional optical and electronic properties. However, their reliance on lead salts raises significant environmental concerns, necessitating a thorough examination of their ecological footprint. Key findings from the study highlight the risks associated with lead leakage and bioavailability, the potential of tin-based alternatives, and the broader implications for sustainable commercialization. One of the primary environmental risks of lead-based halide perovskites is their propensity for leakage during the lifecycle of solar devices. The lead salts used in PSCs are relatively water-soluble, making them highly bioavailable and accessible to plants and other organisms. Experimental evidence indicates that lead from halide perovskites can enter plants approximately ten times more effectively than lead contaminants from other human activities. To assess the impact of lead contamination, the study utilized plants such as mint (Mentha spicata) grown in soil contaminated with varying concentrations of lead. Even low levels of lead contamination significantly increased lead concentrations within plants, with roots, leaves, and stems exhibiting heightened toxicity. In cases of higher contamination (250 mg $kg^{-1}$, as permitted under Chinese agricultural regulations), plants displayed severe symptoms such as blackening, rotting, and extreme lead toxicity. Furthermore, halide perovskites demonstrated higher bioavailability compared to other lead-based contaminants (e.g., $PbI_2$), suggesting a unique chemical interaction between perovskites and soil/pH that facilitates lead uptake by plants. Despite current regulations deeming the lead content in PSCs as minimal (0.1% by weight), the study underscores that this threshold is insufficient to address the disproportionate environmental impact and bioavailability of lead from perovskites. In response to these challenges, the study explores tin-based perovskites as a safer, lead-free alternative for PSCs. Tin exhibits significantly lower bioavailability than lead-based perovskites due to its tendency to oxidize and its low solubility in water. Experiments with plants grown in tin-contaminated soil revealed tin levels within safe limits defined by the FAO (Food and Agriculture Organization), highlighting the potential of tin-based perovskites to mitigate environmental risks. However, the adoption of tin-based alternatives requires further research to address challenges related to stability and efficiency. The broader implications of the study emphasize the need for stricter safety thresholds for lead in PSCs compared to other lead-containing electronics, as well as the importance of substituting lead with inert, environmentally friendly metals like tin. Researchers advocate for systematic screening of the environmental impact of various perovskite compositions to guide their application toward a sustainable market. Without such comprehensive studies, the risks associated with lead from halide perovskites remain underestimated, potentially hindering their safe and sustainable commercialization [120].

In 2023, Yang et al. [79] reported a novel approach to reducing lead toxicity in perovskite solar cells by incorporating a built-in supramolecular complex. The study demonstrated that embedding a cross-linking supramolecular complex composed of 2-hydroxypropyl β-cyclodextrin (HPβCD) and 1,2,3,4-butane tetracarboxylic acid (BTCA) could effectively inhibit lead leakage from severely damaged PSCs. The devices retained 97% of their initial efficiency after 522 h of dynamic water scouring, with lead contamination in water measuring less than 14 ppb. Toxicity evaluations indicated that the chelation between the HPβCD-BTCA complex and lead-bearing perovskites could reduce the toxicity of lead-bearing PSCs to levels comparable to or even lower than their lead-free counterparts. Moreover, the incorporation of the HPβCD-BTCA complex simultaneously improved the stability and reproducibility of PSCs. This innovative strategy paves a new avenue for sustainable PSCs and brings them closer to commercial implementation [121].

In conclusion, while challenges related to the environmental and health risks of materials like lead persist, ongoing efforts to develop lead-free perovskite compositions, enhance protective measures, and implement recycling frameworks demonstrate a clear pathway toward sustainable mass production. Combining these advancements with regulatory support and cost optimization will enable solar technologies to transition into a new era of environmentally responsible and competitive renewable energy solutions.

## References


1. T.J. Seebeck, Magnetische Polarisation der Metalle und Erze durch Temperatur-Differenz (Magnetic polarization of metals and minerals by temperature differences), in *Abhandlungen der Königlichen Akademie der Wissenschaften zu Berlin*, p. 1825
2. M.H. Boyé, J.C. Booth, On the identity of edwardsite and monazite, by Gustav Rose, extracted from the 1st no. of poggendorf's 'annalen der physik und der chemie' for 1840. J. Franklin Inst. **29**(5) (1840). https://doi.org/10.1016/s0016-0032(40)91279-0
3. E. Maciá-Barber, *Thermoelectric Materials: Advances and Applications* (2015). https://doi.org/10.4032/9789814463539

4. J.C.A. Peltier, Nouvelles expériences sur la caloricité des courants électrique (New experiments on the heat effects of electric currents). Annales de Chimie et de Physique **56** (1834)
5. W. Thomson, 4. On a mechanical theory of thermo-electric currents. Proc. R. Soc. Edinb. **3**, 91–98 (1857). https://doi.org/10.1017/S0370164600027310
6. M. Zebarjadi, Heat management in thermoelectric power generators. Sci. Rep. **6**(1), 20951 (2016). https://doi.org/10.1038/srep20951
7. Z. Zhang, Z. Peng, P. Jiang, Y. Sin Ang, C. Zhang, Z. Ma, Thomson effect in thermionic refrigeration: enhanced performance of graphene/2D-semiconductor/graphene heterostructure cooler. J. Appl. Phys. **133**(24) (2023). https://doi.org/10.1063/5.0153026
8. A.C. Ehrlich, S.A. Wolf, Chapter 4 Military applications of enhanced thermoelectrics (2001), pp. 117–124. https://doi.org/10.1016/S0080-8784(01)80138-1
9. Y. Sun et al., Performance boost for bismuth telluride thermoelectric generator via barrier layer based on low Young's modulus and particle sliding. Nat. Commun. **14**(1), 8085 (2023). https://doi.org/10.1038/s41467-023-43879-8
10. G.J. Snyder, A.H. Snyder, Figure of merit ZT of a thermoelectric device defined from materials properties. Energy Environ. Sci. **10**(11), 2280–2283 (2017). https://doi.org/10.1039/C7EE02007D
11. G. Rogl, P. Rogl, Skutterudites, a most promising group of thermoelectric materials. Curr. Opin. Green Sustain. Chem. **4**, 50–57 (2017). https://doi.org/10.1016/j.cogsc.2017.02.006
12. H. Zhu et al., Half-Heusler alloys as emerging high power density thermoelectric cooling materials. Nat. Commun. **14**(1), 3300 (2023). https://doi.org/10.1038/s41467-023-38446-0
13. Z.-G. Chen, G. Han, L. Yang, L. Cheng, J. Zou, Nanostructured thermoelectric materials: current research and future challenge. Prog. Nat. Sci.: Mater. Int. **22**(6), 535–549 (2012). https://doi.org/10.1016/j.pnsc.2012.11.011
14. X.-L. Shi et al., $SrTiO_3$-based thermoelectrics: progress and challenges. Nano Energy **78**, 105195 (2020). https://doi.org/10.1016/j.nanoen.2020.105195
15. D.A. Muller, N. Nakagawa, A. Ohtomo, J.L. Grazul, H.Y. Hwang, Atomic-scale imaging of nanoengineered oxygen vacancy profiles in $SrTiO_3$. Nature **430**(7000), 657–661 (2004). https://doi.org/10.1038/nature02756
16. V. Vulchev, L. Vassilev, S. Harizanova, M. Khristov, E. Zhecheva, R. Stoyanova, Improving of the thermoelectric efficiency of $LaCoO_3$ by double substitution with nickel and iron. J. Phys. Chem. C **116**(25), 13507–13515 (2012). https://doi.org/10.1021/jp3021408
17. C.R. Park, *Advanced Thermoelectric Materials* (Scrivener Publishing LLC, 2019). Accessed 30 May 2025. https://doi.org/10.1002/9781119407348
18. *Thermoelectric Materials and Devices* (Elsevier, 2021). https://doi.org/10.1016/C2018-0-02472-X
19. V.V. Deshmukh et al., Structure, morphology and electrochemical properties of SrTiO3 perovskite: photocatalytic and supercapacitor applications. Environ. Chem. Ecotoxicol. **3**, 241–248 (2021). https://doi.org/10.1016/j.enceco.2021.07.001
20. A. Khan, J. Flitcroft, J. Skelton, *Thermoelectric Properties of n-Type $CaTiO_3$, $SrTiO_3$ and $BaTiO_3$* (2023). https://doi.org/10.26434/chemrxiv-2023-q81v4
21. S. Dahbi, N. Tahiri, O. El Bounagui, H. Ez-Zahraouy, Electronic, optical, and thermoelectric properties of perovskite $BaTiO_3$ compound under the effect of compressive strain. Chem. Phys. **544**, 111105 (2021). https://doi.org/10.1016/j.chemphys.2021.111105
22. N. Kuganathan, R. Srikaran, P.C.M. Fossati, A. Chroneos, Theoretical modeling of defects, dopants, and diffusion in the mineral ilmenite. Minerals **9**(10), 610 (2019). https://doi.org/10.3390/min9100610
23. M. Culebras, R. Torán, C.M. Gómez, A. Cantarero, $La_{1-x}Ca_x MnO_3$ semiconducting nanostructures: morphology and thermoelectric properties. Nanoscale Res. Lett. **9**(1), 415 (2014). https://doi.org/10.1186/1556-276X-9-415
24. O. Phadke, J. Sakhuja, V. Saraswat, U. Ganguly, Exploiting the electrothermal timescale in $PrMnO_3$ RRAM for a compact, clock-less neuron exhibiting biological spiking patterns. Semicond. Sci. Technol. **36**(11), 114001 (2021). https://doi.org/10.1088/1361-6641/ac24e8

25. D. O'Flynn, M.R. Lees, G. Balakrishnan, Magnetic susceptibility and heat capacity measurements of single crystal $TbMnO_3$. J. Phys.: Condens. Matter **26**(25), 256002 (2014). https://doi.org/10.1088/0953-8984/26/25/256002
26. N. Kanas et al., Tuning the thermoelectric performance of $CaMnO_3$-based ceramics by controlled exsolution and microstructuring. ACS Appl. Energy Mater. **5**(10), 12396–12407 (2022). https://doi.org/10.1021/acsaem.2c02012
27. X.Y. Yu et al., The enhanced thermoelectric properties of $BiMnO_3$ ceramics by Sr-doped. Mater. Res. Express **5**(4), 045902 (2018). https://doi.org/10.1088/2053-1591/aabb0b
28. A. Mabrouki, A. Bougoffa, A. Trabelsi, E. Dhahri, K. Khirouni, Optical and thermoelectric properties of $NaNbO_3$ thin film deposed by spray pyrolysis: experimental and DFT study. Eur. Phys. J. Plus **137**(6), 748 (2022). https://doi.org/10.1140/epjp/s13360-022-02963-6
29. M. Yaseen et al., Phase transition and thermoelectric properties of cubic $KNbO_3$ under pressure: DFT approach. J. Market. Res. **11**, 2106–2113 (2021). https://doi.org/10.1016/j.jmrt.2021.02.017
30. A.I. Lebedev, Ferroelectric properties of $RbNbO_3$ and $RbTaO_3$. Phys. Solid State **57**(2), 331–336 (2015). https://doi.org/10.1134/S1063783415020237
31. Power generation material, power generation element, and power generation system, US20160225973A1 (2014)
32. N.T. Thuy, Thermoelectric properties of $LaFeO_3$ system doped with Ti, Co, and Cu. Hue Univ. J. Sci.: Nat. Sci. **126**(1B), 139–144 (2017)
33. S. Shabbir, B. Khalid, H. Sehrish, M.T. Iqbal, N. Morley, H. Anwar, Exploring the structural, morphological, optical, and dielectric properties, along with photocatalytic performance of La-doped $SrFeO_3$ nanofibers. Mater. Res. Bull. **179**, 112970 (2024). https://doi.org/10.1016/j.materresbull.2024.112970
34. J. Wang, M. Saccoccio, D. Chen, Y. Gao, C. Chen, F. Ciucci, The effect of A-site and B-site substitution on $BaFeO_{3-\delta}$: an investigation as a cathode material for intermediate-temperature solid oxide fuel cells. J. Power. Sources **297**, 511–518 (2015). https://doi.org/10.1016/j.jpowsour.2015.08.016
35. T. Yokota, R. Aoyagi, M. Gomi, Thermoelectric properties of Co-doped $BiFeO_3$ and $Bi_{24}CoO_{37}$–$BiFeO_3$ compound systems. J. Ceram. Soc. Jpn. **121**(1416), 675–678 (2013). https://doi.org/10.2109/jcersj2.121.675
36. A.J. Ahmed et al., Significant improvement in electrical conductivity and figure of merit of nanoarchitectured porous $SrTiO_3$ by La doping optimization. ACS Appl. Mater. Interfaces **12**(25), 28057–28064 (2020). https://doi.org/10.1021/acsami.0c01869
37. M. Upadhyay Kahaly, U. Schwingenschlögl, Thermoelectric performance enhancement of $SrTiO_3$ by Pr doping. J. Mater. Chem. A Mater. **2**(27), 10379 (2014). https://doi.org/10.1039/c3ta13947f
38. R. Kabir et al., Improvement in the thermoelectric properties of $CaMnO_3$ perovskites by W doping. J. Mater. Sci. **49**(21), 7522–7528 (2014). https://doi.org/10.1007/s10853-014-8459-x
39. H. Kamata, Y. Yonemura, J. Mizusaki, H. Tagawa, K. Naraya, T. Sasamoto, High temperature electrical properties of the perovskite-type oxide $La_{1-x}Sr_xMnO_{3-d}$. J. Phys. Chem. Solids **56**(7), 943–950 (1995). https://doi.org/10.1016/0022-3697(95)00019-4
40. I. El-Kassab, A.M. Ahmed, P. Mandal, K. Bärner, A. Kattwinkel, U. Sondermann, Heat conductivity of $La_{1-x}Sr_xMnO_3$ surface layers. Phys. B Condens. Matter. **305**(3–4), 233–241 (2001). https://doi.org/10.1016/S0921-4526(01)00618-4
41. F. Delorme et al., Thermoelectric properties of oxygen deficient $(K_{0.5}Na_{0.5})NbO_3$ ceramics. Mater. Lett. **162**, 24–27 (2016). https://doi.org/10.1016/j.matlet.2015.09.124
42. B. Kim et al., High-performance $(Na_{0.5}K_{0.5})NbO_3$ thin film piezoelectric energy harvester. J. Am. Ceram. Soc. **98**(1), 119–124 (2015). https://doi.org/10.1111/jace.13238
43. J. Qi, N. Ma, X. Ma, R. Adelung, Y. Yang, Enhanced photocurrent in $BiFeO_3$ materials by coupling temperature and thermo-phototronic effects for self-powered ultraviolet photodetector system. ACS Appl. Mater. Interfaces **10**(16), 13712–13719 (2018). https://doi.org/10.1021/acsami.8b02543

44. R. Williams, Becquerel photovoltaic effect in binary compounds. J. Chem. Phys. **32**(5), 1505–1514 (1960). https://doi.org/10.1063/1.1730950
45. M. Guarnieri, More light on information [historical]. IEEE Ind. Electron. Mag. **9**(4), 58–61 (2015). https://doi.org/10.1109/MIE.2015.2485182
46. M.D. Archer, Photovoltaics and photoelectrochemistry: similarities and differences. Phys. E Low Dimens. Syst. Nanostruct. **14**(1–2), 61–64 (2002). https://doi.org/10.1016/S1386-9477(02)00450-2
47. T. Prajapati, A. Priyam, *A Review on Photovoltaic Cells* (2023), pp. 497–512. https://doi.org/10.1007/978-981-19-4971-5_36
48. *NREL Best Research-Cell Efficiency Chart. Photovoltaic Research* (NREL, 2025). https://www.nrel.gov/pv/cell-efficiency. Accessed 30 May 2025
49. J.F. Geisz et al., Six-junction III–V solar cells with 47.1% conversion efficiency under 143 suns concentration. Nat. Energy **5**(4), 326–335 (2020). https://doi.org/10.1038/s41560-020-0598-5
50. S. Moon, K. Kim, Y. Kim, J. Heo, J. Lee, Highly efficient single-junction GaAs thin-film solar cell on flexible substrate. Sci. Rep. **6**(1), 30107 (2016). https://doi.org/10.1038/srep30107
51. C. Ballif, F.-J. Haug, M. Boccard, P.J. Verlinden, G. Hahn, Status and perspectives of crystalline silicon photovoltaics in research and industry. Nat. Rev. Mater. **7**(8), 597–616 (2022). https://doi.org/10.1038/s41578-022-00423-2
52. T.D. Lee, A.U. Ebong, A review of thin film solar cell technologies and challenges. Renew. Sustain. Energy Rev. **70**, 1286–1297 (2017). https://doi.org/10.1016/j.rser.2016.12.028
53. A. Hagfeldt, G. Boschloo, L. Sun, L. Kloo, H. Pettersson, Dye-sensitized solar cells. Chem. Rev. **110**(11), 6595–6663 (2010). https://doi.org/10.1021/cr900356p
54. J.Y. Kim, J.-W. Lee, H.S. Jung, H. Shin, N.-G. Park, High-efficiency perovskite solar cells. Chem. Rev. **120**(15), 7867–7918 (2020). https://doi.org/10.1021/acs.chemrev.0c00107
55. J. Yi, G. Zhang, H. Yu, H. Yan, Advantages, challenges and molecular design of different material types used in organic solar cells. Nat. Rev. Mater. **9**(1), 46–62 (2023). https://doi.org/10.1038/s41578-023-00618-1
56. S. Singh, Z.H. Khan, M.B. Khan, P. Kumar, P. Kumar, Quantum dots-sensitized solar cells: a review on strategic developments. Bull. Mater. Sci. **45**(2), 81 (2022). https://doi.org/10.1007/s12034-022-02662-z
57. J. Li, K. Sun, X. Yuan, J. Huang, M.A. Green, X. Hao, Emergence of flexible kesterite solar cells: progress and perspectives. npj Flex. Electron. **7**(1), 16 (2023). https://doi.org/10.1038/s41528-023-00250-7
58. O.M. Saif, A.H. Zekry, M. Abouelatta, A. Shaker, A comprehensive review of tandem solar cells integrated on silicon substrate: III/V versus perovskite. Silicon **15**(15), 6329–6347 (2023). https://doi.org/10.1007/s12633-023-02466-8
59. A.M. Bakhshayesh, H. Abdizadeh, M. Mirhosseini, N. Taghavinia, Designing highly stable yet efficient solar cells based on a new triple-cation quasi-2D/3D hybrid perovskites family. Ceram. Int. **45**(16), 20788–20795 (2019). https://doi.org/10.1016/j.ceramint.2019.07.065
60. A. Kumar et al., Enhanced efficiency and stability of electron transport layer in perovskite tandem solar cells: challenges and future perspectives. Sol. Energy **266**, 112185 (2023). https://doi.org/10.1016/j.solener.2023.112185
61. N. Rai, S. Rai, P.K. Singh, P. Lohia, D.K. Dwivedi, Analysis of various ETL materials for an efficient perovskite solar cell by numerical simulation. J. Mater. Sci.: Mater. Electron. **31**(19), 16269–16280 (2020). https://doi.org/10.1007/s10854-020-04175-z
62. D.B. Mitzi, C.A. Feild, W.T.A. Harrison, A.M. Guloy, Conducting tin halides with a layered organic-based perovskite structure. Nature **369**(6480), 467–469 (1994). https://doi.org/10.1038/369467a0
63. A. Kojima, K. Teshima, Y. Shirai, T. Miyasaka, Organometal halide perovskites as visible-light sensitizers for photovoltaic cells. J. Am. Chem. Soc. **131**(17), 6050–6051 (2009). https://doi.org/10.1021/ja809598r
64. N.-G. Park, M. Grätzel, T. Miyasaka (eds.), *Organic-Inorganic Halide Perovskite Photovoltaics* (Springer International Publishing, Cham, 2016). https://doi.org/10.1007/978-3-319-35114-8

65. T. Nie, Z. Fang, X. Ren, Y. Duan, S. Liu, Recent advances in wide-bandgap organic-inorganic halide perovskite solar cells and tandem application. Nanomicro Lett. **15**(1), 70 (2023). https://doi.org/10.1007/s40820-023-01040-6
66. K. Liang, D.B. Mitzi, M.T. Prikas, Synthesis and characterization of organic−inorganic perovskite thin films prepared using a versatile two-step dipping technique. Chem. Mater. **10**(1), 403–411 (1998). https://doi.org/10.1021/cm970568f
67. S. Pratap et al., Out-of-equilibrium processes in crystallization of organic-inorganic perovskites during spin coating. Nat. Commun. **12**(1), 5624 (2021). https://doi.org/10.1038/s41467-021-25898-5
68. J.E. Bishop, J.A. Smith, D.G. Lidzey, Development of spray-coated perovskite solar cells. ACS Appl. Mater. Interfaces **12**(43), 48237–48245 (2020). https://doi.org/10.1021/acsami.0c14540
69. A. Tara et al., Inkjet-printed $FASn_{1-x}Pb_xI_3$-based perovskite solar cells. ACS Appl. Mater. Interfaces **16**(46), 63520–63527 (2024). https://doi.org/10.1021/acsami.4c12477
70. C. Motta, F. El-Mellouhi, S. Sanvito, Charge carrier mobility in hybrid halide perovskites. Sci. Rep. **5**(1), 12746 (2015). https://doi.org/10.1038/srep12746
71. B. Saparov, D.B. Mitzi, Organic–inorganic perovskites: structural versatility for functional materials design. Chem. Rev. **116**(7), 4558–4596 (2016). https://doi.org/10.1021/acs.chemrev.5b00715
72. J. Bing, L.G. Caro, H.P. Talathi, N.L. Chang, D.R. Mckenzie, A.W.Y. Ho-Baillie, Perovskite solar cells for building integrated photovoltaics—glazing applications. Joule **6**(7), 1446–1474 (2022). https://doi.org/10.1016/j.joule.2022.06.003
73. X. Cheng et al., Mixed organic cation in chiral two-dimensional organic-inorganic hybrid metal halides—an ab initio study of nonlinear optical (NLO) Properties. J. Phys. Chem. C **128**(27), 11392–11400 (2024). https://doi.org/10.1021/acs.jpcc.4c02520
74. E. Aktas et al., Challenges and strategies toward long-term stability of lead-free tin-based perovskite solar cells. Commun. Mater. **3**(1), 104 (2022). https://doi.org/10.1038/s43246-022-00327-2
75. L. Jonathan et al., Hybrid organic-inorganic perovskite halide materials for photovoltaics towards their commercialization. Polymers (Basel) **14**(5), 1059 (2022). https://doi.org/10.3390/polym14051059
76. M.P.U. Haris, S. Kazim, S. Ahmad, Low-temperature-processed perovskite solar cells fabricated from presynthesized $CsFAPbI_3$ powder. ACS Appl. Energy Mater. **4**(3), 2600–2606 (2021). https://doi.org/10.1021/acsaem.0c03160
77. Y. Gao et al., Flexible perovskite solar cells: from materials and device architectures to applications. ACS Energy Lett. **7**(4), 1412–1445 (2022). https://doi.org/10.1021/acsenergylett.1c02768
78. N. Rolston et al., Effect of cation composition on the mechanical stability of perovskite solar cells. Adv. Energy Mater. **8**(9) (2018). https://doi.org/10.1002/aenm.201702116
79. J.P. Hirth, The influence of grain boundaries on mechanical properties. Metall. Trans. **3**(12), 3047–3067 (1972). https://doi.org/10.1007/BF02661312
80. Z. Dai, N.P. Padture, Challenges and opportunities for the mechanical reliability of metal halide perovskites and photovoltaics. Nat. Energy **8**(12), 1319–1327 (2023). https://doi.org/10.1038/s41560-023-01378-6
81. J. Zhao et al., Strained hybrid perovskite thin films and their impact on the intrinsic stability of perovskite solar cells. Sci. Adv. **3**(11) (2017). https://doi.org/10.1126/sciadv.aao5616
82. Y. Gao, D. Lin, P. Liu, T. Shi, W. Xie, Interaction mechanism between water molecules and perovskites. Mater. Chem. Front. **8**(3), 785–799 (2024). https://doi.org/10.1039/D3QM00969F
83. P. Docampo, J.M. Ball, M. Darwich, G.E. Eperon, H.J. Snaith, Efficient organometal trihalide perovskite planar-heterojunction solar cells on flexible polymer substrates. Nat. Commun. **4**(1), 2761 (2013). https://doi.org/10.1038/ncomms3761
84. Y. Li et al., High-efficiency robust perovskite solar cells on ultrathin flexible substrates. Nat. Commun. **7**(1), 10214 (2016). https://doi.org/10.1038/ncomms10214

85. K. Li et al., An efficient, flexible perovskite solar module exceeding 8% prepared with an ultrafast $PbI_2$ deposition rate. Sci. Rep. **8**(1), 442 (2018). https://doi.org/10.1038/s41598-017-18970-y
86. S. Pisoni et al., Tailored lead iodide growth for efficient flexible perovskite solar cells and thin-film tandem devices. NPG Asia Mater. **10**(11), 1076–1085 (2018). https://doi.org/10.1038/s41427-018-0099-1
87. D. Zhao et al., Efficient two-terminal all-perovskite tandem solar cells enabled by high-quality low-bandgap absorber layers. Nat. Energy **3**(12), 1093–1100 (2018). https://doi.org/10.1038/s41560-018-0278-x
88. L. Gao et al., Flexible, transparent nanocellulose paper-based perovskite solar cells. npj Flex. Electron. **3**(1), 4 (2019). https://doi.org/10.1038/s41528-019-0048-2
89. Y.Y. Kim et al., Roll-to-roll gravure-printed flexible perovskite solar cells using eco-friendly antisolvent bathing with wide processing window. Nat. Commun. **11**(1), 5146 (2020). https://doi.org/10.1038/s41467-020-18940-5
90. L. Li et al., Flexible all-perovskite tandem solar cells approaching 25% efficiency with molecule-bridged hole-selective contact. Nat. Energy **7**(8), 708–717 (2022). https://doi.org/10.1038/s41560-022-01045-2
91. W. Xu et al., Multifunctional entinostat enhances the mechanical robustness and efficiency of flexible perovskite solar cells and minimodules. Nat. Photonics **18**(4), 379–387 (2024). https://doi.org/10.1038/s41566-023-01373-z
92. Z. Li et al., Boosting mechanical durability under high humidity by bioinspired multisite polymer for high-efficiency flexible perovskite solar cells. Nat. Commun. **16**(1), 1771 (2025). https://doi.org/10.1038/s41467-025-57102-3
93. Y. Jung, K. Hwang, Y. Heo, J. Kim, D. Vak, D. Kim, Progress in scalable coating and roll-to-roll compatible printing processes of perovskite solar cells toward realization of commercialization, Adv. Opt. Mater. **6**(9) (2018). https://doi.org/10.1002/adom.201701182
94. H.C. Weerasinghe et al., The first demonstration of entirely roll-to-roll fabricated perovskite solar cell modules under ambient room conditions. Nat. Commun. **15**(1), 1656 (2024). https://doi.org/10.1038/s41467-024-46016-1
95. X. Wang et al., Ultrathin ($\sim$30 $\mu$m) flexible monolithic perovskite/silicon tandem solar cell. Sci. Bull. (Beijing) **69**(12), 1887–1894 (2024). https://doi.org/10.1016/j.scib.2024.04.022
96. K. Lin et al., Perovskite light-emitting diodes with external quantum efficiency exceeding 20 percent. Nature **562**(7726), 245–248 (2018). https://doi.org/10.1038/s41586-018-0575-3
97. Z. Liu et al., Gas-solid reaction based over one-micrometer thick stable perovskite films for efficient solar cells and modules. Nat. Commun. **9**(1), 3880 (2018). https://doi.org/10.1038/s41467-018-06317-8
98. G. Zheng et al., Manipulation of facet orientation in hybrid perovskite polycrystalline films by cation cascade. Nat. Commun. **9**(1), 2793 (2018). https://doi.org/10.1038/s41467-018-05076-w
99. Z. Chen et al., Thin single crystal perovskite solar cells to harvest below-bandgap light absorption. Nat. Commun. **8**(1), 1890 (2017). https://doi.org/10.1038/s41467-017-02039-5
100. L. Lee, J. Baek, K.S. Park, Y.-E. Lee, N.K. Shrestha, M.M. Sung, Wafer-scale single-crystal perovskite patterned thin films based on geometrically-confined lateral crystal growth. Nat. Commun. **8**(1), 15882 (2017). https://doi.org/10.1038/ncomms15882
101. Y. Lei et al., A fabrication process for flexible single-crystal perovskite devices. Nature **583**(7818), 790–795 (2020). https://doi.org/10.1038/s41586-020-2526-z
102. V. Babu et al., Improved stability of inverted and flexible perovskite solar cells with carbon electrode. ACS Appl. Energy Mater. **3**(6), 5126–5134 (2020). https://doi.org/10.1021/acsaem.0c00702
103. Z. Li et al., Hyperbranched polymer functionalized flexible perovskite solar cells with mechanical robustness and reduced lead leakage. Nat. Commun. **14**(1), 6451 (2023). https://doi.org/10.1038/s41467-023-41931-1
104. J. Jin et al., Spontaneous bifacial capping of perovskite film for efficient and mechanically stable flexible solar cell. Nat. Commun. **16**(1), 90 (2025). https://doi.org/10.1038/s41467-024-55652-6

105. M. Kaltenbrunner et al., Flexible high power-per-weight perovskite solar cells with chromium oxide–metal contacts for improved stability in air. Nat. Mater. **14**(10), 1032–1039 (2015). https://doi.org/10.1038/nmat4388
106. S. Kang et al., Ultrathin, lightweight and flexible perovskite solar cells with an excellent power-per-weight performance. J. Mater. Chem. A Mater. **7**(3), 1107–1114 (2019). https://doi.org/10.1039/C8TA10585E
107. A. Grabowska, R. Fuentes Pineda, P. Spinelli, G. Soto Pérez, F.A. Vinocour Pacheco, V. Babu, Development of lightweight and flexible perovskite solar cells on ultrathin colorless polyimide foils. ACS Appl. Mater. Interfaces **16**(36), 48676–48684 (2024). https://doi.org/10.1021/acsami.4c11355
108. S. Khatoon et al., Perovskite solar cell's efficiency, stability and scalability: a review. Mater. Sci. Energy Technol. **6**, 437–459 (2023). https://doi.org/10.1016/j.mset.2023.04.007
109. Y. Wang et al., Printing strategies for scaling-up perovskite solar cells. Natl. Sci. Rev. (2021). https://doi.org/10.1093/nsr/nwab075
110. S. Tao, L. Polavarapu, P. Vivo, Perovskite photovoltaics: stability and scalability. Sci. Rep. **13**(1), 4370 (2023). https://doi.org/10.1038/s41598-023-31512-z
111. T.E. Rardin, R. Xu, Printing processes used to manufacture photovoltaic solar cells. J. Technol. Stud. **37**(2), 62–68 (2011). https://doi.org/10.21061/jots.v37i2.a.1
112. B.R. Hunde, A.D. Woldeyohannes, 3D printing and solar cell fabrication methods: a review of challenges, opportunities, and future prospects. Results Opt. **11**, 100385 (2023). https://doi.org/10.1016/j.rio.2023.100385
113. P. Holzhey, M. Prettl, S. Collavini, N.L. Chang, M. Saliba, Toward commercialization with lightweight, flexible perovskite solar cells for residential photovoltaics. Joule **7**(2), 257–271 (2023). https://doi.org/10.1016/j.joule.2022.12.012
114. L. McGovern, E.C. Garnett, S. Veenstra, B. van der Zwaan, A techno-economic perspective on rigid and flexible perovskite solar modules. Sustain. Energy Fuels **7**(21), 5259–5270 (2023). https://doi.org/10.1039/D3SE00828B
115. C.E. Torrence, C.S. Libby, W. Nie, J.S. Stein, Environmental and health risks of perovskite solar modules: case for better test standards and risk mitigation solutions. iScience **26**(1), 105807 (2023). https://doi.org/10.1016/j.isci.2022.105807
116. A.S. Thomas, A review on antimony-based perovskite solar cells. Equilib. J. Chem. Eng. **6**(2), 75 (2022). https://doi.org/10.20961/equilibrium.v6i2.64322
117. S.M. Jain, T. Edvinsson, J.R. Durrant, Green fabrication of stable lead-free bismuth based perovskite solar cells using a non-toxic solvent. Commun. Chem. **2**(1), 91 (2019). https://doi.org/10.1038/s42004-019-0195-3
118. J. Chen et al., Efficient tin-based perovskite solar cells with trans-isomeric fulleropyrrolidine additives. Nat. Photonics **18**(5), 464–470 (2024). https://doi.org/10.1038/s41566-024-01381-7
119. Z. Yue, H. Guo, Y. Cheng, Toxicity of perovskite solar cells. Energies (Basel) **16**(10), 4007 (2023). https://doi.org/10.3390/en16104007
120. J. Li et al., Biological impact of lead from halide perovskites reveals the risk of introducing a safe threshold. Nat. Commun. **11**(1), 310 (2020). https://doi.org/10.1038/s41467-019-13910-y
121. M. Yang et al., Reducing lead toxicity of perovskite solar cells with a built-in supramolecular complex. Nat. Sustain. **6**(11), 1455–1464 (2023). https://doi.org/10.1038/s41893-023-01181-x

# Chapter 7
# Integrating Perovskites into Wearables

The integration of perovskite materials into wearable electronics demands a fundamental reconsideration of conventional synthesis approaches. While traditional perovskite synthesis methods such as solid-state reactions, hydrothermal processing, and high-temperature chemical vapor deposition, have proven effective for laboratory-scale device fabrication, they are relatively incompatible with the temperature-sensitive flexible substrates and scalability requirements of wearable applications. Thus, the synthesis techniques discussed in this chapter are dedicatedly applicable for the formation of wearable structures equipped with perovskite materials, and then the integration methods to seamlessly combine perovskite materials into wearable structures are presented.

## 7.1 Synthesis Methods

The development of perovskite-based wearable devices centers around the ability of material scientists to synthesize high-quality materials under conditions compatible with flexible and often temperature-sensitive substrates. Traditional perovskite fabrication methods, while effective in bulk or rigid-device applications, frequently involve high temperatures, toxic solvents, or multi-step purification processes that are impractical for wearable platforms. This section introduces synthesis techniques specifically adapted for wearable applications—focusing on approaches that offer material tunability, environmental stability, and integration compatibility with soft, flexible substrates such as polymers, cellulose and fibers.



M. Darab et al., *Perovskites for Next-Generation Wearable Technology*, Engineering Materials, https://doi.org/10.1007/978-3-032-03931-6_7

### 7.1.1 Modified Hot Injection

The modified hot injection method is a widely employed synthetic route for producing halide perovskite nanocrystals (NCs), particularly well-suited for applications requiring controlled morphology and tunable optoelectronic properties. This method is a refinement of the classical hot-injection technique, which typically involves the rapid injection of a cation precursor into a hot solution containing metal halides and organic ligands under an inert atmosphere. The instantaneous supersaturation that follows this injection initiates rapid nucleation, while the short residence time at elevated temperatures allows precise control over crystal growth. The modified version of this approach introduces carefully optimized steps such as staged degassing, tailored ligand compositions, and temperature ramping protocols. These adjustments increase reproducibility, reduce the occurrence of secondary phases, and enhance the uniformity of the resulting nanocrystals [1–3].

This synthetic strategy is particularly advantageous for wearable electronics and photonic devices, which demand not only optical precision but also process compatibility with flexible and often temperature-sensitive substrates. The nanocrystals synthesized via this method exhibit narrow size distributions and emission linewidths, making them suitable for use in wearable light-emitting diodes (LEDs), photodetectors, and optical sensors. Furthermore, the use of organic ligands such as oleylamine and oleic acid during synthesis provides colloidal stability and surface passivation, which improves the environmental stability of the nanocrystals and facilitates their dispersion in polymer matrices [4].

These ligands also aid in integrating the nanocrystals into flexible substrates, including electrospun fibers and elastomeric composites. While the injection step typically occurs at elevated temperatures (120–160 °C), the reaction is quenched rapidly, allowing for low overall thermal stress and compatibility with temperature-sensitive wearable platforms. Another important advantage is the scalability of the method. Once reaction conditions are optimized, the process can be scaled up for larger-batch production, although relatively high temperatures and inert conditions remain challenging for the mass production of perovskite nanocrystals.

A representative application of the modified hot injection method is the synthesis of cesium lead bromide ($CsPbBr_3$) nanocrystals [4], an all-inorganic halide perovskite known for its high photoluminescence quantum yield, color purity, and relative environmental stability. In this procedure, a cesium-oleate precursor is prepared by dissolving cesium carbonate in a mixture of oleic acid and a nonpolar solvent such as octadecene under inert conditions. Separately, a lead bromide solution is prepared by dissolving $PbBr_2$ in octadecene in the presence of oleylamine and oleic acid. Both mixtures are degassed and heated under nitrogen. Once the lead precursor reaches the desired temperature, the cesium-oleate solution is rapidly injected, initiating nanocrystal formation. The reaction is quickly quenched in an ice-water bath to arrest further growth. Ethanol or acetone is then used as an antisolvent to precipitate the nanocrystals, which are collected and redispersed for further processing.

The $CsPbBr_3$ nanocrystals produced by this method exhibit strong green photoluminescence, excellent size and shape uniformity, and high compatibility with flexible materials. These properties make them particularly attractive for incorporation into wearable light-emitting fabrics, energy-harvesting nanogenerators, and stretchable photodetectors. The step-by-step synthesis process for $CsPbBr_3$ nanocrystals, as described above, is schematically illustrated in Fig. 7.1, which outlines the experimental workflow including precursor preparation, hot injection, quenching, and final nanocrystal collection.

An important advantage of the modified hot injection method lies in its contribution to enhanced stability of the resulting perovskite nanocrystals. By carefully tuning the synthesis conditions, particularly the ligand environment and thermal quenching steps, the method produces nanocrystals with improved resistance to degradation from moisture, oxygen, and thermal fluctuations [3, 5]. This increased environmental robustness is particularly critical for wearable applications, where devices are frequently subjected to dynamic and often harsh operational conditions such as perspiration, body heat, and ambient humidity. The enhanced structural and chemical stability afforded by this synthesis route may potentially extend the operational lifespan of perovskite-based components in real-world wearable systems.

Controlled synthesis is achieved through adjustments in the process, such as varying solvents or applying encapsulation techniques, which enable precise control over nanocrystal size, shape, and composition, thereby enhancing their performance in flexible and wearable applications [1, 5].

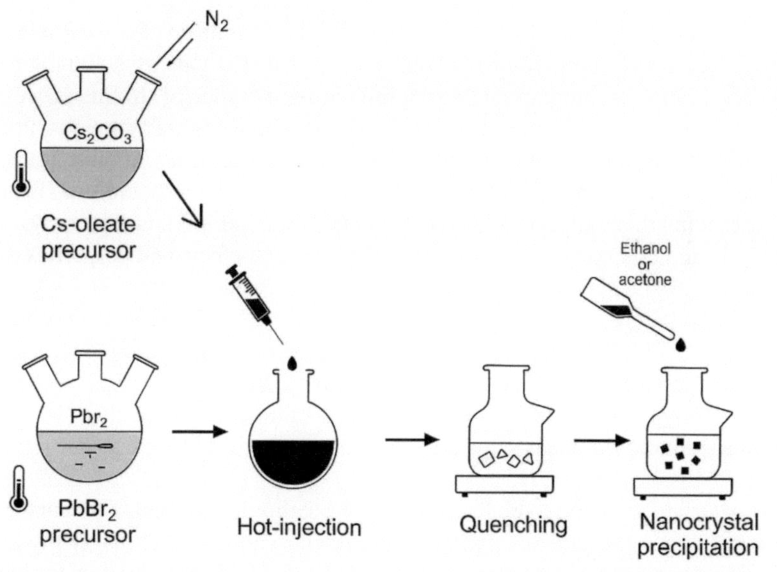


**Fig. 7.1** $CsPbBr_3$ nanocrystals step-by-step synthesis route using modified hot injection method adapted from [4]

The modified hot-injection method has also been successfully adapted for the synthesis of lead-free perovskites, such as $Cs_2SnI_6$ [2], offering a safer and more environmentally friendly alternative to traditional lead-based systems. These materials maintain proper optoelectronic properties while significantly reducing toxicity—an essential consideration for wearable electronics, where prolonged skin contact and environmental exposure are common. The ability to tailor reaction conditions for the incorporation of alternative metal cations broadens the material landscape and also aligns with the growing demand for sustainable and biocompatible components in the next-generation wearable devices.

While modified hot injection techniques deliver exceptional nanocrystal quality with superior size uniformity and crystalline structure, their industrial scalability remains a significant limitation. For mass production, the method requires sophisticated instrumentation including precision injection systems, controlled heating apparatus, and specialized nozzles or spray functions to maintain the critical temperature and timing parameters necessary for optimal nucleation and growth. These stringent requirements for proper control over reaction conditions, combined with the need for expensive relatively high temperature equipment, present substantial barriers to large-scale manufacturing and economic viability in commercial production environments.

### *7.1.2 Low Temperature Solution Process*

One widely used approach for fabricating perovskite thin films for wearable applications is the room temperature solvent techniques that can be a foundation for further integration techniques such as spin-coating method [6]. In this process, the entire synthesis is carried out at room temperature and regularly involves precipitating perovskite nanocrystals or films by adding an antisolvent (like toluene or chlorobenzene) to a perovskite precursor solution at room temperature. There are reports showing that with careful solvent controls room temperature, perovskite such as $MAPbI_3$ demonstrate comparable structural and optoelectronic properties to high temperature crystals owing to similar single crystal characteristics [7].

Low temperature processing is crucial for wearables because it may produce high-quality perovskite materials without thermal stress, preserves the integrity of temperature-sensitive substrates, and can be easily controlled to tune particle size and morphology. The mild conditions make it compatible with flexible polymer substrates from the synthesis stage and also has the potential to mass production and scale up.

For materials prepared to be fed into a spin-coating integration, a precursor solution containing a mixture of organic or inorganic cations (such as methylammonium, formamidinium, or cesium) and metal halides (typically lead or tin halides) is dissolved in polar aprotic solvents like dimethylformamide (DMF) and dimethyl sulfoxide (DMSO).

The solution is then deposited onto a prepared substrate by spinning (or spin-coating in the laboratory vocabulary), where centrifugal force spreads the liquid uniformly. This will be described in detail in the Integration section, however the details for the precursor preparations are offered in this section.

An example of this technique is the fabrication of methylammonium lead halide perovskites, such as $CH_3NH_3PbI_3$ ($MAPbI_3$) [6]. Methylammonium lead halides perovskites such as $CH_3NH_3PbI_3$ are posed as promising photoactive materials mainly because of their long charge–carrier diffusion length (greater than 1 μm [8]) strong absorption coefficient (around $10^5$ $cm^{-1}$) [9] and the possibility of band gap tunability from 1.1 to 2.5 eV [10] by alteration in chemical composition (See also Chap. 3, Band Gap Engineering). These materials have been reported to be suitable for colored emitters [11] energy harvesting [12] and photodetectors [13, 14].

Figure 7.2 shows the steps taken for the synthesis of a variant of $MAPbI_3$ perovskites where A and B sites are replaced by functionalized compounds to be spin-coated on a UV-treated substrate as a foundation for a fiber-based single layer graphene (SLG)/$CH_3NH_3PbI_3$ perovskite (PVK) photodetector.

The researchers synthesized a mixed-cation, mixed-halide perovskite with the composition $(Cs_{0.06}FA_{0.79}MA_{0.15})Pb(I_{0.85}Br_{0.15})_3$, where FA and MA were used alongside cesium (Cs) to enhance stability and optoelectronic properties. The perovskite precursor solution was prepared by dissolving $PbI_2$, FAI, MABr, and $PbBr_2$ in a 4:1 volume mixture of dimethylformamide (DMF) and dimethyl sulfoxide (DMSO)—a solvent combination that improves film uniformity by slowing crystallization. To further tune the bandgap and phase stability, 5 vol% of a CsI stock solution (1.5 M in DMSO) was added, introducing small $Cs^+$ ions into the perovskite lattice to suppress phase segregation. The solution was then diluted with additional DMF:DMSO (4:1) as a move to optimize viscosity for spin-coating. This type of synthesis should be entirely performed in an inert environment to prevent moisture-induced degradation, as water disrupts perovskite crystallization and promotes $PbI_2$ formation. In the case of the report in [6], all the steps were carried out in a nitrogen-filled glovebox.

The final perovskite film is often deposited by two-step spin-coating with chlorobenzene anti-solvent dripping to rapidly remove excess solvents, ensuring

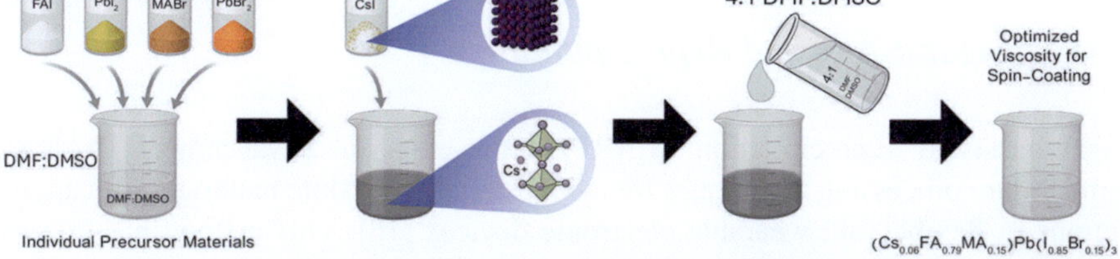


**Fig. 7.2** Low temperature solution $MAPbI_3$ synthesis steps adopted from [6]

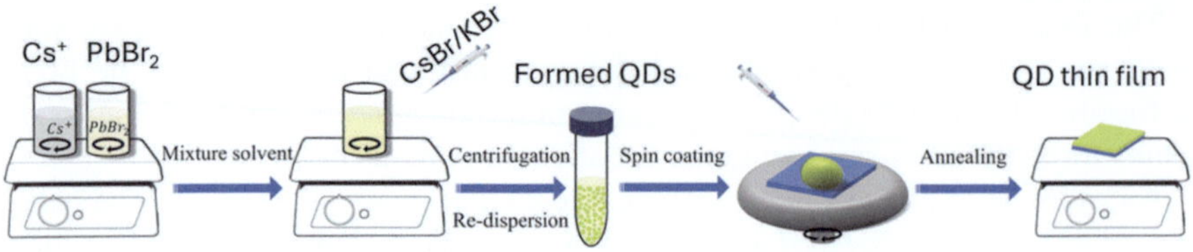


**Fig. 7.3** Synthesis at room-temperature and structural characterizations of CsBr/KBr assisted $CsPbBr_3$ QDs. Reproduced from [15] published by Wiley–VCH GmbH. Licensed under Creative Commons Attribution 4.0

a dense, pinhole-free morphology to be discussed in the Integration section in this chapter.

All-inorganic cesium lead bromide ($CsPbBr_3$) quantum dots thin films have been synthesized through a room temperature antisolvent precipitation method as self-powered wearable photodetectors [15]. In this method, a strategic mixed-cation approach was utilized for enhanced performance in wearable electronics. The synthesis protocol schematically shown in Fig. 7.3 involves the preparation of distinct $Cs^+$ and $PbBr_2$ precursors that are subsequently mixed with a binary solvent system of hexane and 2-propanol, inducing immediate yellow coloration indicative of perovskite formation within seconds.

An innovation in this approach is the controlled injection of a CsBr/KBr precursor mixture, where the molar fraction $x = [Cs^+/K^+]/([Cs^+/K^+] + [Cs^+])$ is systematically varied from 0.025 to 0.3 to optimize quantum dot properties. This mixed-cation strategy enables precise tuning of the optoelectronic characteristics while maintaining the structural integrity of the perovskite lattice. The CsBr/KBr precursor is deliberately prepared in an identical solvent mixture to the $PbBr_2$ precursor to eliminate ligand interference and ensure homogeneous nucleation conditions. Following synthesis, the colloidal $CsPbBr_3$ quantum dots undergo rigorous purification through centrifugation cycles to remove residual solvents and impurities, after which they are re-dispersed in appropriate solvents for device fabrication. The final processing involves spin coating deposition followed by controlled thermal annealing to promote film formation and enhance crystalline quality while maintaining compatibility with flexible substrates.

### *7.1.3 Ligand-Assisted Reprecipitation*

Ligand-assisted reprecipitation (LARP) represents an advancement in low-temperature processing techniques for synthesizing perovskite materials with applications in flexible and wearable electronic devices [16]. This method has proven particularly effective for the controlled synthesis of $CsPbI_3$ perovskite quantum dots, with researchers implementing various automated approaches and process optimizations to enhance synthetic efficiency and reproducibility. The LARP technique exploits differential solubility principles, where precursor materials exhibit high solubility in polar aprotic solvents while the desired perovskite products demonstrate poor solubility in non-polar solvents, enabling facile precipitation and isolation of phase-pure materials under ambient conditions.

The synthetic versatility of LARP is exemplified in the preparation of bismuth-based perovskite variants, including $Cs_3Bi_2I_9$ and $CsBi_3I_{10}$ [17], which can be selectively obtained through stoichiometric control of precursor ratios. In a typical synthesis, cesium iodide (CsI) and bismuth triiodide ($BiI_3$) are dissolved in N,N-dimethylformamide (DMF) at predetermined molar ratios—3:2 for $Cs_3Bi_2I_9$ or 1:3 for $CsBi_3I_{10}$—followed by rapid injection into toluene under continuous stirring. Figure 7.4 schematically shows the synthesis process where the stoichiometric tuning can lead to accurate band gap engineering.

The immediate color transition from vermilion to orange indicates successful nucleation and growth of the desired perovskite phase, with the precipitation process completed within 30 min at room temperature. Subsequent centrifugation, washing with toluene, and thermal drying at 80 °C yield pure crystalline products with distinct structural and optical properties that can be systematically tuned through precursor stoichiometry, demonstrating the precision and flexibility inherent in the LARP methodology.

LARP synthesis route with modified middle steps to produce high-quality $CsPbBr_3$ perovskite nanocrystals with enhanced reaction efficiency and environmental compatibility is also reported [18]. By dissolving stoichiometric CsBr and $PbBr_2$ with reduced concentrations of oleic acid (OA) and saturated alkylamines (e.g., octylamine) in DMF, and injecting this precursor into ethyl acetate (EA) instead of the traditional toluene, the process achieves rapid nucleation and growth under ambient

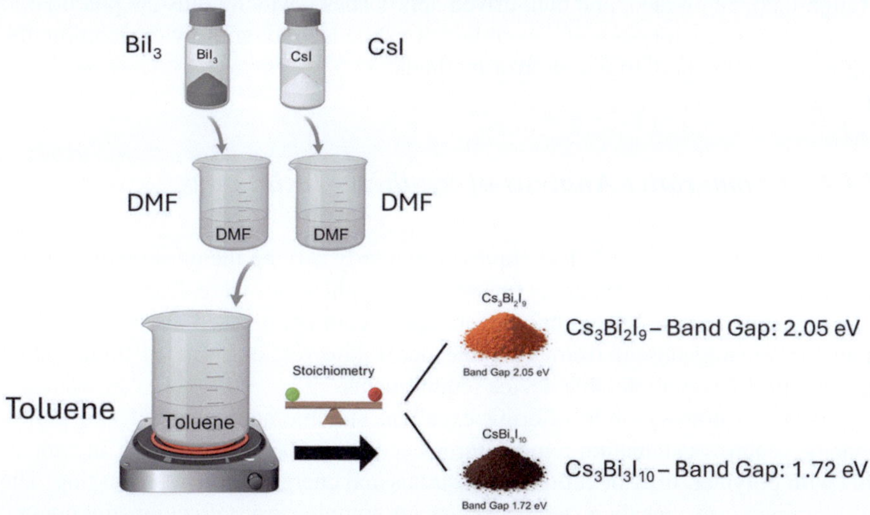


**Fig. 7.4** LARP process with stoichiometry-dependent selective products adapted from [17]

conditions. This substitution enables a dramatic improvement in product yield from ~18% (TOL-LARP) to ~70% (EA-LARP), while reducing ligand use by 60% and solvent waste by 62.5%. The nanocrystals exhibit excellent optical properties with photoluminescence quantum yields exceeding 80% after post-synthetic surface passivation using short-chain alkylammonium bromide ligands. Furthermore, the method supports scalable synthesis up to 0.5 L batch sizes without compromising nanocrystal quality, and allows for efficient halide exchange ($Cl^-$ or $I^-$) to tune emission across the visible spectrum. These advancements make the EA-LARP route a promising and scalable strategy for producing perovskite nanocrystals tailored for wearable and optoelectronic applications.

Another variant of LARP technique, enhanced through robotic automation, is presented in the study by Sanchez et al. [19]. The synthesis route presented in this work is based on a robot-assisted automated ligand-assisted reprecipitation (LARP) technique designed for high-throughput, precise, and reproducible fabrication of $CsPbBr_3$ perovskite nanocrystals. The method utilizes a liquid-handling robotic platform to systematically vary precursor concentrations, ligand types, and antisolvent volumes, enabling a highly controlled reaction environment. Precursors of CsBr and $PbBr_2$ are dissolved in a polar solvent along with ligands such as oleic acid and oleylamine, and then rapidly injected into a non-polar antisolvent (e.g., toluene) by the robotic system. This automation allows fine-tuning of nanocrystal size, morphology, and photoluminescence characteristics with minimal human intervention. The key achievement of this study is the demonstration of an optimized, reproducible synthesis protocol that yields perovskite nanocrystals with high photoluminescence quantum yields (PLQY) and stability, while also mapping out detailed parameter–property relationships across a large design space. This work exemplifies a future trend in perovskite nanocrystal synthesis for wearables—the integration of high-tech, automated, and data-driven approaches—which holds the potential to accelerate the development of wearable perovskite technologies by overcoming the limitations of manual, trial-and-error methods.

### *7.1.4 Comparative Analysis of Synthesis Techniques*

In this section, the synthesis techniques particularly serving the purpose of creating perovskite-based active materials for wearable applications were discussed. Several conventional perovskite synthesis techniques commonly used in bulk applications are notably absent from wearable perovskite research due to fundamental incompatibilities with flexible device requirements.

Sol–gel synthesis, while offering excellent stoichiometric control and homogeneity, requires calcination temperatures exceeding 500 °C, making it incompatible with polymer, fiber or cellulose substrates and energy-efficient processing. The high-temperature annealing step necessary for complete crystallization and organic component removal would damage flexible substrates and consume excessive energy in roll-to-roll manufacturing processes.

Co-precipitation methods, though capable of producing phase-pure perovskites, typically require subsequent high-temperature sintering steps to achieve adequate crystallinity and density. The aqueous processing conditions can also lead to hydrolysis of moisture-sensitive perovskite precursors, particularly for hybrid organic–inorganic compositions. Additionally, the resulting powders require further processing into films or coatings, adding complexity to the integration process.

Sonochemical synthesis, while offering rapid reaction kinetics and good mixing, faces challenges in terms of reproducibility and scale-up for continuous perovskites manufacturing processes. The acoustic cavitation effects are difficult to control uniformly across large areas, leading to inhomogeneous particle distributions. The high local temperatures and pressures generated during sonication can also cause degradation of organic components in hybrid perovskites.

Microwave-assisted synthesis, although rapid and energy-efficient, suffers from poor penetration depth and non-uniform heating in large-scale applications. The technique is also limited by the electromagnetic properties of the precursor solutions and may not be suitable for continuous processing methods required for wearable device manufacturing.

Solid-state synthesis methods, including mechanochemical approaches, are fundamentally incompatible with wearable applications due to their inability to produce thin films directly and the high-energy ball milling or grinding processes required. These methods also typically require high-temperature post-processing steps to achieve phase purity and crystallinity.

The synthesis methods discussed in this section then included modified hot injection, low temperature solution processes and LARP.

Modified hot injection synthesis offers exceptional control over nanocrystal size and morphology through precise temperature and precursor concentration management, typically operating at 150–200 °C. This technique produces highly crystalline perovskite nanocrystals with narrow size distributions, which is crucial for achieving consistent optical and electronic properties in wearable applications. The method allows for compositional flexibility, enabling the incorporation of different halide compositions and cation substitutions to tune bandgaps and stability. However, the high temperature requirements increase energy consumption and processing complexity, while the reliance on toxic organic solvents such as oleylamine and octadecene raises environmental and safety concerns. The complex purification process involving multiple washing steps with organic solvents adds to the overall processing time and cost, and the requirement for inert atmosphere conditions necessitates specialized equipment.

Low temperature solution processes represent a more sustainable approach to perovskite synthesis, operating at temperatures below 100 °C and often at room temperature. This technique offers significant advantages in terms of energy efficiency and compatibility with flexible substrates that cannot withstand high processing temperatures. The method typically employs polar solvents such as dimethylformamide or dimethyl sulfoxide, which dissolve the precursors effectively at low temperatures. The reduced thermal budget makes this approach particularly attractive for roll-to-roll processing and large-scale manufacturing. However, low temperature synthesis often results in poor crystallinity and higher defect densities compared to high-temperature methods. The slower kinetics at reduced temperatures can lead to broader size distributions and less controlled morphologies. Additionally, the resulting perovskite films may exhibit reduced stability and photoluminescence quantum yields due to incomplete crystallization and higher trap state densities.

LARP provides a unique approach to perovskite synthesis by utilizing the controlled precipitation of perovskite nanocrystals from solution through the addition of an antisolvent. This technique operates at room temperature and produces colloidal perovskite nanocrystals with proper size control and high photoluminescence quantum yields. The method is particularly advantageous for its simplicity and speed, requiring only mixing of precursor solutions with an antisolvent under ambient conditions. The presence of ligands during precipitation helps stabilize the nanocrystals and prevents agglomeration, while the rapid precipitation kinetics can lead to relatively high crystalline quality. However, LARP suffers from challenges related to reproducibility, as the precipitation process is sensitive to mixing conditions, temperature fluctuations, and precursor purity. The technique often produces polydisperse size distributions compared to hot injection methods, and the long-term stability of the resulting colloidal solutions can be problematic due to ligand desorption and particle aggregation. Simultaneously, the new achievements in the automation and robotic-enhancement of LARP has promises for overcoming the barriers in the commercialization of the technique.

The selection of an appropriate synthesis technique for wearable perovskite applications requires careful consideration of multiple competing factors. While modified hot injection offers superior crystal quality, the high processing temperatures limit its compatibility with flexible substrates. Conversely, low temperature solution processes provide excellent substrate compatibility but compromise material quality. LARP presents a middle ground with moderate crystal quality and processing simplicity, though reproducibility remains a concern.

Table 7.1 summarizes the key parameters and trade-offs among these three synthesis approaches, highlighting the complex decision matrix that researchers must navigate when developing perovskite materials for wearable applications.

The absence of conventional high-temperature synthesis methods from this comparison underscores the unique constraints imposed by wearable device requirements, where substrate compatibility and processing energy considerations often outweigh the pursuit of optimal crystal quality.

Furthermore, the integration methods to be discussed in the next section represent a decisive factor in determining which synthesis approach researchers should pursue, as the compatibility between synthesis output and subsequent processing steps is critical for ensuring smooth fabrication workflows and maintaining material integrity throughout the device assembly process.

**Table 7.1** Comparative analysis of synthesis techniques for wearable perovskite applications

| Parameter | Modified hot injection | Low temperature solution | Ligand-assisted reprecipitation (LARP) |
|---|---|---|---|
| Processing temperature | 150–200 °C | <100 °C | Room temperature |
| Crystal quality | Excellent | Poor to moderate | Good |
| Size distribution | Very narrow | Broad | Moderate |
| Processing time | Hours | Minutes to hours | Seconds to minutes |
| Substrate compatibility | Moderate | Excellent | Excellent |
| Environmental impact | High (toxic solvents) | Moderate | Low to moderate |
| Scalability | Moderate | High | High |
| Reproducibility | High | Moderate | Low |
| Equipment complexity | High | Low | Very low |
| Capability for automation and AI integration | Low | Moderate | High |
| Cost | High | Low to moderate | Low |

## 7.2 Integration Techniques

While the synthesis techniques discussed previously enable the preparation of high-quality perovskite materials, the successful implementation of these materials in wearable applications requires their effective integration into flexible structures. The synthesized perovskites, whether in the form of nanocrystals, precursor solutions, or colloidal dispersions, must be deposited onto or incorporated within flexible substrates such as polymer films, cellulose-based materials, and various types of fibers. This integration step presents unique challenges different from conventional rigid electronics, as the deposition processes must maintain the mechanical flexibility of the substrate while ensuring uniform film formation, adequate adhesion, and preservation of the perovskite's functional properties.

The mechanical properties of flexible substrates impose significant constraints on the integration process, requiring processing temperatures typically below 150 °C to prevent thermal degradation of polymer or fiber chains and maintain substrate integrity. Additionally, the surface energy mismatch between hydrophobic substrates and polar perovskite precursor solutions necessitates careful consideration of wetting behavior and adhesion mechanisms. The dynamic mechanical stresses experienced during bending and stretching of wearable devices further complicate the integration process, as the deposited perovskite films must maintain electrical continuity and optical performance under mechanical deformation.

Several integration techniques have been developed to be particularly suitable for wearable perovskite applications, each offering certain advantages in terms of scalability, film quality, and substrate compatibility. This section examines these methods and discusses the advantages and disadvantages of each one with the prospect of lab and industrial scale fabrications of wearable perovskites.

### *7.2.1 Inkjet Printing*

Digital inkjet printing technology, evolved from conventional graphic printing methods, has been evolved into a suitable manufacturing approach for wearable device fabrication [20]. The technique's inherent precision, efficient material usage, and production scalability make it exceptionally fitting for creating flexible, low-weight components essential in wearable energy harvesting and health monitoring systems. Through computer-guided deposition mechanisms, optoelectronic materials can be precisely positioned onto non-planar surfaces, fabric substrates, and various unconventional materials commonly utilized in wearable designs. This capability facilitates the creation of customized device patterns that conform to the distinctive contours and dynamic movements characteristic of wearable applications.

Contrasting with conventional rigid device manufacturing processes, inkjet printing delivers optimal material consumption, cost reduction benefits, improved production scalability for personalized manufacturing, and remarkable design adaptability for seamless integration into garments and flexible accessories. Nevertheless, the rapid solidification of printed inks can generate fragmented perovskite layers with substantially elevated defect concentrations, potentially undermining the structural integrity and longevity demanded by wearable device applications [21].

Dual-stage methodology developed by Pesch et al. [22] specifically targets the requirements of wearable photovoltaic systems. This approach begins with precisely controlled metal containing precursor vapor deposition onto bendable substrates, subsequently followed by accurate inkjet application of organic precursor materials. By fine-tuning critical printing variables such as droplet distribution patterns and substrate thermal conditions, the researchers achieved high energy conversion performance while retaining the mechanical adaptability crucial for wearable device functionality. This methodology successfully prevents formation of drying-related structural flaws that could impair flexibility and eliminates hazardous solvents that pose concerns in consumer product development. The incorporation of dimethyl sulfoxide vapor processing further improved material blending efficiency, yielding reliable performance levels reaching 18.2% efficiency while maintaining the mechanical characteristics necessary for conformable and flexible device operation.

All-inkjet printing was used as a comprehensive fabrication technique for flexible perovskite solar cells, enabling the deposition of every functional layer—including electrodes, transport layers, and the perovskite absorber—directly onto a flexible polyethylene naphthalate (PEN) substrate [23]. Reported by Gao and Meng [23], this method capitalized on the digital, non-contact nature of inkjet printing to achieve precise material placement, high film uniformity, and minimal waste, making it particularly suitable for large-area, lightweight, and mechanically compliant photovoltaic devices. Ethyl acetate vapor annealing was employed to enhance crystallization and improve the morphological quality of the $CH_3NH_3PbI_3$ films, resulting in dense, uniform grains with low surface roughness—key characteristics for reliable performance under mechanical stress in wearable applications. This fully printed approach yielded devices with an active area of 120 $cm^2$ and a record-high power conversion efficiency of 16.78%, the highest reported for PeSCs fabricated solely by inkjet printing. Moreover, the devices demonstrated strong environmental stability, with only minor efficiency losses after six months in ambient conditions when encapsulated with fluorinated thermoplastic polyurethane. This work highlights the viability of all-inkjet printing as a scalable and efficient integration method for next-generation flexible and wearable solar energy systems.

Ink engineering represents a critical challenge in inkjet printing integration techniques, requiring precise formulation of perovskite precursor solutions to achieve optimal viscosity, surface tension, and wetting properties for reliable droplet formation and deposition. Despite these formulation complexities, there have been advances in ink engineering [24] that enhanced inkjet printing capabilities for wearable perovskite devices by enabling the production of smoother, more uniform films with substantially reduced defect densities. The key breakthrough lies in developing multi-component ink systems that incorporate crystallization modulators, surface energy modifiers, and rheological additives to control the drying and solidification processes during and after droplet impact. By precisely controlling crystallization kinetics through optimized solvent selection, anti-solvent incorporation, and thermal management during the printing process, researchers [25] have achieved improvements in film quality, grain structure uniformity, and overall device stability. These engineered inks enable better control over perovskite grain growth, minimize coffee ring effects, reduce surface roughness, and enhance film adhesion to flexible substrates, ultimately resulting in wearable optoelectronic devices with improved performance consistency and operational longevity under mechanical stress and environmental exposure conditions.

Inkjet printing as an established integration technique for wearable perovskite devices, offers significant advantages including precise material deposition on non-planar and fabric substrates, optimal material utilization with reduced waste, cost-effective scalability for personalized manufacturing, and exceptional design flexibility for seamless garment integration. However, the technique faces critical challenges primarily related to rapid ink solidification that can produce fragmented perovskite layers with elevated defect concentrations, potentially compromising structural integrity and device longevity. Additionally, ink engineering complexities require precise formulation control of viscosity, surface tension, and wetting properties, along with sophisticated management of crystallization kinetics and solvent systems to minimize coffee ring effects and surface roughness while ensuring adequate film adhesion to flexible substrates under mechanical stress conditions.

### 7.2.2 Vapor-Assisted Solution Processing

The Vapor-Assisted Solution Process (VASP) has become a compelling method for integrating perovskite materials into photovoltaic systems, with potential applications in wearable devices. It is believed that VASP is an integration method nearing commercial viability [26, 27].

By combining features of vapor-phase and solution-based processing, VASP enables the formation of high-quality perovskite films with excellent grain uniformity, full surface coverage, and microscale grain structures [28]. The low-temperature process begins with the deposition of a lead halide precursor, such as $PbI_2$, onto a substrate—commonly fluorine-doped tin oxide (FTO) coated with a compact $TiO_2$ layer [28]. This is followed by annealing in a methylammonium iodide (MAI) vapor environment under nitrogen at temperatures around 150 °C [29]. The process results in a complete transformation of the inorganic layer into a perovskite phase, typically $MAPbI_3$, characterized by high crystallinity, reduced surface roughness, and improved optoelectronic properties [28].

For wearable device integration, VASP presents several potential advantages. Its moderate processing temperature is compatible with flexible substrates, though the specific compatibility with thermally sensitive substrates such as polymers and textiles requires careful consideration of substrate thermal stability limits [30]. The gradual, thermodynamically driven conversion process allows uniform film formation, which could potentially be adapted for non-planar surfaces [29]. The resulting films exhibit tightly packed grains and minimized pinholes, which enhance film quality and reduce charge recombination [28]. However, the mechanical robustness under bending and stretching conditions specific to wearable applications requires further investigation.

Unlike conventional one-step solution processes or vacuum-based techniques, VASP separates the deposition of organic and inorganic components, leading to dual-source vapor deposition (DSVD) [31]. This decoupling of the reaction pathway enables more controlled film growth and reduces defect density typically introduced by rapid crystallization [29]. The method addresses key limitations found in earlier fabrication techniques, such as uncontrolled reaction rates and degradation during solution processing steps.

The primary limitations of VASP include the relatively slow kinetics of the gas–solid reaction, which extends processing times and may pose challenges for high-throughput production [29]. Additionally, while the method shows promise for flexible substrates, comprehensive studies on mechanical durability under repeated flexing and environmental stability specific to wearable applications remain limited. Despite these challenges, the method's ability to produce high-quality films with structural uniformity positions it as a potential pathway for integrating perovskite optoelectronics into next-generation flexible and potentially wearable systems [32].

A newer generation technique among the vapor deposition integration approaches is surface energy-assisted patterning vapor deposition (SEAPVD) [32], which includes a significant advancement in direct fabrication of patterned perovskite arrays through innovative surface engineering approaches. The researchers [32] developed this sophisticated method (schematically shown in Fig. 7.5) for fabricating all-inorganic $CsPbBr_3$ perovskite film arrays with precise spatial control, beginning with strategic surface functionalization of the target substrate where they modulate surface energy across a wide range from 25.8 to 75.4 mJ $m^{-2}$ through controlled self-assembly of octadecyl-trichlorosilane. This surface energy variation creates distinct growth and non-growth regions on the substrate, with high surface energy areas promoting heterogeneous nucleation by reducing the Gibbs free energy barrier for perovskite crystal formation.

The vapor deposition process utilizes a tubular furnace system where perovskite powder serves as the evaporation source, with the material being carried by argon stream to the functionalized substrate positioned downstream. The integration process is governed by three critical parameters: surface energy, growth temperature, and growth time. Surface energy primarily determines the array properties including resolution, pixel dimensions, and pattern shapes, while growth temperature and time collectively control the film morphology characteristics such as grain size, thickness, and pinhole formation.

The key achievement of this work lies in demonstrating exceptional pattern selectivity of 96.7% while maintaining high-quality film properties. The method successfully produced perovskite arrays with remarkable resolution capability of 423 dpi, indicating strong potential for high-density optoelectronic device integration. The researchers achieved precise compositional control by regulating halogen ratios in the evaporation source, enabling fabrication of $CsPbX_3$ films with tunable bandgaps ranging from 2.98 to 1.74 eV. The resulting perovskite films exhibited excellent optical properties with photoluminescence peaks at 531 nm and absorption edges at 537 nm, consistent with high-quality $CsPbBr_3$ material. The films demonstrated dual-component photoluminescence lifetimes of 2.52 ns and 12.4 ns, attributed to surface and bulk recombination processes respectively.

Perhaps most significantly, the integrated perovskite photodetector arrays demonstrated outstanding optoelectronic performance with an on/off ratio of 13 887 and responsivity of 47.5 A $W^{-1}$, coupled with exceptional chemical, environmental, and operational stability. The successful integration of these arrays into photoplethysmography sensing systems for pulse monitoring represents a major advancement in wearable optoelectronics, particularly given their ability to function effectively under ultra-low light conditions. This achievement establishes the SEAPVD method as a viable pathway for creating long-term stable, high-performance perovskite-based wearable optoelectronic systems with applications extending beyond traditional display and sensing technologies.

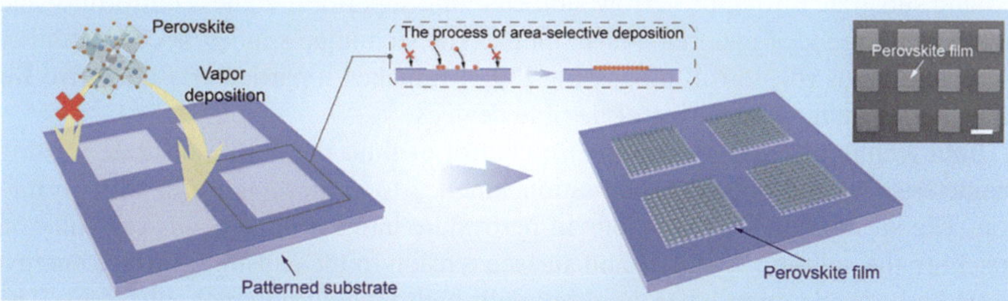


**Fig. 7.5** Vapor deposition of perovskite films array. **a** Schematic diagram of vapor deposition of perovskite films array. The inset is SEM images of the as-fabricated perovskite films array. The scale bar is 100 μm. Reproduced from [32] published by Wiley–VCH GmbH. Licensed under Creative Commons Attribution 4.0

Through systematic optimization, the researchers identified optimal conditions of 430 °C growth temperature and 30-min deposition time, which yielded compact, pinhole-free films with excellent selectivity. While these specific parameters are tailored for $CsPbBr_3$ systems, the growth temperature and processing conditions for other wearable applications or different perovskite compositions would require individual optimization depending on the thermal stability, decomposition kinetics, and nucleation characteristics of each material system. However, the fundamental principles established in this research—particularly the quantitative relationships between surface energy modulation, nucleation thermodynamics, and selectivity control—provide a systematic framework that can streamline the optimization process for diverse perovskite compositions. The parametric dependencies identified offer predictive guidelines that could be integrated with machine learning algorithms to accelerate parameter optimization for new material systems. This systematic approach transforms what has traditionally been an empirical trial-and-error process into a more methodical optimization strategy, potentially enabling AI-driven materials discovery and process development for next-generation wearable optoelectronic devices.

### 7.2.3 *Blade and Slot-Die Coating*

Blade coating and slot-die coating represent two complementary solution-based deposition techniques that have gained prominence [33] in general perovskite device fabrication [34–36] and specifically flexible devices [34, 37, 38] due to their compatibility with large-area processing and low-temperature requirements. Blade coating operates through the controlled spreading of a precursor solution across a substrate using a stationary blade, where film thickness is primarily governed by the blade-to-substrate gap and coating velocity. This technique offers excellent parameter tunability and remains widely adapted in research environments due to its operational simplicity and minimal equipment requirements. Slot-die coating, conversely, employs a precision-engineered die head that dispenses solution directly onto a moving substrate through a narrow slot, enabling superior thickness uniformity and significantly reduced material consumption. The technique's inherent compatibility with continuous roll-to-roll processing [39, 40] makes it particularly attractive for industrial-scale manufacturing of flexible devices.

Both techniques rely on the precise control of fluid rheology, substrate wetting characteristics, and solvent evaporation kinetics to achieve uniform film formation. The success of these methods in perovskite integration depends critically on matching the solution viscosity and surface tension to the substrate surface energy, particularly challenging when working with hydrophobic polymer substrates. The controlled shear environment in both techniques can influence perovskite crystallization kinetics, potentially affecting grain size and orientation, which directly impacts the optoelectronic properties of the resulting films.

The practical implementation of these techniques for wearable perovskite integration is exemplified by the work of Kokaba et al. [41], who demonstrated a fully ambient processing approach for flexible perovskite photovoltaics. Their strategy employed blade coating for the $SnO_2$ electron transport layer, where the incorporation of phenyltrimethylammonium chloride enhanced particle-to-particle interactions and improved film conformity on textured polyethylene terephthalate (PET)/ indium-doped tin oxide (ITO) substrates. The subsequent slot-die coating of the $MAPbI_3$ perovskite layer utilized lead acetate and chloride precursors to promote crystallization under ambient conditions, eliminating the need for controlled atmosphere processing. This combined approach achieved power conversion efficiencies of 17.6%, demonstrating the viability of these techniques for high-performance flexible perovskite devices while maintaining scalability for wearable applications.

The work by Xue et al. [42] demonstrates the application of blade coating for fabricating high-quality, flexible hole transport layers (HTLs) in perovskite solar cells designed for wearable applications. They developed a blade-coatable HTL ink with optimized viscosity and mechanical flexibility, enabling uniform film formation on plastic substrates under ambient processing conditions. This integration approach ensured intimate interfacial contact between the HTL and adjacent layers while preserving the flexibility and electrical conductivity essential for mechanically robust devices. The processed HTL films exhibited excellent electrical properties and mechanical resilience, which proved critical for maintaining device integrity under repeated mechanical deformation. The resulting flexible PSCs achieved a power conversion efficiency of 19.7% and retained over 85% of their initial efficiency after 5000 bending cycles, demonstrating the effectiveness of blade coating as a scalable technique for wearable perovskite integration. This work exemplifies how ink formulation, rheological optimization, and low-temperature deposition via blade coating can be synergistically combined to satisfy the dual requirements of high performance and mechanical durability in wearable electronics.

In addition to its widespread utility in wearable perovskite solar cells, blade coating has also emerged as an effective integration technique for other wearable optoelectronic applications, such as photodetectors. In Tong's study [43], the authors demonstrate a fully ambient, scalable fabrication route for cesium-doped triple cation perovskite photodetector. A schematic representation of the integration process is shown in Fig. 7.6.

This research demonstrates an innovative integration approach for manufacturing perovskite photodetectors by combining multiple roll-to-roll (R2R) printing techniques. The team successfully integrated R2R micro-gravure printing, R2R slot-die coating, and blade-coating methods to fabricate fully-printed, flexible cesium-doped triple cation perovskite photodetectors on polyethylene terephthalate (PET) substrates. This multi-technique integration approach enabled the fabrication of high-quality perovskite films through a two-step solution process under ambient conditions, representing a significant advancement in scalable manufacturing of flexible optoelectronic devices. The combination of these complementary printing and coating methods allowed for precise control over film formation while maintaining the advantages of each technique.

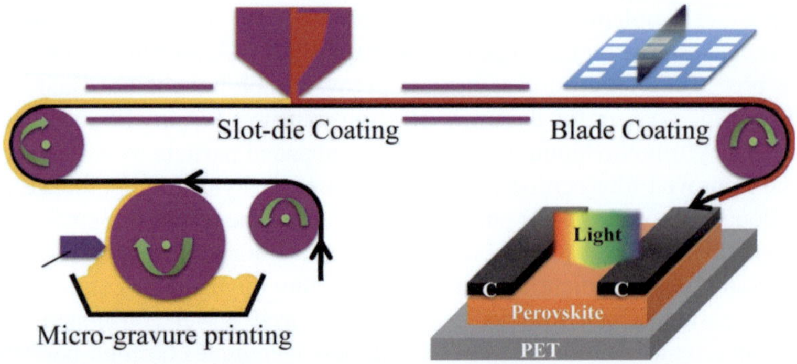


**Fig. 7.6** Schematic illustration of fully-printing process for Cs-doped triple cation perovskite photodetector. Reproduced with permission from [43] © 2019 Elsevier Ltd.

The integrated manufacturing approach yielded remarkable performance achievements that validate the effectiveness of their multi-method strategy. The resulting photodetectors exhibited exceptional optoelectronic properties including a broad photoresponse spectrum from ultraviolet to near-infrared, a high responsivity of 1.62 A $W^{-1}$, an outstanding light-to-dark current ratio of $1.83 \times 10^4$, and excellent detectivity reaching $7.77 \times 10^{12}$ Jones. Beyond performance metrics, the integration methods produced devices with exceptional mechanical durability, maintaining 80% of original performance after 2000 bending cycles, and notable ambient stability, retaining 80% performance after 700 h of exposure to unencapsulated ambient conditions.

Critically, this integrated approach eliminated the need for traditional encapsulation methods—a breakthrough that simplifies the manufacturing process and reduces production costs while maintaining device integrity. The combination of these complementary printing and coating methods allowed for precise control over film formation. The large-scale uniformity achieved through their integrated approach resulted in highly reproducible photodetector arrays promising for commercial manufacturing applications.

Blade coating and slot-die coating represent highly complementary techniques that bridge laboratory research and industrial manufacturing for integrating perovskites into wearable electronic structures. Blade coating excels in both flexible perovskite solar cells (PSCs) and photodetectors by delivering high efficiency combined with exceptional mechanical flexibility and stability, making it particularly valuable for laboratory optimization and prototype development where precise control over film thickness and morphology is critical.

Slot-die coating, on the other hand, offers superior scalability and manufacturing throughput while maintaining high responsivity, stability, and mechanical flexibility across flexible optoelectronic devices, positioning it as the preferred technique for industrial-scale production of wearable electronics. The complementary nature of these techniques becomes evident in their integration: blade coating provides the foundation for understanding perovskite film formation mechanisms and optimizing device architectures at the research level, while slot-die coating translates these insights into continuous, high-volume manufacturing processes essential for commercial wearable applications.

Together, they enable a seamless transition from laboratory proof-of-concept to industrial production, with blade coating serving as the research and development workhorse for material optimization and slot-die coating providing the scalable manufacturing pathway necessary for integrating high-performance perovskite optoelectronics into flexible, durable wearable devices that can withstand the mechanical stresses and environmental conditions of real-world applications.

### *7.2.4 Spin Coating*

Spin coating is a widely utilized technique for fabricating perovskite materials, including those intended for wearable applications [44]. In this section, we examine spin coating in relation to the previously discussed synthesis methods, where the resulting perovskite formulations—optimized for concentration and rheological behavior—are subsequently processed into thin films through spin coating.

One method introduced in the Synthesis Methods section was the preparation of perovskites with solution processing for spin-coating integration into flexible substrates that can be used in wearable devices. The example given was graphene-perovskite fibers for wearable photodetector application [6]. It was described that the prepared perovskite solution was deposited onto a substrate by spin-coating. This substrate was already treated with UV to induce hydrophilicity that helps anchoring the perovskites crystals more strongly on the surface. In the spin-coating operation, the centrifugal force spreads the liquid uniformly all over the surface.

During spinning, an antisolvent—typically a nonpolar solvent such as chlorobenzene—is dripped onto the film to rapidly induce crystallization. This is followed by thermal annealing at moderate temperatures to drive off residual solvents and promote complete crystal formation. The entire process is often carried out in an inert atmosphere to prevent degradation from oxygen or contamination ingress including moisture. This method allows for fine control over film thickness, grain size, and uniformity, making it suitable for optoelectronic devices and wearable applications.

Figure 7.7 shows the step-by-step processes for the integration of SLG on fiber before the final integration of other layer including perovskite, for a wearable structure according to [6]. In this approach, a mixed-cation, mixed-halide perovskite of composition $(Cs_{0.06}FA_{0.79}MA_{0.15})Pb(I_{0.85}Br_{0.15})_3$ were integrated onto graphene-coated fiber substrates for use in wearable photodetectors. The synthesized perovskite synthesized by solution processing was spin-coated onto UV-treated substrates and antisolvent-assisted crystallization using chlorobenzene. The resulting films were thermally annealed and showed proper environmental stability and optoelectronic performance, confirming the method's suitability for flexible and wearable optoelectronic devices.

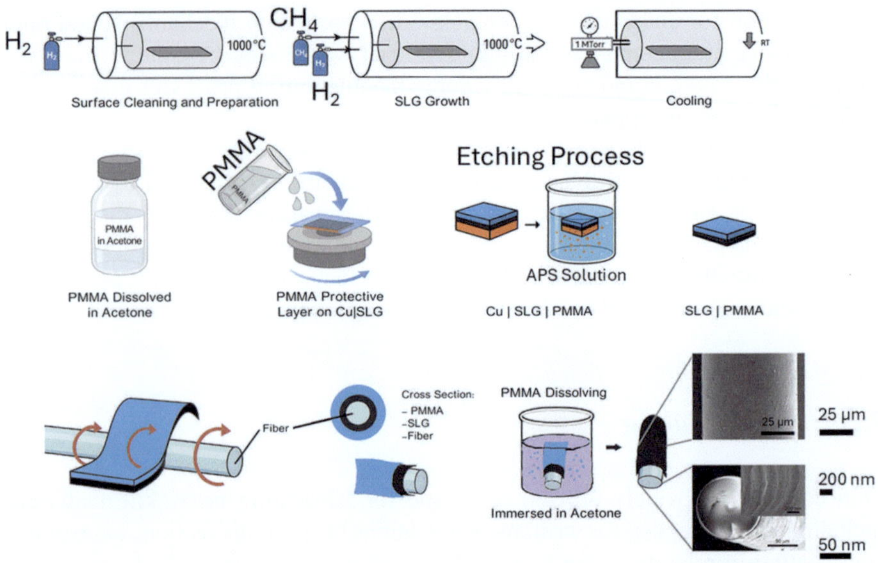


**Fig. 7.7** Step-by-step process of integration of SLG on fiber adopted from [6] published by Wiley–VCH GmbH. Licensed under Creative Commons Attribution 4.0

The two-step spin-coating process (2000 rpm for 10 s followed by 6000 rpm for 30 s) was employed to achieve optimal perovskite film morphology and crystallinity on the graphene-coated fiber substrate. The initial low-speed spin ensures complete wetting of the hydrophobic graphene surface and uniform distribution of the precursor solution, preventing dewetting and coffee-ring effects. This step is particularly crucial for conformal coating on the non-planar fiber geometry, where capillary forces and surface energy gradients can lead to non-uniform coverage.

The subsequent high-speed spin employs centrifugal force to rapidly thin the film and remove excess solvent, while simultaneously promoting controlled solvent evaporation. This dual mechanism—mechanical removal via centrifugal acceleration and enhanced evaporation due to increased surface area and air flow—creates optimal conditions for perovskite crystallization. The timed anti-solvent (chlorobenzene) quenching, typically introduced in the final seconds of the high-speed step, induces abrupt supersaturation of the precursor solution, dramatically increasing nucleation density while limiting crystal growth time to produce compact, pinhole-free films.

This sequential approach is specifically adapted for the unique challenges of fiber-based device fabrication, including substrate curvature effects that alter fluid dynamics, interfacial energy considerations between the hydrophobic graphene surface and polar perovskite precursors, and the need to maintain conformal coating around the cylindrical geometry. The method yields perovskite films with large crystalline domains, minimal grain boundaries, and optimal interfacial contact with the underlying graphene layer—all essential for efficient charge transport and collection in the photodetector architecture. The process combines established principles of solution processing and crystallization kinetics with substrate-specific optimizations that address the fundamental materials science challenges of creating high-quality perovskite films on curved, hydrophobic graphene surfaces.

Figure 7.8 shows the schematic representation of the integration step outcomes of the assembled fiber-based SLG/PVK photodetectors for wearable applications.

The process begins with a cleaned silica fiber substrate serving as the foundation, followed by the precise wrapping of CVD single-layer graphene around the fiber to form a conductive gate electrode with optimized sheet resistance. An innovation in this work is in the dual-dielectric integration strategy, where 150 nm $Al_2O_3$ is deposited via atomic layer deposition for superior dielectric properties, followed by 200 nm parylene C through physical vapor deposition to provide mechanical robustness and prevent crack-induced failures under fiber bending conditions. The channel formation involves rolling another layer of CVD SLG onto the dielectric-coated fiber, creating the active transistor structure. The final integration step employs spin-coating to deposit the perovskite layer directly onto the rolled graphene channel,

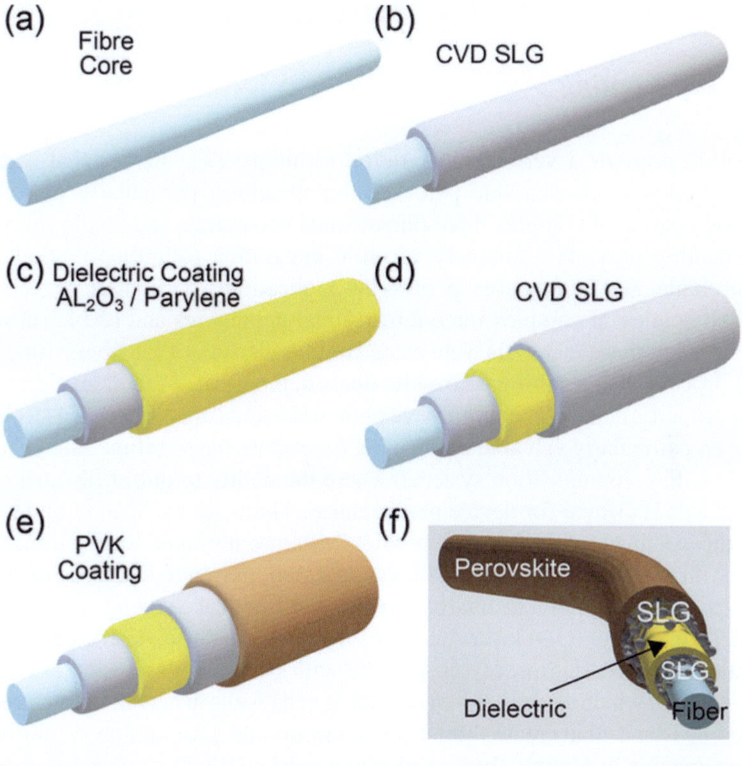


**Fig. 7.8** Fiber-based SLG/PVK PDs fabrication process. **a** Silica fiber substrate. **b** NR = 6 around the fiber as a gate electrode. **c** Gate dielectric layer composed of Al2O3 and parylene C. **d** Rolled NR = 6 channel. **e** Spin-coated PVK on rolled CVD SLG channel. **f** Assembled fiber-based SLG/PVK PD. Reproduced from [6] published by Wiley–VCH GmbH. Licensed under Creative Commons Attribution 4.0

forming a hybrid photoactive interface that combines graphene's excellent charge transport properties with perovskite's strong light absorption capabilities. This multi-technique integration approach successfully addresses the mechanical challenges of fiber-based devices while maintaining excellent electrical performance, ultimately achieving a flexible, fiber-integrated photodetector.

Spin coating remains one of the most accessible and controllable techniques for integrating perovskite materials, offering high film uniformity, tunable thickness, and excellent reproducibility under well-optimized conditions. Its compatibility with a broad range of perovskite precursor formulations makes it particularly useful for lab-scale exploration of wearable devices. However, the method also presents limitations when applied to wearable applications. Spin coating is inherently wasteful, with significant material loss, and is largely restricted to small, flat substrates, making it poorly suited for complex geometries or large-area flexible systems. Additionally, its scalability for industrial production is limited compared to printing or roll-to-roll processes. Despite these limitations, spin coating is a valuable tool for prototyping and fundamental research in wearable perovskite technologies.

### 7.2.5 *Dip-Coating*

Dip-coating presents a versatile integration technique [33, 45–48] that addresses specific challenges in wearable perovskite applications, particularly excelling in conformal coating of complex three-dimensional geometries and textile substrates.

Dip-coating presents a uniquely versatile integration technique that addresses specific challenges in wearable perovskite applications, particularly excelling in conformal coating of complex three-dimensional geometries and textile substrates. Layer-by-layer, these materials were coated onto transparent PET fibers using a dip-coating process. The resulting perovskite fibers demonstrated excellent electroluminescent properties, alongside impressive photodetection capabilities. This technique has proven particularly valuable for creating perovskite-based bifunctional fibers for wearable LiFi communication systems, where the ability to uniformly coat curved fiber surfaces is critical for device performance. The work by Shan et al. exemplifies the dip-coating for fabrication of perovskite light-emitting/detecting bifunctional fibers for these systems [49]. Based on this research, the team developed an innovative dip-coating integration technique that addresses critical challenges in fabricating uniform quantum dot (QD) layers on fiber substrates. The researchers identified that conventional dip-coating processes struggle with perovskite QDs due to their low surface tension, which causes uneven spreading and results in network structures with isolated QD islands that create short-circuit currents and poor luminance uniformity. To overcome this limitation, they developed a hybrid QD ink system incorporating poly(triarylamine) (PTAA) and 1,3,5-tri[(3-pyridyl)-phen-3-yl]benzene (TmPyPB) alongside the perovskite QDs. This hybrid approach strategically increases the ink's viscosity and surface tension, enabling better control over the film-forming process through enhanced capillary forces and the cooperative effects of gravity, viscous drag, and liquid–vapor surface tension.

The team's achievement represents a breakthrough in fiber-based optoelectronics, successfully creating ultra-smooth perovskite QD films with remarkably low surface roughness of just 1.9 nm compared to over 10 nm for pure QD films. This dramatic improvement in surface quality enables the fabrication of flexible quantum dot light-emitting fibers (QLEFs) with homogeneous photoluminescence properties that can be woven into complex patterns and textile structures. The versatility of their technique is demonstrated through successful application on various substrates including PET fibers, silver wires, and zinc wires, with the resulting fibers maintaining bright and uniform emission under UV illumination. Their approach may open new possibilities for wearable photonic devices and LiFi communication systems, as evidenced by their ability to create multicolor emission patterns and integrate multiple fibers into functional textile-based optical communication networks.

Adnan and Lee developed [50] a sequential dip-coating integration technique that addresses manufacturing limitations associated with conventional spin-coating methods for perovskite solar cells. Their approach utilizes an aqueous lead nitrate precursor system instead of toxic organic solvents, incorporating a ZnO interfacial layer on mesoporous $TiO_2$ substrates to facilitate lead precursor adsorption from water-based solutions. The ZnO layer enables significant adsorption of the non-halide lead precursor, which does not occur effectively on bare $TiO_2$ surfaces. This integration technique offers practical advantages for large-area processing and reduces environmental concerns associated with hazardous solvent use, making it potentially suitable for wearable electronics manufacturing where safety and scalability are important considerations.

The integration methodology consists of two complementary processes: Successive Sequential Deposition (SSD) and Successive Solid-State Ion Exchange and Reaction (SSIER). Figure 7.9 shows the integration process where pictures of the as-prepared electrodes after a number of SSIER repetitions followed by annealing at room temperature are also deposited.

The SSD process involves repeated cycles of lead precursor deposition followed by methylammonium iodide treatment to build up the perovskite layer and improve surface coverage. The SSIER technique addresses film stability by controlling solid-state reactions that eliminate unreacted lead precursor, preventing decomposition during processing. Through systematic optimization, the researchers determined that three SSD cycles followed by three SSIER cycles provide optimal film morphology and crystallinity, resulting in perovskite layers with reduced pin-hole formation and improved uniformity compared to single-step deposition methods.

The dip-coating technique demonstrated practical scalability through successful fabrication on large substrates (780 $cm^2$) while achieving a power conversion efficiency of 12.41% under standard test conditions. The method's compatibility with aqueous processing eliminates the need for high-temperature annealing steps required by some alternative approaches and reduces material costs compared to spin-coating techniques. The water-based processing approach also presents opportunities for integration with flexible substrates that may be sensitive to organic

**Fig. 7.9** Schematic description of the preparation of $MAPbI_3$ perovskite films using sequential dipping of ZnO covered m-$TiO_2$/c-$TiO_2$/FTO substrates in aqueous $Pb(NO_3)_2$ and MAI solutions. Reproduced from [50] published by Nature. Licensed under Creative Commons Attribution 4.0

solvents, though the authors note that film thickness uniformity remains a challenge for large-area applications. These characteristics make the technique a viable option for manufacturing perovskite devices where conventional spin-coating methods are impractical or cost-prohibitive.

Furthermore, dip-coating has shown remarkable potential in addressing one of perovskite's greatest challenges for wearable applications: moisture sensitivity. Although not necessarily wearable, scientists from China and Australia [51] have designed a water-resistant perovskite solar cell that can operate in a humid environment and maintain its efficiency over three weeks. The technique's ability to create protective surface modifications and ion-exchange layers potentially makes it indispensable for developing environmentally stable wearable perovskite devices. In textile-based applications, dip-coating enables the integration of perovskite materials directly onto fabric substrates, creating truly wearable electronic textiles that maintain both the mechanical properties of the base textile and the optoelectronic functionality of the perovskite layer, thus representing a critical enabling technology for next-generation smart textiles and wearable electronics.

### 7.2.6 *Spray-Coating*

Spray coating has gained recognition as a promising technique for integrating perovskite materials into flexible and wearable electronics, thanks to its compatibility with low-temperature processing, scalability, and ability to uniformly coat large or non-planar surfaces [52–54]. Unlike spin coating, which is constrained to small-area, flat substrates, spray coating enables conformal deposition on flexible films, textiles, and curved surfaces—key features for next-generation wearable devices. Variants such as ultrasonic, airbrush, and electrostatic spray methods provide certain advantages in terms of droplet control, film uniformity, and processing throughput.

The method also supports ambient processing and is adaptable to roll-to-roll manufacturing platforms, making it suitable for large-scale, cost-effective production. While continued optimization of film morphology and crystallization is necessary, several studies have already demonstrated the suitability of spray coating for flexible perovskite device fabrication, highlighting its relevance to wearable applications.

One such study introduces an ultrasonic spray-coating technique for depositing a low-temperature, compact $TiO_2$ electron transport layer (ETL) onto flexible polymer substrates—specifically ITO-coated PEN—forming the foundation for perovskite absorber layers. Pre-synthesized $TiO_2$ nanoparticles ($\leq$10 nm) were dispersed in ethanol at varying concentrations and spray-coated in "wet-film" and "dry-film" regimes, achieving thicknesses between 30–150 nm. To enhance film quality, a small amount of titanium diisopropoxide bis(acetylacetonate) (TAA) was incorporated. Deposition parameters—such as solution flow rate, carrier gas pressure, nozzle-to-substrate distance, and raster speed—were optimized to ensure uniform, pinhole-free films at processing temperatures below 150 °C. The compact $TiO_2$ layer was subsequently combined with a spin-coated $MAPbI_3$ perovskite and transferred to flexible ITO-PEN substrates, enabling all-low-temperature processing compatible with wearable electronics.

The resulting flexible perovskite solar cells (PSCs) achieved a power conversion efficiency (PCE) of 10.87% on ITO-PEN substrates and 14.32% on ITO-glass, marking a strong performance for spray-coated flexible devices. Importantly, the optimized "wet-film" regime produced a dense $TiO_2$ film with low roughness, ensuring good electron transport and mechanical flexibility. The devices also exhibited promising mechanical endurance, maintaining functional integrity when bent to a curvature radius of 50 mm. This demonstrates that ultrasonic spray-coating is a viable integration method for flexible perovskites, offering scalable, ambient-compatible deposition with precise control over film thickness and quality.

### *7.2.7 Electrospinning*

Electrospinning is a technique used as simultaneous synthesis and integration method in which a high-voltage field draws polymer solutions containing nanomaterials into fine fibers. This method involves using a high electric field to draw fibers from a polymer solution, resulting in continuous fibers with diameters ranging from nanometers to micrometers [55–57]. A recent example is [58] electrospinning synthesis of lead-free perovskite nanofiber-based triboelectric used for wearable fall alarm for elderly citizens. The researchers in this case employed electrospinning process to fabricate $CsBi3I_{10}$ (CBI)/PVDF nanofiber composite films, where perovskite precursors (CsI and $BiI_3$) were dissolved alongside PVDF in a mixed solvent system of DMF and acetone. During electrospinning, the rapid evaporation of organic solvents facilitated simultaneous nucleation and uniform growth of lead-free bismuth perovskite crystals within the PVDF polymer matrix, creating homogeneous nanofiber structures. This technique enabled the CBI perovskite to act as a nucleating agent that enhanced β-phase formation in PVDF, resulting in improved surface charge density and superior triboelectric performance with a maximum power density of 7.11 W $m^{-2}$ for wearable sensing applications [58].

Polyvinylidene fluoride (PVDF) and polyethylene oxide (PEO) were utilized to formulate precursor solutions with precise compositional control. The synthesis route to achieve the perovskite precursor is similar to low-temperature solution techniques presented in the previous section with the exception that the perovskite solution will be mixed with polymer solution before further integration. In the research by Wu et al. [59] preparation of two distinct precursor solutions is reported: a perovskite solution containing equimolar precursor salts (CsBr and $PbBr_2$ or $PbI_2$) dissolved in dimethyl sulfoxide (DMSO), and a polymer solution comprising a 9:1 PVDF:PEO blend. The incorporation of polyethylene oxide (PEO) serves a dual function, providing enhanced spinnability for the polymer solution while acting as a local moisture absorber during solvent evaporation, which facilitates controlled perovskite nucleation and crystallization. Figure 7.10 shows the steps taken to synthesize and integrate the perovskite decorated polymer stainless-steel yarn.

The polymerization occurs during the electrospinning process itself, where the electric field-driven liquid jet undergoes rapid solvent evaporation under controlled environmental conditions (16 ± 5% relative humidity, 40 ± 5 °C). This creates an ideal environment for in situ perovskite crystal formation within the polymer matrix through solvent engineering principles. The PVDF polymer chains undergo conformational changes during this process, with the presence of perovskite crystals

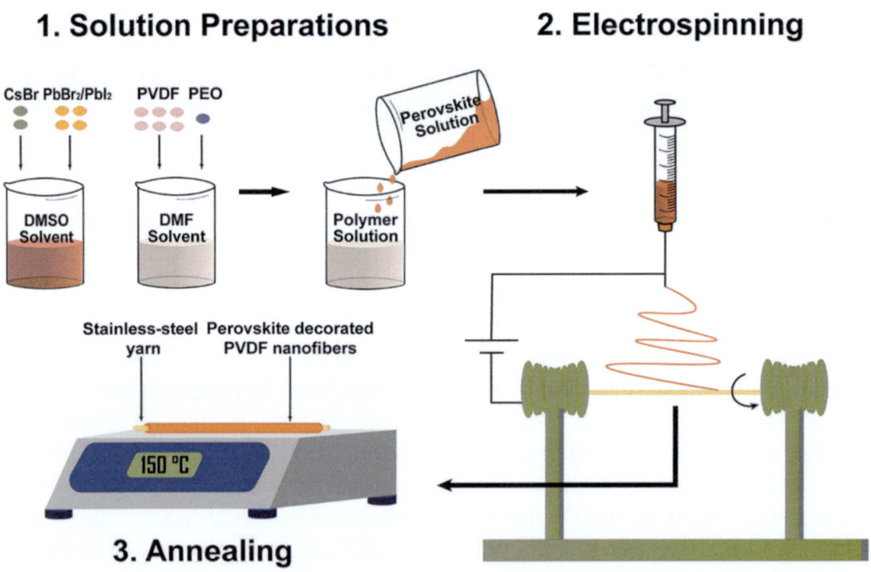


**Fig. 7.10** Schematic of preparation steps for perovskite decorated PVDF nanofibers on the stainless-steel yarn. Reproduced from [59] Copyright © 2023 The Authors. Published by American Chemical Society. Licensed under Creative Commons Attribution 4.0

potentially promoting β-phase formation, which is crucial for piezoelectric applications. The simultaneous evaporation of DMSO solvent drives both the crystallization of halide perovskites and the solidification of the polymer matrix, creating intimate interfacial contact between the two phases.

Post-synthesis treatment involves thermal annealing at 150 °C for 10 min followed by vacuum treatment at 50 °C for 15 h, which serves to remove residual solvents and enhance the crystallization of both PVDF and perovskite components. This comprehensive fabrication approach demonstrates the unique dual nature of electrospinning as both a synthesis and integration technique for wearable perovskite devices, where crystal formation and matrix incorporation occur in a single-step process, offering several key advantages including continuous fiber production, scalable manufacturing potential, and excellent mechanical flexibility suitable for textile integration. However, challenges remain in achieving uniform perovskite distribution within the polymer matrix, controlling crystal size and orientation during the rapid electrospinning process, and maintaining long-term stability of the perovskite-polymer interface under mechanical stress. The technique's success in creating functional wearable sensors with high power densities while maintaining flexibility positions electrospinning as a potentially promising pathway for next-generation flexible electronics, particularly for applications requiring conformal integration into textiles and garments.

### *7.2.8 Comparative Analysis of Integration Techniques for Wearable Perovskites*

The integration of perovskite materials into wearable electronics represents a critical technological challenge that demands careful evaluation of processing techniques capable of balancing efficiency, scalability, mechanical flexibility, and manufacturing feasibility. Contemporary research in this field has identified seven primary integration methodologies that warrant comprehensive comparative analysis: inkjet printing, vapor assisted solution processing, spray-coating, blade and slot-die coating, spin-coating, dip-coating, and electrospinning. Immature or less reproducible lab scale techniques such as brush coating on paper substrate [60] were not discussed in this chapter.

#### 7.2.8.1 Performance and Quality Considerations

Among the integration techniques examined, vapor-assisted solution processing (VASP) and its advanced variant surface energy-assisted patterning vapor deposition (SEAPVD) demonstrate superior film quality characteristics. VASP achieves exceptional grain uniformity, full surface coverage, and microscale grain structures through its thermodynamically driven conversion process, while SEAPVD achieves remarkable pattern selectivity of 96.7% with high-resolution capability of 423 dpi. These vapor-based methods produce films with tightly packed grains and minimized pinholes, resulting in reduced charge recombination and enhanced optoelectronic performance.

Spin coating, despite its widespread laboratory adoption, offers excellent control over film thickness and uniformity under optimized conditions. The technique enables fine-tuning of grain size and crystallinity through precise control of spinning parameters, antisolvent timing, and thermal treatment. However, its performance advantages are primarily realized on small, flat substrates under controlled environmental conditions, limiting its applicability to complex wearable geometries.

Solution-based coating methods present a mixed performance profile. Blade coating and slot-die coating have demonstrated impressive achievements, with blade-coated devices reaching 19.7% power conversion efficiency while maintaining performance in a longer period of times. Similarly, slot-die coated photodetectors exhibit high responsivity and exceptional detectivity. Inkjet printing has achieved notable success with all-printed devices reaching 16.78% efficiency over 120 $cm^2$ active areas, though rapid ink solidification can produce fragmented layers with elevated defect concentrations.

#### 7.2.8.2 Scalability and Manufacturing Viability

Manufacturing scalability varies significantly across integration techniques, with clear distinctions between laboratory-scale methods and industrially viable processes. Slot-die coating emerges as the most scalable technique, offering inherent compatibility with continuous roll-to-roll processing and superior thickness uniformity with reduced material consumption. Its precision-engineered die head system enables high-throughput production essential for commercial wearable device manufacturing.

Inkjet printing presents excellent scalability potential through its digital, non-contact deposition mechanism, enabling efficient material usage and production scalability for personalized manufacturing. The technique's computer-guided precision allows for customized device patterns that conform to wearable application requirements. However, ink engineering complexities require sophisticated formulation control and crystallization management systems.

Spray coating methods, particularly ultrasonic variants, offer decent scalability with ambient processing compatibility and adaptability to roll-to-roll manufacturing platforms. The technique enables large-area coating with uniform deposition on non-planar surfaces, making it suitable for cost-effective production. Conversely, spin coating faces significant scalability limitations due to its inherent material waste, restriction to small substrates, and batch processing nature.

Electrospinning presents unique scalability characteristics as a continuous fiber production method, offering potential for textile integration applications. However, achieving uniform perovskite distribution and controlling crystallization during rapid processing remain challenging for large-scale implementation.

#### 7.2.8.3 Mechanical Durability and Flexibility

Mechanical robustness under bending and stretching conditions represents a critical requirement for wearable applications. Blade coating demonstrates exceptional mechanical durability, with devices maintaining over 85% performance after 5000 bending cycles. This outstanding flexibility results from optimized ink formulation and rheological control that maintains intimate interfacial contact between layers under mechanical stress.

Dip-coating excels in conformal coating of complex three-dimensional geometries and textile substrates, producing ultra-smooth films with remarkably low surface roughness (1.9 nm). This technique's ability to create uniform coatings on curved fiber surfaces makes it particularly valuable for textile-based wearable applications, as demonstrated in bifunctional light-emitting/detecting fibers for LiFi communication systems.

Slot-die coated devices show impressive mechanical endurance, maintaining 80% performance after 2000 bending cycles while demonstrating notable ambient stability with 80% performance retention after 700 h of unencapsulated exposure. The technique's controlled deposition environment contributes to proper film adhesion and mechanical integrity.

VASP and SEAPVD methods, while producing high-quality films, require further investigation regarding mechanical durability under repeated flexing conditions specific to wearable applications. The techniques offer relatively high compatibility with flexible substrates, though comprehensive mechanical durability studies remain limited.

#### 7.2.8.4 Processing Complexity and Environmental Requirements

Processing complexity and environmental requirements significantly impact manufacturing feasibility and cost-effectiveness. Solution-based methods generally offer advantages in processing simplicity and environmental compatibility. Dip-coating with aqueous processing eliminates toxic organic solvents and reduces environmental concerns, making it suitable for safer manufacturing environments.

Blade coating and slot-die coating enable fully ambient processing approaches, eliminating the need for controlled atmosphere processing while maintaining high performance. This ambient compatibility reduces manufacturing complexity and associated costs. Similarly, spray coating supports ambient processing with adaptability to existing manufacturing infrastructure.

Vapor-based methods, while producing superior film quality, require more sophisticated processing environments with controlled atmosphere conditions and precise temperature management. VASP processing involves specialized furnace systems with inert gas environments, consequently increasing equipment complexity and operational costs.

Spin coating demands controlled environmental conditions for optimal results, requiring inert atmosphere processing to prevent degradation and contamination. The technique's sensitivity to environmental factors limits its manufacturing flexibility compared to ambient-compatible alternatives.

#### 7.2.8.5 Application-Specific Suitability

Different integration techniques demonstrate varying suitability for specific wearable applications. For high-performance photovoltaic applications requiring maximum efficiency, VASP and SEAPVD methods offer superior film quality, while blade coating provides an optimal balance of performance and mechanical durability for flexible solar cells.

Textile-based applications benefit significantly from dip-coating and electrospinning techniques, which enable direct integration onto fabric substrates while maintaining both mechanical properties and optoelectronic functionality. These methods are particularly valuable for smart textiles and truly wearable electronic textiles.

Large-area applications favor slot-die coating and spray coating for their scalability and uniform coverage capabilities, while customized devices may benefit from inkjet printing's digital precision and design flexibility.

The selection of appropriate integration techniques must consider the specific requirements of each wearable application, including performance targets, mechanical demands, manufacturing scale, and cost constraints. While no single technique excels in all categories, the combination of complementary methods or hybrid approaches may provide optimal solutions for complex wearable perovskite systems.

#### 7.2.8.6 Future Perspectives and Hybrid Approaches

The future development of wearable perovskite integration techniques likely lies in hybrid approaches that combine the strengths of multiple methods. For instance, the integration of blade coating for research optimization with slot-die coating for manufacturing scalability represents a promising pathway from laboratory development to commercial production.

Similarly, the combination of vapor-based methods for critical layers with solution-based techniques for supporting layers could optimize both performance and manufacturability. Advanced process control systems incorporating machine learning algorithms may enable real-time optimization of integration parameters, transforming empirical approaches into systematic, predictive manufacturing processes.

The development of environmentally stable integration techniques remains crucial for practical wearable applications. Methods like dip-coating with protective surface

**Table 7.2** Comparative analysis of integration techniques for wearable perovskite applications

| Integration technique | Scalability | Mechanical durability | Processing complexity | Substrate compatibility | Material efficiency | Maturity |
|---|---|---|---|---|---|---|
| Inkjet printing | High (digital, non-contact) | Moderate (fragmented layers possible) | Moderate (ink engineering required) | Excellent (non-planar, fabric) | Excellent (minimal waste) | Medium–high (established printing technology, advancing ink formulations) |
| VASP/SEAPVD | Moderate (specialized equipment) | Under investigation | High (controlled atmosphere, precise temperature) | Good (150 °C compatible) | Moderate | Medium (nearing commercial viability per document) |
| Blade coating | Moderate (research scale) | Excellent | Low (ambient processing) | Good (polymer substrates) | Good | Medium (widely adapted in research, proven performance) |
| Slot-die coating | Excellent (R2R compatible) | Excellent | Low (ambient processing) | Excellent (continuous processing) | Excellent | High (R2R compatible, industrial scalability demonstrated) |
| Spin coating | Poor (batch processing, small area) | Good (under controlled conditions) | Moderate (controlled environment) | Limited (flat substrates only) | Poor (high material waste) | High (mature laboratory technique, limited industrial relevance) |
| Dip-coating | Moderate (sequential processing) | Excellent (conformal coating) | Low (aqueous processing available) | Excellent (3D geometries, textiles) | Good | Medium (proven scalability, environmental benefits) |

(continued)

**Table 7.2** (continued)

| Integration technique | Scalability | Mechanical durability | Processing complexity | Substrate compatibility | Material efficiency | Maturity |
|---|---|---|---|---|---|---|
| Spray coating | High (R2R adaptable) | Good (conformal on curved surfaces) | Low (ambient compatible) | Excellent (large area, non-planar) | Good | Medium–high (established industrial technique, adapting to perovskites) |
| Electrospinning | Moderate (continuous fiber) | Excellent (fiber flexibility) | Moderate (controlled environment) | Excellent (textile integration) | Poor | Medium (emerging for perovskites, established for polymers) |

modifications and hybrid polymer-perovskite systems show promise for addressing moisture sensitivity challenges while maintaining integration flexibility.

As the field advances, the focus on sustainable and scalable integration techniques will likely drive further innovation in ambient processing methods, reduced-toxicity precursor systems, and energy-efficient manufacturing processes. The ultimate success of wearable perovskite technologies will depend on achieving optimal integration approaches that satisfy the complex requirements of performance, durability, scalability, and sustainability. Table 7.2 briefly compares the integration techniques for wearable perovskites discussed in the chapter.

## References


1. T.C.A. da Silva, C. Fernández-Saiz, R.S. Sánchez, A.F. Gualdrón-Reyes, I. Mora-Seró, B. Julián-López, A soft-chemistry route to prepare halide perovskite nanocrystals with tunable emission and high optical performance. J. Sol Gel Sci. Technol. **114**(1), 131–138 (2025). https://doi.org/10.1007/s10971-023-06171-1
2. T. Koyanagi et al., Hot-injection and ultrasonic irradiation syntheses of Cs2SnI6 quantum dot using Sn long-chain amino-complex. J. Nanopart. Res. **22**(3) (2020). https://doi.org/10.1007/s11051-020-04787-w
3. J. H. Yoo et al., Simultaneous encapsulation of mixed-halide perovskite on synthesis step using a thermally expandable lamellar capsule to fabricate the multi-conversion layer. Adv. Opt. Mater. **11**(8) (2023). https://doi.org/10.1002/adom.202202393
4. J. Chen, Q. Jing, F. Xu, Z. Lu, Y. Lu, High-sensitivity optical-fiber-compatible photodetector with an integrated CsPbBr_3–graphene hybrid structure. Optica **4**(8), 835 (2017). https://doi.org/10.1364/OPTICA.4.000835
5. H. Li, W. Lu, G. Zhao, W. Dong, G. Han, B. Song, Evolution and mechanism of cesium lead bromide nanostructures in oleylamine-rich system by hot-injection method. Adv. Mater. Interfaces **10**(1) (2023). https://doi.org/10.1002/admi.202201916
6. S. Akhavan et al., Graphene-perovskite fibre photodetectors. Adv. Mater. **36**(35) (2024). https://doi.org/10.1002/adma.202400703
7. R. Gupta et al., Room temperature synthesis of perovskite (MAPbI3) single crystal by anti-solvent assisted inverse temperature crystallization method. J. Cryst. Growth **537**, 125598 (2020). https://doi.org/10.1016/j.jcrysgro.2020.125598

8. S.D. Stranks et al., Electron-hole diffusion lengths exceeding 1 micrometer in an organometal trihalide perovskite absorber. Science (1979) **342**(6156), 341–344 (2013). https://doi.org/10.1126/science.1243982
9. S. De Wolf et al., Organometallic halide perovskites: sharp optical absorption edge and its relation to photovoltaic performance. J. Phys. Chem. Lett. **5**(6), 1035–1039 (2014). https://doi.org/10.1021/jz500279b
10. N.J. Jeon et al., Compositional engineering of perovskite materials for high-performance solar cells. Nature **517**(7535), 476–480 (2015). https://doi.org/10.1038/nature14133
11. J. Xing et al., Color-stable highly luminescent sky-blue perovskite light-emitting diodes. Nat. Commun. **9**(1), 3541 (2018). https://doi.org/10.1038/s41467-018-05909-8
12. D.P. McMeekin et al., A mixed-cation lead mixed-halide perovskite absorber for tandem solar cells. Science (1979) **351**(6269), 151–155 (2016). https://doi.org/10.1126/science.aad5845
13. F. Li et al., Ambipolar solution-processed hybrid perovskite phototransistors. Nat. Commun. **6**(1), 8238 (2015). https://doi.org/10.1038/ncomms9238
14. L. Dou et al., Solution-processed hybrid perovskite photodetectors with high detectivity. Nat. Commun. **5**(1), 5404 (2014). https://doi.org/10.1038/ncomms6404
15. K. Shen et al., Flexible and self-powered photodetector arrays based on all-inorganic $CsPbBr_3$ quantum dots. Adv. Mater. **32**(22) (2020). https://doi.org/10.1002/adma.202000004
16. S. Zang, J. Chen, Y. Yu, X. Qin, H. Liu, Ligand-assisted reprecipitation of highly luminescent perovskite nanocrystals. ACS Appl. Nano Mater. **8**(7), 3680–3687 (2025). https://doi.org/10.1021/acsanm.5c00412
17. S. Vijaya, J. Subbiah, D.J. Jones, S. Anandan, LARP-assisted synthesis of $CsBi_3I_{10}$ perovskite for efficient lead-free solar cells. RSC Adv. **13**(15), 9978–9982 (2023). https://doi.org/10.1039/D3RA00365E
18. C.K. Ng, W. Yin, H. Li, J.J. Jasieniak, Scalable synthesis of colloidal $CsPbBr_3$ perovskite nanocrystals with high reaction yields through solvent and ligand engineering. Nanoscale **12**(8), 4859–4867 (2020). https://doi.org/10.1039/C9NR10726F
19. S.L. Sanchez, Y. Tang, B. Hu, J. Yang, M. Ahmadi, Understanding the ligand-assisted reprecipitation of CsPbBr3 nanocrystals via high-throughput robotic synthesis approach. Matter **6**(9), 2900–2918 (2023). https://doi.org/10.1016/j.matt.2023.05.023
20. M. Gao, L. Li, Y. Song, Inkjet printing wearable electronic devices. J. Mater. Chem. C Mater. **5**(12), 2971–2993 (2017). https://doi.org/10.1039/c7tc00038c
21. M. Rani et al., Breaking barriers: addressing challenges in perovskite solar cell development. J. Alloys Compd. **1010**, 177648 (2025). https://doi.org/10.1016/j.jallcom.2024.177648
22. R. Pesch et al., Hybrid two-step inkjet-printed perovskite solar cells. Solar RRL **8**(13) (2024). https://doi.org/10.1002/solr.202400165
23. B. Gao, J. Meng, Flexible CH3NH3PbI3 perovskite solar cells with high stability based on all inkjet printing. Sol. Energy **230**, 598–604 (2021). https://doi.org/10.1016/j.solener.2021.10.072
24. Z. Li et al., Ink engineering of inkjet printing perovskite. ACS Appl. Mater. Interfaces **12**(35), 39082–39091 (2020). https://doi.org/10.1021/acsami.0c09485
25. S. Li, A. Kosek, M.N. Jahangir, R. Malhotra, C. Chang, Inkjet printing of perovskites for breaking performance–temperature tradeoffs in fabric-based thermistors. Adv. Funct. Mater. **31**(1) (2021). https://doi.org/10.1002/adfm.202006273
26. J. Liu et al., Crystalline quality control in sequential vapor deposited perovskite film toward high efficiency and large scale solar cells. Sol. Energy Mater. Sol. Cells **233** (2021). https://doi.org/10.1016/j.solmat.2021.111382
27. T. Abzieher et al., Vapor phase deposition of perovskite photovoltaics: short track to commercialization? Energy Environ. Sci. **17**(5), 1645–1663 (2024). https://doi.org/10.1039/d3ee03273f
28. Q. Chen et al., Planar heterojunction perovskite solar cells via vapor-assisted solution process. J. Am. Chem. Soc. **136**(2), 622–625 (2014). https://doi.org/10.1021/ja411509g
29. M.R. Leyden, M.V. Lee, S.R. Raga, Y. Qi, Large formamidinium lead trihalide perovskite solar cells using chemical vapor deposition with high reproducibility and tunable chlorine

concentrations. J. Mater. Chem. A Mater. **3**(31), 16097–16103 (2015). https://doi.org/10.1039/C5TA03577E
30. G.E. Eperon, V.M. Burlakov, P. Docampo, A. Goriely, H.J. Snaith, Morphological control for high performance, solution-processed planar heterojunction perovskite solar cells. Adv. Funct. Mater. **24**(1), 151–157 (2014). https://doi.org/10.1002/adfm.201302090
31. M. Liu, M.B. Johnston, H.J. Snaith, Efficient planar heterojunction perovskite solar cells by vapour deposition. Nature **501**(7467), 395–398 (2013). https://doi.org/10.1038/nature12509
32. Z. Xu et al., Surface energy-assisted patterning of vapor deposited all-inorganic perovskite arrays for wearable optoelectronics. Adv. Sci. **11**(25) (2024). https://doi.org/10.1002/advs.202402635
33. D. Burkitt, J. Searle, D.A. Worsley, T. Watson, Sequential slot-die deposition of perovskite solar cells using dimethylsulfoxide lead iodide ink. Materials **11**(11), 2106 (2018). https://doi.org/10.3390/ma11112106
34. J. Liu et al., Evolutionary manufacturing approaches for advancing flexible perovskite solar cells. Joule **8**(4), 944–969 (2024). https://doi.org/10.1016/j.joule.2024.02.025
35. Y. Ahmed et al., All-scalable CH3NH3PbI3 perovskite solar cells fabricated in ambient air. Solar RRL **7**(15) (2023). https://doi.org/10.1002/solr.202300288
36. J.B. Whitaker et al., Scalable deposition of polycrystalline perovskite thin films towards high-efficiency and large-area perovskite photovoltaics, in *2018 IEEE 7th World Conference on Photovoltaic Energy Conversion Wcpec 2018 A Joint Conference of 45th IEEE Pvsc 28th Pvsec and 34th Eu Pvsec* (2018), pp. 2808–2811. https://doi.org/10.1109/PVSC.2018.8547355
37. P. Wu, F. Zhang, Advancing flexible perovskite solar cells from lab to commercialization: materials and deposition strategies. Adv. Funct. Mater. (2025). https://doi.org/10.1002/adfm.202507644
38. Y. Yerlanuly et al., Elucidating the hysteresis effect in printed flexible perovskite solar cells with SnO2 quantum dot- and PCBM-based electron transport layers. Heliyon **10**(21) (2024). https://doi.org/10.1016/j.heliyon.2024.e39667
39. N. Khambunkoed et al., Slot-die-coated zinc tin oxide film for carbon-based methylammonium-free perovskite solar cells. Surf. Rev. Lett. **28**(11) (2021). https://doi.org/10.1142/S0218625X21501092
40. J. Ciro, M.A. Mejía-Escobar, F. Jaramillo, Slot-die processing of flexible perovskite solar cells in ambient conditions. Sol. Energy **150**, 570–576 (2017). https://doi.org/10.1016/j.solener.2017.04.071
41. M.R. Kokaba et al., Enhanced particle-to-particle interaction of tin oxide electron transporter layer for scalable flexible perovskite solar cells. Solar RRL **8**(7) (2024). https://doi.org/10.1002/solr.202301013
42. T. Xue et al., Mechanically robust and flexible perovskite solar cells via a printable and gelatinous interface. ACS Appl. Mater. Interfaces **13**(17), 19959–19969 (2021). https://doi.org/10.1021/acsami.1c00813
43. S. Tong et al., Fully-printed, flexible cesium-doped triple cation perovskite photodetector. Appl. Mater. Today **15**, 389–397 (2019). https://doi.org/10.1016/j.apmt.2019.03.001
44. N. Zarabinia et al., Method for fabricating flexible solar cell perovskite semiconductors via a sheet of paper applicator soaked in anti-solvent, in *Fleps 2022 IEEE International Conference on Flexible and Printable Sensors and Systems Proceedings* (2022). https://doi.org/10.1109/FLEPS53764.2022.9781479
45. Y. Hui, Y. Ren, J. Song, P. Hu, H. Fan, F. Teng, Preparation of flexible ultraviolet photodetectors based on ZnO film using a dip-coating method on different flexible substrates. J. Phys. D Appl. Phys. **55**(35) (2022). https://doi.org/10.1088/1361-6463/ac71e6
46. Y. Li et al., Flexible perovskite light-emitting diodes: recent progress, applications and challenges. npj Flex. Electron. **9**(1), 32 (2025). https://doi.org/10.1038/s41528-025-00408-5
47. Y. Zhao et al., Low-temperature fabrication of stable black-phase CsPbI3 perovskite flexible photodetectors toward wearable health monitoring. Nano Micro Lett. **17**(1) (2025). https://doi.org/10.1007/s40820-024-01565-4

48. M.R. Ahmadian-Yazdi, F. Zabihi, M. Habibi, M. Eslamian, Effects of process parameters on the characteristics of mixed-halide perovskite solar cells fabricated by one-step and two-step sequential coating. Nanoscale Res. Lett. **11**(1) (2016). https://doi.org/10.1186/s11671-016-1601-8
49. Q. Shan et al., Perovskite light-emitting/detecting bifunctional fibres for wearable LiFi communication. Light Sci. Appl. **9**(1), 163 (2020). https://doi.org/10.1038/s41377-020-00402-8
50. M. Adnan, J.K. Lee, All sequential dip-coating processed perovskite layers from an aqueous lead precursor for high efficiency perovskite solar cells. Sci. Rep. **8**(1), 2168 (2018). https://doi.org/10.1038/s41598-018-20296-2
51. S. Yang, Y. Wang, P. Liu, Y.-B. Cheng, H.J. Zhao, H.G. Yang, Functionalization of perovskite thin films with moisture-tolerant molecules. Nat. Energy **1**(2), 15016 (2016). https://doi.org/10.1038/nenergy.2015.16
52. T. Thornber et al., Nonplanar spray-coated perovskite solar cells. ACS Appl. Mater. Interfaces **14**(33), 37587–37594 (2022). https://doi.org/10.1021/acsami.2c05085
53. D.K. Mohamad, J. Griffin, C. Bracher, A.T. Barrows, D.G. Lidzey, Spray-cast multilayer organometal perovskite solar cells fabricated in air. Adv. Energy Mater. **6**(22) (2016). https://doi.org/10.1002/aenm.201600994
54. J.E. Bishop, T.J. Routledge, D.G. Lidzey, Advances in spray-cast perovskite solar cells. J. Phys. Chem. Lett. **9**(8), 1977–1984 (2018). https://doi.org/10.1021/acs.jpclett.8b00311
55. A. Greiner, J.H. Wendorff, Electrospinning: a fascinating method for the preparation of ultrathin fibers. Angew. Chem. Int. Ed. **46**(30), 5670–5703 (2007). https://doi.org/10.1002/anie.200604646
56. S. Ghosh, P. Roy, D. Lahiri, *Recent Trends in Electrospinning for the Preparation of Ultrathin Plastic and Polymer Fibers for Bio-Medical Applications*, vol. 1–4 (2022). https://doi.org/10.1016/B978-0-12-820352-1.00130-9
57. D. Ji et al., Electrospinning of nanofibres, Nat. Rev. Methods Primers **4**(1) (2024). https://doi.org/10.1038/s43586-023-00278-z
58. H. Su et al., One-step electrospinning synthesis of lead-free perovskite nanofiber-based triboelectric nanogenerator for wearable intelligent fall alarm. J. Colloid Interface Sci. **696**, 137836 (2025). https://doi.org/10.1016/j.jcis.2025.137836
59. S. Wu et al., Cesium lead halide perovskite decorated polyvinylidene fluoride nanofibers for wearable piezoelectric nanogenerator yarns. ACS Nano **17**(2), 1022–1035 (2023). https://doi.org/10.1021/acsnano.2c07320
60. W. Wang et al., Structural, optical and flexible properties of CH3NH3PbI3 perovskite films deposited on paper substrates. Opt. Mater. (Amst) **114**, 110926 (2021). https://doi.org/10.1016/j.optmat.2021.110926

# Chapter 8
# Applications of Perovskite Wearables in Medical Science

In order to meet the needs of an aging population, healthcare must adapt, especially in the area of ongoing, minimally invasive monitoring. The field of medical wearables continues to develop due to materials based on perovskites. Devices such as autonomous pacemakers, conformal photoplethysmography sensors, and gesture-responsive human–machine interfaces are made possible by their remarkable characteristics, which include piezoelectricity, long carrier diffusion lengths, and adjustable bandgaps. Perovskites have opened up new applications for radiological sensing, including power-free X-ray and γ-ray detectors and wearable monitors for real-time PET/CT imaging. These materials have improved mechanical durability, physiological stability, and biocompatibility. Because perovskite-enabled technologies integrate energy harvesting, analytics, and wireless communication, they represent a significant advancement in intelligent, self-sustaining medical wearables.

## 8.1 Introduction

As the population ages, especially for those who cannot take care of themselves, effective and ongoing health monitoring is growing in relevance. Often found in older wearable devices, batteries can be costly and create safety concerns. To improve real-time internal monitoring, theranostics is turning to minimally invasive technologies including ingestible gastrointestinal sensors and osseosurface electronics. Directly bonding osseosurface devices to bone, they use its stable impedance for consistent, long-term sensing, so enabling functions including thermography and strain measurement without batteries. Though localization and clinical integration remain difficult, ingestible capsules offer physiological data from within the GI tract with minimum disturbance. These technologies taken together show a move toward more exact, discreet diagnostic tools within individualized medicine [1, 2]. Modern wearable systems, which incorporate self-powered customizable designs, now feature



M. Darab et al., *Perovskites for Next-Generation Wearable Technology*, Engineering Materials, https://doi.org/10.1007/978-3-032-03931-6_8

far-field Radio Frequency Energy (RFEH) as part of their energy collecting technologies. When linked devices are coupled with wireless communication technologies including Bluetooth, radio-frequency identification (RFID), and Near Field Communications (NFC), they can effectively communicate data and run without batteries. Wearable biosensors have been quite beneficial for tailored medication as well as point-of-care diagnosis. Among the main signals they can track real-time are walking speed, breathing, and activity level. They find use in creative technologies including wound infection sensors free of batteries as well as in traditional medical uses including heart monitoring and detection of gastrointestinal diseases [3–5]. Among the several materials investigated for these uses, perovskite-based compounds have shown especially great promise based on their remarkable optoelectronic, piezoelectric, and mechanical characteristics. Originally praised for their photovoltaic efficiency, hybrid organic–inorganic perovskites and lead-free inorganic analogues are now redefining standards in biomedical wearables, allowing uses spanning biosensing, energy harvesting, radiation monitoring, and intelligent human–machine interaction. The combination of mechanical flexibility and solution-processability endows perovskite materials with tunable bandgaps, extended carrier diffusion lengths and high carrier mobilities, rendering them ideally suited for integration into conformal electronic and optoelectronic devices. These features have been very helpful in building skin-compatible sensors running on photoplethysmography (PPG) and other optical systems. Recent integration of fluorinated organic spacer layers and natural beeswax coatings, combined with encapsulation in biocompatible polymer matrices and dynamic ionic barriers, has markedly enhanced the operational stability and mechanical robustness of perovskite-based sensors in physiological environments [6]. These synergistic modifications confer self-cleaning, cytocompatible interfaces while mitigating moisture ingress and ionic migration, thus preserving carrier transport and high signal-to-noise ratios under continuous flexure. Consequently, flexible perovskite photoplethysmographic and respiratory monitoring devices sustain clinically relevant data fidelity, enabling early cardiovascular disease detection and underpinning preventive healthcare strategies [7, 8]. These novel materials enhance sensor sensitivity and device adaptability in addition to enable continuous bio-signal monitoring, effective physiological data collecting, on-demand drug delivery, and also improve the sensor sensitivity and device flexibility. These devices enable non-invasive early diagnosis by means of environmental and biological indicator detection based on physical and chemical sensor combinations, so identifying early disease. Advances in these disciplines are allowing more reasonably priced, sustainable, intelligent health monitoring devices made possible by new materials and wireless energy collecting. These innovations are supposed to meet the demands of modern healthcare [9–11].

The most amazing application of these materials is in radiological sensing, where their naturally high atomic-weight components make them quite good in absorbing and transforming ionizing radiation. High resistivity and mechanical robustness of flexible detector films combined into durable polymer matrices allow self-powered X-ray detection for uses including dental and chest imaging. Matching or surpassing the performance of conventional photoconductive detectors, advanced device designs using custom, lead-free compositions achieve ultrahigh charge yields and record-low detection limits. These developments draw attention to the great adaptability of this materials class for flexible radiographic sensors of next generation, Fig. 8.1 [12, 13]. Gamma-ray sensors developed from this materials family are quickly becoming rather important in nuclear medicine. Their designed layered architectures track injected radiotracers during PET/CT scans with sub-microcurie sensitivity using both the photoelectric effect and Compton scattering. Apart from improving spatial resolution, these devices can be designed into wearable forms with live feedback on radiotracer distribution and clearance, so enabling real-time therapy kinetic monitoring. Furthermore suggesting a route towards more compact, flexible radiography systems, scintillators derived from the same platform have shown the potential to rival—and in some cases surpass—traditional cesium iodide (CsI) detectors in diagnostic imaging [14].

Beyond passive sensing, perovskite materials convert biomechanical stimuli into highly sensitive and repeatable electrical outputs, so enabling dynamic human–machine interfaces. These interfaces guarantee charge stability and mechanical durability during high-cycle operation by means of hyperbranched polymers (HBPs),

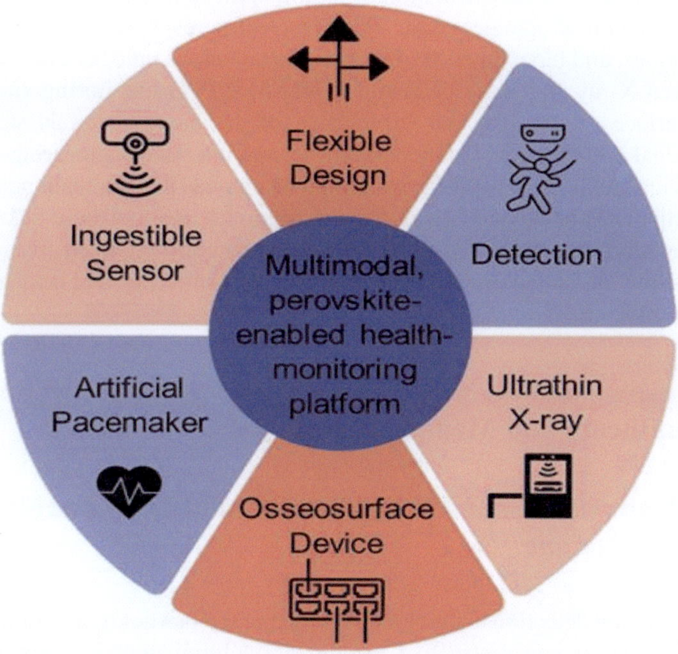


**Fig. 8.1** Schematic of a multimodal, perovskite-enabled health-monitoring platform integrating: **a** an ingestible sensor within the gastrointestinal tract; **b** osseosurface devices adhered to wrist and knee bone surfaces for mechanical and biochemical sensing; **c** an ultrathin, flexible perovskite X-ray detector over the thorax for in situ radiographic monitoring; and **d** a wireless energy-harvesting unit to power and network all modules for continuous diagnostic and therapeutic feedback

dopamine-crosslinked matrices, and stretchable substrates such as PDMS. By integrating these devices with machine learning algorithms, advanced gesture recognition, ambient control, and interactive feedback systems are facilitated, so attaining recognition accuracy as high as 95% [15]. High-performance piezoelectric nanogenerators that harvest biomechanical movement and directly convert it into electrical energy have been developed in response to the unrelenting drive for really autonomous wearables. These perovskite-based piezoelectric materials routinely produce currents above 0.2 mA—enough to sustain implantable medical electronics such as pacemakers and neuromodulators when treated as thin films on flexible polymer supports. Crucially, the acceptance of lead-free compositions solves long-standing biocompatibility problems, so opening the path for continuous, safe operation near biological tissues. Parallel with this, perovskite–polymer composites have been developed into ultrasensitive pressure sensors: they retain performance over long cycling, show rapid response and recovery times, and achieve sensitivities on the order of 10 $kPa^{-1}$. Such multifunctional materials platforms lay the foundation for next-generation diagnostic and therapeutic wearable systems by uniting self-powering harvester ability with real-time physiological monitoring [16, 17].

As mentioned, perovskite wearables mix several sensing capabilities, effective energy harvesting, and easy detection of radiographic and nuclear medical images. Their combination of optoelectronic and piezoelectric responses, robust encapsulation techniques, and biocompatible, lead-free formulations unlocks everything from self-powered X-ray and λ-ray detectors to real-time PPG monitoring and gesture-driven interfaces. Following our exploration of the basic materials strategies—compositional optimization, defect passivation, and interface engineering—we then look at device architectures and encapsulation techniques that maintain performance under mechanical and physiological stress. To demonstrate how these flexible materials are transforming wearable healthcare, we end by outlining fresh areas of use for them including implanted drugs, chronic-condition monitoring, and next-generation diagnostic imaging.

## 8.2 Healthcare and Medical Monitoring

### *8.2.1 Perovskite-Based X-Ray Detectors and Their Imaging Applications*

Ionizing radiation detection technology is crucial in various fields, including medical imaging, industrial non-destructive testing, agricultural nuclide tracking, public safety, and environmental protection. The medical field's growing reliance on radiation for diagnostic purposes has sparked significant concerns regarding the associated cancer risks for patients. The fundamental principle of X-ray detection is based on the interaction of X-ray photons with matter through various mechanisms, including the photoelectric effect, Compton scattering, Rayleigh scattering, and pair production. In the photoelectric effect, X-ray photons are absorbed by atoms, producing photocarriers. Compton scattering involves photons colliding with free electrons, transferring energy, and generating high-energy electrons that can create low-energy free electrons in a semiconductor [18]. X-ray detection technologies operate on two main principles: indirect conversion, which uses scintillating materials to convert X-rays into detectable photons, and direct conversion, where semiconductor materials generate charge carriers directly from X-rays, Fig. 8.2. Direct radiation detectors convert X-ray photons into excess free charges, which are collected by electrodes with an external bias. These detectors often incorporate thin-film transistors (TFTs) and CMOS readout circuits for efficient charge collection and storage. Key performance parameters for X-ray detectors include dark current, sensitivity, detection limit, response time, spatial resolution, and gain factor, with spatial resolution being particularly important for identifying small lesions or fractures [19]. Each method has its advantages and challenges; indirect detection produces UV–visible photons that correspond to the incident radiation but suffers from issues like low image contrast and increased quantum noise. In contrast, direct detection yields superior performance metrics, including higher detective quantum efficiency (DQE) and better contrast-to-noise ratios, making it more suitable for capturing fine anatomical details and potentially allowing for lower X-ray doses [20]. In X-ray imaging, spatial resolution is crucial for differentiating traits that are near together. In many fields, including medicine and industry, the ability to detect millimeter-scale cracks and malignant tumors requires sophisticated detectors with high modulation transfer function (MTF) capabilities. Radiography pictures with higher spatial resolution have a much better chance of detecting complex characteristics when trained using deep learning neural networks [21]. Traditional materials such as CsI and $Gd_2O_2S$ are commonly employed for indirect conversion, while α-Se is predominantly used for direct conversion in commercial flat panel X-ray imagers (FPXIs). The novel idea for X-ray imaging signal processing, titled "CIX—A Detector for Spectral Enhanced X-ray Imaging by Simultaneous Counting and Integrating," makes use of pixelated semiconductor sensors that can integrate charges and count photons all at once. The detector's dynamic range is increased by this dual-mode operation, enabling it to successfully handle both low and high X-ray fluxes. Operating in both modes simultaneously allows the system to measure the average energy of incoming photons, which provides spectrum data that improves picture contrast. Medical applications such as X-ray computed tomography greatly benefit from these characteristics, since they allow for precise tissue distinction and imaging at varied exposure levels. The constraints of amorphous selenium have made it difficult to integrate pixelated readout circuits with X-ray converters [22].

Metal halide perovskite (MHP) detectors, especially lead-based variants, combine high bulk resistivity and integrated circuit compatibility with exceptional X-ray attenuation—scaling as $\alpha \propto Z^4/AE^3$ (where Z is the average atomic number, A is the atomic mass, and E is the X-ray photon energy)—to achieve ultra-sensitive detection at low dose rates, making them ideal for medical diagnostics, safety inspections, and scientific research. Single-crystal and nanocrystal systems—such as methylammonium

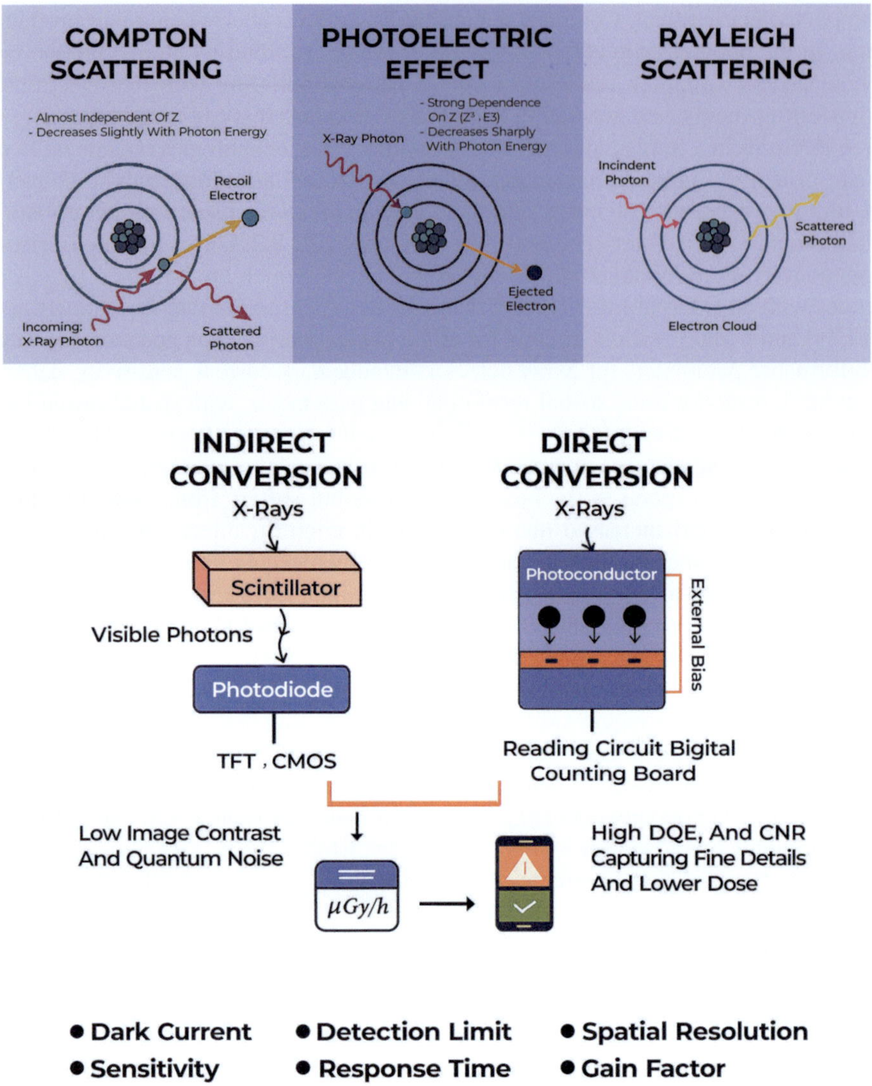


**Fig. 8.2** Overview of X-ray interaction processes and detector technologies. **a** Compton scattering, **b** The photoelectric effect, **c** Rayleigh scattering, and **d** X-ray detector modalities

lead iodide ($MAPbI_3$), cesium silver bismuth bromide ($Cs_2AgBiBr_6$), and cesium lead bromide ($CsPbBr_3$)—exhibit large mobility–lifetime products, long carrier diffusion lengths and intrinsic defect tolerance. In particular, $CH_3NH_3PbBr_3$ single crystals outperform CdTe in both carrier mobility and lifetime, yielding enhanced X-ray response [19, 23, 24]. Excessive dark current has, however, constrained further gains; by employing statistical models to link dark and photocurrents and adopting dark-current-shunting architectures, researchers have achieved record-low leakage, ultra-high sensitivity and high-resolution imaging in realistic device configurations [25–27]. Looking ahead, low-dimensional, large-band-gap perovskites (e.g. $Cs_3Bi_2I_xBr_{9-x}$) promise near-noise-free detection, and while 3D scintillators excel via low exciton binding energies, 2D analogues—with higher binding energies—can boost light yield and speed response [28]. Metal halide perovskites—whether as polycrystalline films or all-inorganic nanocrystals—have supplanted conventional scintillators and direct-conversion semiconductors for X-ray detection owing to their high-Z composition, strong photon attenuation and tunable bandgaps. Two-dimensional organic–inorganic hybrids, comprising alternating bulky organic cations and inorganic halide sheets, form natural quantum wells that confine excitons, boosting radioluminescence yield, accelerating decay kinetics and eliminating afterglow for high-frame-rate, high-resolution imaging. For example, Li-doped phenethylammonium lead bromide ($PEA_2PbBr_4$:Li) in microcapillary arrays achieves spatial resolution and contrast on par with—or superior to—CsI:TI and α-Se detectors. The pronounced anisotropy, adjustable bandgap, high photoluminescence quantum yield and enhanced moisture resistance of these 2D perovskites—combined with low-temperature, solution-based processing enabled by large organic spacers—facilitate scalable, cost-effective fabrication. Continued compositional optimization and targeted doping further elevate scintillation brightness and operational stability, positioning low-dimensional perovskite hybrids as leading candidates for next-generation X-ray imaging [29].

Detector figures of merit depend critically on both material properties and device architecture. In scintillators, efficient optical coupling, minimal self-absorption and uniform light transport are achieved through optimized layer thicknesses, large Stokes shifts and high-quality interfaces. A substantial separation between radioluminescence and absorption bands virtually eliminates self-absorption, yielding high-intensity light output and precise, real-time X-ray imaging. Although lead-halide perovskite scintillators deliver exceptional performance, their intrinsic lead toxicity limits practical deployment. Improved performance and the elimination of lead-related health risks have led researchers to create lead-free perovskite scintillators, which are non-toxic and make better use of Bi and Sn elements. Direct radiation detector manufacture has made use of lead-free perovskites, the majority of which outperform α-Se in terms of sensitivity and have lower detection limits. Pixeled $Cs_2AgBiBr_6$ perovskite array X-ray detectors, as an example, have a high X-ray sensitivity of about $1.91 \times 10^3$ μC $Gy_{air}^{-1}$ $cm^{-2}$ [30]. In a recent investigation, researchers systematically optimized the X-ray scintillation properties of the double perovskite $Cs_2Ag_{0.6}Na_{0.4}In_{0.85}Bi_{0.15}Cl_6$ by modulating its $Bi^{3+}$ content. Substituting $In^{3+}$ with higher-atomic-number $Bi^{3+}$ ions significantly increases the material's photoelectric absorption cross section, thereby enhancing the capture of incident X-ray photons. Concurrent enhancements in radiative recombination efficiency and photoluminescence quantum yield further convert a larger fraction of the absorbed energy into visible emission. The concurrent formation of self-trapped excitons and a pronounced Stokes shift between absorption and emission spectra minimizes in-material reabsorption losses. As a result, the optimized perovskite achieves a light yield of $(3.9 \pm 0.7) \times 10^4$ photons/MeV. Importantly, thermal annealing and prolonged X-ray exposure do not impair its luminescent performance, demonstrating both the stability and practical viability of this composition for durable X-ray detection applications, Fig. 8.3 [31]. Moreover, photoferroelectric perovskites enhance X-ray detector performance by leveraging their intrinsic polarization to drive spontaneous separation and efficient transport of photogenerated charge carriers. Notably, a high-performance X-ray detector was developed using a perovskite single crystal, achieving impressive sensitivity levels. Building on this, a new study introduces a wearable, flexible X-ray detector made from a lead-free photoferroelectric perovskite membrane. The device processes signals from X-ray exposure through a charge amplifier and transmits data to an electronic counter, allowing for real-time monitoring of radiation levels in various environments. This innovation marks a significant step towards the creation of wearable X-ray dosimeters [32]. Another membrane, created by embedding a hybrid metal halide perovskite, $[NH_3(CH_2)_4NH_3]BiI_5$, within a porous nylon polymer, is designed to withstand bending stresses, making it ideal for integration into clothing. The detector achieved a high resistivity of $2.0 \times 10^{12}$ Ω cm, reducing dark current and improving the signal-to-noise ratio (SNR). The charge-carrier mobility and lifetime product of lead-free photoferroelectric perovskite single crystals (LFPPM) were found to be $1.3 \times 10^{-3}$ $cm^2$ $V^{-1}$, enhancing charge transport and collection. The detector's radiation response was assessed through current–voltage measurements, revealing a radiation photovoltage of 0.98 V and significant sensitivity at a 50 V bias. It also performed well in self-driven mode, with a photocurrent density of approximately 20 pA at a dose rate of 6.8 $\mu Gy_{air}$ $s^{-1}$. The flexible X-ray detector's operational stability was tested under cumulative X-ray exposure, like those of X-ray chest evaluation, indicating its potential for direct X-ray imaging applications. LFPPM detectors maintained consistent performance during bending tests, demonstrating their mechanical stability [33].

While single crystal X-ray detectors made from metal halides demonstrate significantly higher sensitivity compared to commercial α-Se detectors, they come with drawbacks such as biotoxicity and compatibility issues with flexible substrates. Metal-free perovskites are organic materials maintaining a perovskite-like lattice but substituting protonated amine-based cations for conventional metal cations. Using octahedral coordination, strong intermolecular interactions, and organic orbital overlap, the $ABX_3$ structure in metal-free perovskites replicates the behavior of conventional perovskites. Unlike conventional metal-based perovskites, these materials preserve the corner-sharing $BX_6$ octahedral structure, necessary for charge transport and electronic structure. Similar to the electronic band structure seen in conventional semiconductors, organic components including π-conjugated molecular orbitals allow charge carrier delocalization. These properties allow metal-free perovskites to show photoresponse, radioluminescence, and direct bandgaps, therefore facilitating their perfect application in optoelectronics and X-ray detection free of heavy metal requirements. These materials' structural adaptability helps to increase the thermal and chemical stability of electronics, therefore enhancing their lifetime [34]. However, challenges in their preparation and processing persist. To improve

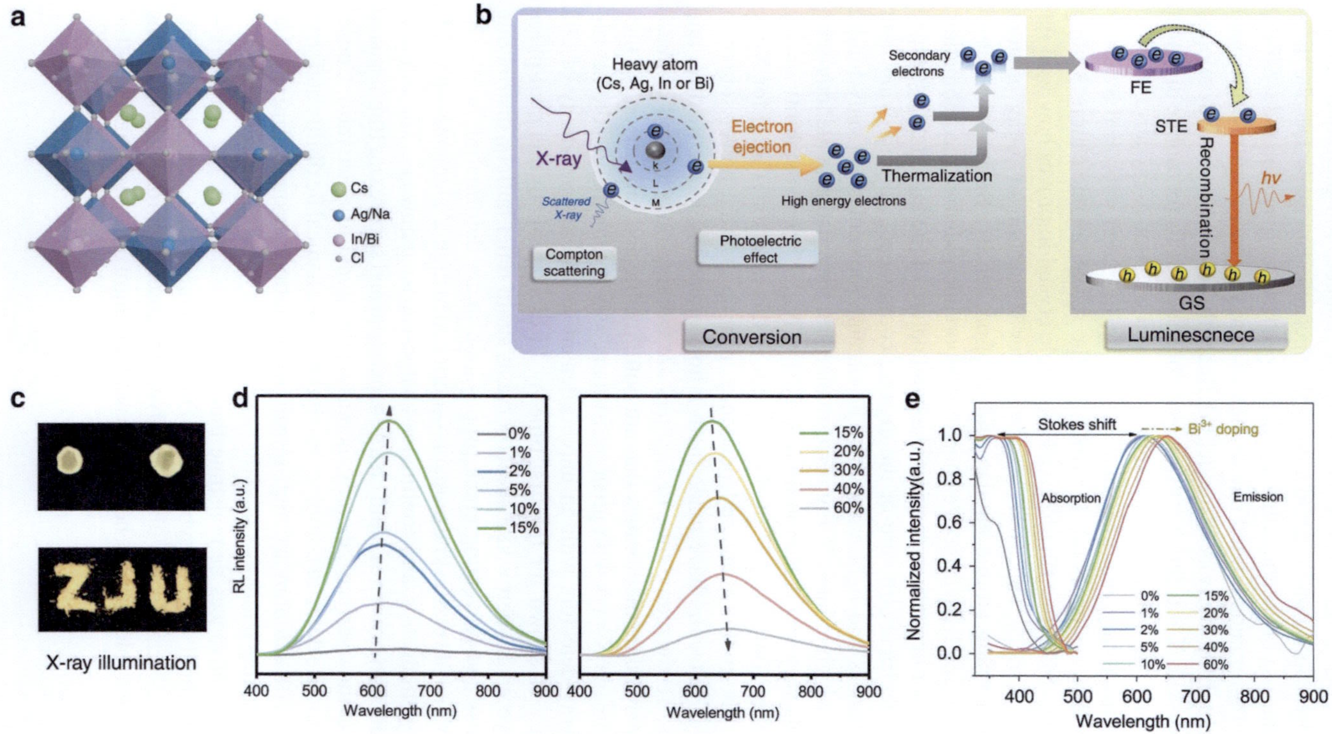


**Fig. 8.3** **a** Crystal structure of the double perovskite. **b** Proposed mechanism of X-ray scintillation in a lead-free halide double-perovskite scintillator. **c** Photographs of $Cs_2Ag_{0.6}Na_{0.4}In_{0.85}Bi_{0.15}Cl_6$ single crystals and powder under X-ray illumination (dose rate: 189 $\mu Gy_{air}$ $s^{-1}$, voltage: 50 kV). **d** RL spectra of $Cs_2Ag_{0.6}Na_{0.4}In_{1-y}Bi_{iy}Cl_6$ powder with different $Bi^{3+}$ contents under X-ray excitation with a dose rate of 189 $\mu Gy_{air}$ $s^{-1}$ at a voltage of 50 kV. **e** Stokes shift of $Cs_2Ag_{0.6}Na_{0.4}In_{1-y}Bi_{iy}Cl_6$ with different $Bi^{3+}$ contents. Reproduced from Ref. [31] published by Light: Science & Applications. Licensed under Creative Commons Attribution 4.0

their performance, researchers suggest substituting halogen elements with pseudohalides (such as $BF_4^-$, $PF_6^-$, and $ClO_4^-$), which enhances ion migration energy, intermolecular hydrogen bonding, and adjustable electronic bandgaps. Despite these advancements, the use of pseudohalide metal-free perovskites (pseudohalide–MFPs) in X-ray detection remains limited due to lack of heavy elements, ultra-wide band gap, and unreliable biotoxicity assessment [35]. Recently, researchers have developed a wearable perovskite X-ray detector using pseudohalide–MFPs for mobile diagnostics. The detector, which has been initially hampered by poor X-ray absorption, a wide band gap, instability, and uncertain biotoxicity, has been improved by introducing heavy atom covalent bonds (C–Br/Cl) on the A-site organic cation. Adding $NH_4(BF_4)_3$ to the (MDABCOBr)-$NH_4(BF_4)_3$ single crystal improves how well the material absorbs X-rays. This is so because better absorption of X-ray photons results from both rising atomic number and electron density. Better scintillation performance, resulting from effective charge transport and radiative recombination, comes from a higher crystalline density. The band nature of the material is changed to enhance the electrical structure of the organic components, so improving their X-ray responsiveness and light emission. The organic molecules not only inhibit ionic migration, which reduces X-ray performance, but they also provide long-term stability. With its very low detection limit of 50.1 $nGy_{air}\ s^{-1}$, great stability during operation, and sensitivity of 2377 $\mu C\ Gy_{air}^{-1}\ cm^{-2}$, this detector is ideal for X-ray detection [34]. The mobility and independence from an external power source of self-driven X-ray detectors have made them a hot commodity in the medical and industrial fields. To overcome this, researchers are working on a method to mass-produce thick perovskite membranes that are both flexible and self-driven, so they can detect and image X-rays with great performance. A flexible perovskite heterojunction for X-ray detectors with an unprecedented self-driven sensitivity of 292 $\mu C\ Gy_{air}^{-1}\ cm^{-2}$ is shown for the first time by Li et al. [36]. In this dual-phase architecture, layered 2D perovskite sheets serve as strain-relief buffers, distributing mechanical stress and preventing crack propagation under deformation, while the vertically integrated 3D perovskite domains—enriched in high-Z elements—provide enhanced X-ray attenuation and continuous charge-carrier pathways. The intimate 2D/3D heterojunction promotes rapid charge separation and suppresses trap-assisted recombination at the interfaces, thereby increasing the signal-to-noise ratio. Consequently, the composite detector combines exceptional mechanical flexibility and scalable fabrication with autonomous, high-sensitivity operation—requiring no external power source—making it ideally suited for wearable X-ray imaging application [36]. Collectively, these advances demonstrate that perovskite-based X-ray detectors—through compositional tuning, hybrid 2D/3D architectures, and scalable, flexible device engineering—can deliver ultra-sensitive, low-dose, high-resolution imaging in autonomous, wearable formats, heralding a new paradigm in medical and industrial radiography.

### 8.2.2 *Perovskite-Based Radiotracer Detectors for Medical Imaging*

One potential difficulty with radiopharmaceutical injections is the risk of extravasation events, which occur when the radiotracer leaks into neighboring tissue. During medical treatments, gamma scintillation sensors are used to improve the detection of radioactive material flow. Radiotracer injections for PET/CT scans are monitored by three sensors that are strategically positioned above the tumors, injection arms, and control arms. Wearable perovskite technology provides an online monitoring option for radiotracers in nuclear medicine, Fig. 8.4 [37]. Lead-halide perovskites, especially low-dimensional hybrids, combine high-atomic-number constituents with intrinsic radiation and defect tolerance, yielding suppressed dark currents (via minimal self-doping) and excellent carrier mobilities. This enables high detection sensitivity even at low dose rates. At the perovskite–electrode interface, however, energy-level misalignment and ion migration can induce leakage currents and degrade stability. To mitigate these effects, interface engineering strategies—such as diffusion barriers, interlayers for energy alignment, and surface passivation—are employed to stabilize charge-transport channels [38]. High-resolution gamma- and X-ray detection further requires perovskite films of substantial thickness (hundreds of micrometers), low defect density, and high crystallinity to maximize photon attenuation, charge generation, and to suppress nonradiative recombination. Fabrication methods—including spin-coating, blade-coating, mechanical pressing, sintering, and additive printing—have been optimized to produce such thick, uniform films with elevated sensitivity. Finally, chemical passivation, fullerene-based interlayers, and organic halide salt additives have been demonstrated to enhance long-term stability and performance, paving the way for clinically viable, high-resolution perovskite detectors [39, 40].

Additionally, addressing ion migration through the use of low-dimensional perovskites and optimizing the choice of organic cations has shown promise in improving long-term stability. By using the inherent optoelectronic properties of 2D layered $PEA_2PbBr_4$ perovskite thin films, wearable detectors built from them may detect gamma-ray emissions from nuclear medicine radiotracers in real-time. The main component of the detecting mechanism is the interaction between high-energy gamma rays and the perovskite lattice. High atomic number (Z) lead (Pb) atoms in the perovskite allow efficient ionizing radiation absorption. The photoelectric effect and Compton scattering are the main mechanisms by which these interactions produce electron–hole pairs inside the crystal structure. Effective charge transfers and collection following production is made possible by the layered perovskite structure using the organic spacer cation ($PEA^+$), which provides spatial confinement. This procedure produces a quantifiable photocurrent that varies exactly with the flux of entering gamma rays. Conformal and fit for wearable use, the detector's 2D layered construction guarantees mechanical flexibility while preserving outstanding charge mobility and low dark current for steady and sensitive signal collecting. Constant monitoring of gamma-ray activity during radiotracer application is made feasible by this real-time detection capacity. The detector turned out to be quite accurate, with a maximum

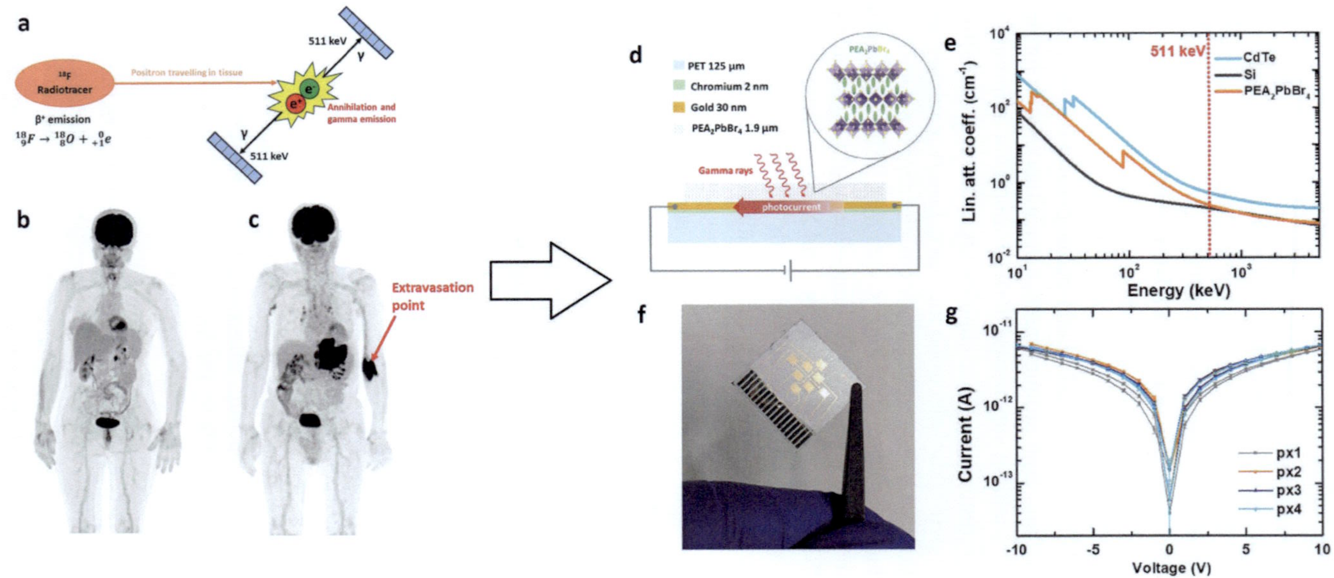


**Fig. 8.4** Wearable perovskite-based radiotracer detectors enable real-time gamma monitoring for the distribution and identification of radiotracer extravasation, featuring wireless data transmission for online monitoring. **a** A schematic illustration of the radioactive decay of 18F, succeeded by positron–electron annihilation and the subsequent emission and detection of gamma rays. **b** PET imaging for the secure administration of radiopharmaceuticals. **c** PET imaging of an extravasation event (denoted by the red arrow). **d** Schematic depiction of the device architecture and the configuration of PEA2PbBr4 perovskite constituting the active layer of the detector. **e** The linear attenuation coefficient as a function of photon energy for PEA2PbBr4 perovskite and two reference materials for gamma-ray detectors, namely Si and CdTe. The dashed orange line denotes the characteristic energy of gamma-ray emission resulting from positron annihilation subsequent to 18F decay. **f** Image of the perovskite detector being examined. **g** Graph of the current against voltage measurements for the four pixels. The error bars are derived from the instrument's precision. Reproduced from Ref. [41] published by Advanced Materials Technologies. Licensed under Creative Commons Attribution 4.0

accuracy variation of around 8%. Its ability to record real-time events, such as injections and extravasation, shows its potential to enhance clinical nuclear medicine operations in terms of patient safety and treatment accuracy [41]. Another report presents a novel wearable gamma ray detector that utilizes hybrid halide perovskite methyl ammonium lead bromide (MAPB). This flexible, paper-like device works at room temperature and detects gamma rays by changing its resistance when it comes into contact with radiation, thanks to electron–hole pairs created in the perovskite when exposed to gamma rays. Featuring interdigitated electrodes, it enhances current response and boasts a maximum calibrated sensitivity of 5.26 $\mu$C $Gy^{-1}$ $cm^{-2}$, along with reasonable flexibility. The detector is capable of functioning across a broad range of gamma photon energies and can identify radiation levels as low as 0.01 $\mu$Ci. With a shelf life exceeding six months, rapid response time, and low power consumption, it is expected to have applications in healthcare and point-of-use quick radiation detection [42]. However, commercialization faces challenges like ion migration, fermi level pinning, electrochemical reactions, and integration difficulties. The design of metal electrodes is crucial for achieving effective energy band alignment and ensuring device longevity. Solutions to challenges like Fermi level pinning and redox reactions at the electrode/perovskite interface are vital for enhancing the overall performance of wearable perovskite gamma-ray detectors. By overcoming these hurdles, wearable perovskite detectors could revolutionize radiation detection in various applications, from medical imaging to environmental monitoring.

## 8.3 Human–Machine Interfaces (HMI)

### *8.3.1 Evolution of HMIs*

The key technical systems allowing communication and interaction between people and machines are human–machine interfaces. Acting as essential translational platforms, HMIs enable the smooth passage of commands, data, and feedback across computational and biological domains. Ullah et al. [43] highlight that wearable sensor-based HMIs should minimize cognitive load and physical exertion to provide an intuitive, low-friction user experience, thereby enhancing adoption and operational accuracy. HMIs' intellectual and technical development closely reflects more general changes in digital technologies and materials science. Designed mostly for military and industrial uses, the first significant phase—which ran the 1940s through the 1970s—dependent on crude mechanical control systems. Graphical user interfaces (GUIs) in personal computers emerged in the 1980s and 1990s, following which devices including the keyboard, mouse, and visual displays became central to human–machine interaction [44, 45]. In the early 2000s, the emergence of capacitive touchscreens, kinesthetic haptic actuators, and voice-recognition modules defined the third generation of human–machine interfaces. In the 2020s, a fourth paradigm has arisen—driven by the infusion of machine-learning and AI algorithms into edge-computing nodes, the deployment of ubiquitous low-latency wireless networks (e.g., IoT, 5G/6G), and the advent of stretchable, biocompatible electronic materials. Together, these innovations yield multimodal, context-aware HMI platforms that adapt in real time to user intent while minimizing cognitive effort and enabling seamless, conformal integration with the human body, Fig. 8.5 [46]. Modern human–machine interface research prioritizes biointegration by seamlessly aligning mechanical, physiological, and cognitive interactions with electronic systems. Innovations in flexible electronics, wearable sensors, transparent electrodes, stretchable substrates, and biocompatible hydrogels underpin this transformation, while the transition from Industry 4.0 to human-centric Industry 5.0 reinforces design paradigms that balance technical performance with adaptable, emotionally aware, and ethically grounded human–robot collaboration [43, 44, 47].

Among the new technologies influencing the next generation of HMIs, soft, functional materials—including hybrid nanomaterials and perovskites—stand out as especially transforming. These materials provide a special mix of lightweight mechanical qualities, breathability, flexibility, and natural self-healing capacity that helps to create interfaces able to monitor real-time physiology and provide adaptive feedback [48]. These dynamic HMIs are now intelligent ecosystems able to interpret complicated user intent, emotional states, and ambient environmental cues rather than fixed conduits for commands. Brain-computer interfaces, adaptive machine learning algorithms, multi-sensory feedback systems, and self-powered sensing technologies are continuously merging to challenge what qualifies as human–machine interaction [49]. Now central in many important fields including healthcare, education, manufacturing, virtual and augmented reality, prosthesis, and remote robotics are HMIs. This cross-disciplinary penetration emphasizes the critical need of ongoing innovation in materials engineering, user-centric system design, and ethical frameworks to guarantee that future HMI systems not only improve human capabilities but also maintain human dignity, autonomy, and social values [47]. In a nutshell technology undergoes a basic paradigm change from rigid, mechanical, device-centric interaction models to flexible, adaptive, human-centric ecosystems. HMIs are poised to become the cornerstone of a future in which the line separating human cognition from machine intelligence disappears into a seamless continuum as they progressively mix organic and synthetic systems.

Human–machine interaction for assistant devices refers to a range of interfaces meant to satisfy several user needs and application environments. Usually through gestures like tapping or swiping, the touch-based interface—which lets users interact with devices by touching a screen—is among the most often used forms. Because of their simplicity, tablets and cellphones extensively feature these interfaces. Likewise, voice-activated interfaces have grown in popularity since they let hands-free operation by spoken commands. These interfaces allow virtual assistants such as Amazon Alexa and Google Assistant to provide smooth, voice-based interaction. Gesture-based interfaces translate physical motions—such as hand or full-body movements—into control signals that enhance immersion in virtual and augmented reality, whereas brain–computer interfaces (BCIs) establish direct neural pathways

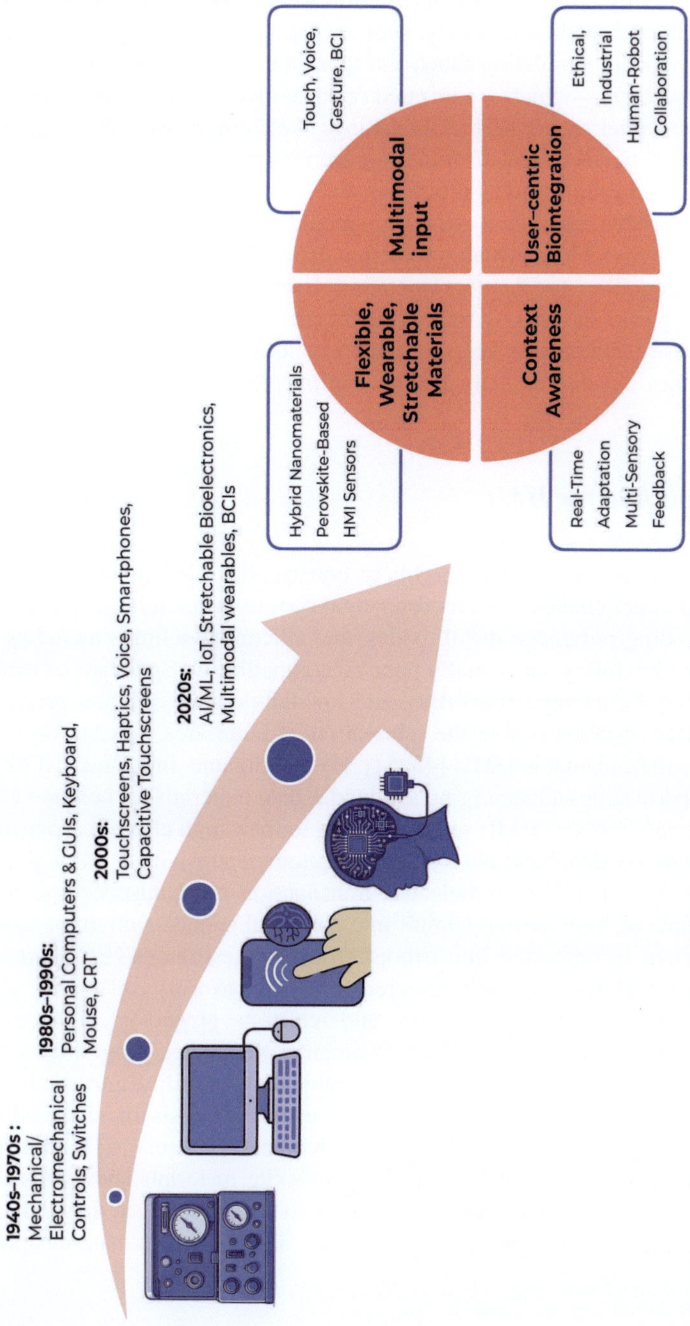


**Fig. 8.5** Evolution of Human–Machine Interfaces (HMIs)

to external devices, enabling purely thought-driven operation. For those with disabilities, BCIs especially help to enable interaction without physical movement. Often used in applications needing accuracy, such medical simulations, haptic interfaces give tactile feedback, simulating touch via vibration or force. On the other hand, non-contact interfaces—which use infrared or capacitive sensors—detect user input without physical touch, hence perfect for settings like public kiosks where hygiene or contactless interaction is crucial. Combining several input sources—touch, voice, and gestures—the rise of multi-modal interfaces gives consumers flexibility and adaptation. At last, wearable interfaces—which are included into smartwatches and fitness trackers—offer a portable, constant interaction experience that frequently combines several input modalities for flawless operation. Every one of these interface forms is designed for particular uses, so improving the usability and functionality of assistant tools. These technologies keep making user interactions more appealing and accessible as they develop in many different settings [50].

### *8.3.2 Perovskite for HMI*

Human–machine interfaces call for highly responsive, flexible, durable, and capable of functioning under challenging environmental conditions materials. Conventional materials including polymers, metal oxides, and silicon have limits including brittleness, limited flexibility, and usually poor biocompatibility. By means of mechanical compliance, lightweight properties, and low-temperature solution processing, perovskite-based systems enable the fabrication of large-area, flexible devices fit for wearable and implantable HMIs [48, 51]. Hybrid organic–inorganic (HOIP) and all-inorganic perovskites are emerging as foundational materials for next-generation human–machine interface (HMI) sensors, owing to their high charge-carrier mobilities, minimal defect densities, and exceptional photoresponsivities. Through piezoelectric, photodetection, and pyroelectric transduction mechanisms, they convert mechanical, optical, and thermal stimuli into electrical signals with unprecedented sensitivity. When incorporated into triboelectric nanogenerator (TENG) architectures, these perovskites yield self-powered sensing platforms capable of gesture recognition, wearable device operation, and real-time physiological monitoring in precision- and energy-critical AI environments. Metal-free perovskite variants further enhance this approach by providing mechanical flexibility, intrinsic biocompatibility, and low toxicity—thus establishing a sustainable basis for advanced wearable electronics, electronic skins, and implantable health sensors [52]. Particularly metal-free variants like MDABCO-$NH_4I_3$, perovskite materials show remarkable piezoelectric properties, which make them quite fit for HMIs since they can effectively transform mechanical energy into electrical signals. These materials provide major benefits in terms of biocompatibility, self-powered functionality, and flexibility, so allowing the creation of low-toxicity, highly sensitive, durable wearable and implantable devices. Offering a sustainable and flexible solution for next-generation technologies, the combination of perovskite-based piezoelectric nanogenerators (PENGs) into HMI systems promises to improve applications including gesture recognition, smart health monitoring, and electrical stimulation therapies [53]. These materials' unusual crystal structure lets them bend and stretch without upsetting their electronic characteristics, so preserving effective electron flow even under deformation. By means of a well-organized lattice that reduces flaws and evenly distributes strain, this mechanical compliance is obtained, so preserving electronic performance. Furthermore, perovskite solar cells' great power conversion efficiency allows energy harvesting straight from ambient light, so enabling devices to run free from large batteries. Perovskites are positioned as the best solution for next-generation HMIs in health monitoring and augmented reality systems thanks to biocompatibility, exceptional sensing capabilities, scalable, low-cost fabrication techniques [54]. Advances in flexible electroluminescent devices for HMIs could also be driven by perovskite-based systems, so enabling high-performance, self-powered displays free from large batteries. These materials show amazing electroluminescent qualities, providing clear, effective light emission while preserving mechanical compliance and flexibility—qualities absolutely essential for building bendable or stretchable interfaces [55].

Especially in flexible photodetectors, piezoelectric nanogenerators, and pressure sensors—essential parts for real-time human–machine communication—perovskites have shown outstanding performance [56]. Particularly underlined in recent studies is perovskite adaptation to environmental stimuli. Development of HMIs depends mostly on advanced perovskite nanoarrays, especially in tele-perception and tele-operated robotic systems. Applications requiring exact, dependable, and extended performance depend on increased sensitivity and long-term stability, which these nanoarrays have been designed to provide. Three main techniques—ligand engineering, encapsulation, and metal cation dopant introduction—help to improve the performance of these nanoarrays. Ligand engineering changes the surface chemistry of perovskite nanocrystals so enhancing their stability and preventing degradation over time. Encapsulation methods shield the nanocrystals from environmental elements including moisture and oxygen, which might otherwise lower their efficacy. By strengthening the crystal structure and properties of the perovskite material, metal cation dopants help to increase its stability and sensitivity. These developments allow perovskite nanoarrays to remain highly appropriate for integration into non-contact sensing technologies used in HMIs since they enable their functionality for prolonged periods [48]. Duan et al. [57] recently look at including perovskite-based materials into flexible composites with improved environmental stimulus responsiveness. The work presents a composite material formed from trimethylchloromethyl ammonium (TMCM)–$MnCl_3$ microrods embedded within a styrene-ethylene-butylene-styrene (SEBS) matrix. High tensile strain capabilities (>1300%) and improved mechanical and optical qualities follow from this integration into a flexible composite. Including monopole-monopole interactions, halogen bonding, and van der Waals forces, the molecular interactions between TMCM-$MnCl_3$ guarantee homogeneous dispersion of the ferroelectric microrods and help to provide the composite's increased flexibility and sensitivity. Using a piezoelectric response to mechanical deformation and photoluminescence under UV light, the composite shows dual responsiveness and can be used in advanced anticountersfeiting and encryption technologies. By means of pattern recognition and piezoelectric responses for anticountersfeiting and anti-tampering functionalities, perovskite material improves its versatility and qualifies for applications requiring both mechanical flexibility and optical reactivity, such high-security password systems and tamper-evident tags [57].

Moreover, the capacity to precisely control their bandgap and electronic characteristics helps perovskites to be tailored for particular sensing applications inside HMIs, such brain-computer interfacing or gesture recognition. Because of their unique ionotronic properties, perovskite nickelates—especially rare-earth nickelates ($ReNiO_3$, where Re is a rare-earth ion, a class of quantum materials whose electrical properties are essentially dominated by the strong interactions among electrons in them)—like $NdNiO_3$ and $SmNiO_3$ have emerged as promising materials for BMIs. Protons and other small ions can be reversibly accepted and released by these materials, so modulating their electronic conductivity in response to ionic doping. This quality is essential for designing devices that interact with biological systems, where ionic signals constitute the main means of communication. In BMIs, perovskite nickelates detect and react to neurotransmitter release. When interfaced with brain tissue, for example, these materials can measure changes in their electrical resistance to identify the release of dopamine, a neurotransmitter linked with movement and reward. This resistance change results from the protonation of the nickelate surface by dopamine, made possible by the enzyme accelerating protons from dopamine to the nickelate lattice. These interactions provide real-time monitoring of neurotransmitter dynamics in situ, so offering important new perspectives on neural activity and maybe allowing closed-loop control in neuroprosthetic uses. Moreover, perovskite nickelates have been used to replicate fundamental learning and memory processes in the brain by means of synaptic behaviors including potentiation and depression. Applying electric pulses allows the proton distribution within the nickelate lattice to be controlled, producing tree-like conductance states resembling the synaptic weight fluctuations seen in biological systems. Development of neuromorphic systems able to learn and adapt in a way similar to the human brain depends on this capacity to replicate synaptic plasticity. A major first toward building more natural and effective interfaces between electronic systems and the nervous system is the inclusion of perovskite nickelates into BMIs. Their capacity to replicate synaptic functions and transduce ionic signals into electronic responses define them as a fundamental material in the progress of brain-machine interface technologies, Fig. 8.6 [58, 59].

In the field of intelligent gesture recognition, a significant advancement is the construction of conductive elastomeric sensors made of polydimethylsiloxide (PDMS) coupled with nanoparticles. By raising sensitivity and operational range, this novel design solves the constraints of conventional flexible strain sensors. These sophisticated sensors are positioned deliberately at finger joints such that real-time

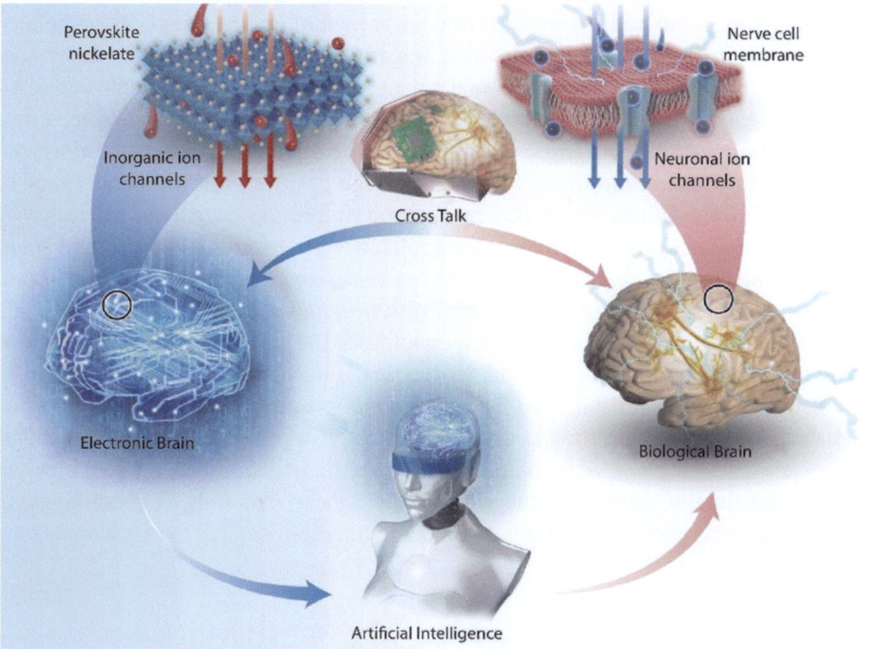


**Fig. 8.6** Perovskite nickelate-based brain-machine interface (BMI) for ionic-electronic transduction at the brain interface. From Ref. [59] published by Materials Lab. Licensed under Creative Commons Attribution 4.0

finger motion monitoring is feasible. Together with specialized hardware and software, the sensors enable precise gesture recognition. For example, machine learning approaches with an accuracy of up to 97% have been used to identify American Sign Language gestures, so illustrating the possibilities of these systems in enabling communication for those with hearing problems. Including such intelligent gesture recognition gloves into HMI systems presents a sensible way to create more simple and natural interfaces. Over many uses including assistive technologies, virtual reality, and robotics, these gloves can improve user interaction using creative sensor technologies and machine learning algorithms [60]. Recently Wu et al. [61] report a wearable sensor based on perovskite-based materials for shadow recognition, so enabling ambient and non-obtrusive human–computer interaction (HCI). Combining a composite film of perovskite crystals with liquid crystal (LC) oligmers, the sensor promotes the ordered nucleation and growth of perovskite crystals at the micrometer scale. Under ambient light conditions (~500 lx), this structural design greatly improves the optoelectronic characteristics of the sensor, producing a high on–off ratio exceeding 100 and fast millisecond response times. After thousands of bending cycles and little degradation over over 1200 h of storage, the sensor's performance is remarkably constant with almost 90% of its photocurrent intensity maintained. The sensor also shows negligible lead ion ($Pb^{2+}$) leakage with concentrations as low as 1.02 μg/L, so addressing typical environmental and health concerns related to perovskite materials. Where stability, sensitivity, and environmental safety are critical in HCI, these characteristics make the sensor a constant candidate for wearable applications. Run through a machine learning algorithm, the wearable sensor picks hand-cast shadows as input motions, so enabling non-contact interaction with objects. With a maximum of 95% and an average gesture recognition accuracy of 88%, this sensor provides users with a perfect and simple approach to run devices with just basic hand motions. Particularly for applications in wearable devices and smart environments, the development of this perovskite-based shadow recognition sensor signals a significant turn toward the realization of more natural, unobtrusive, and efficient human–computer interaction technologies [61]. Still, despite these encouraging qualities, issues including lead toxicity and environmental damage have to be resolved if we are to fully realize the commercial possibilities of perovskite-based HMI technologies [51]. Growing demand for adaptive, intelligent interfaces drives the convergence of perovskite material science with artificial intelligence, bioelectronics, and flexible device engineering to place these materials as a fundamental basis for the next evolution of human–machine interaction.

### *8.3.3 Innovations in Flexible Perovskite Technologies for Human–Machine Interfaces*

By means of strategic substrate integration and material engineering, flexible perovskite architectures have become essential innovations in wearable electronics and human–machine interfaces (HMIs), so overcoming the inherent brittleness of conventional materials. Devices achieve high sensitivity while conforming to dynamic surfaces like human skin by embedding perovskite layers onto elastic substrates such as polydimethylsilane (PDMS), polyurethane, or breathable textiles, so enabling next-generation health-monitoring patches, smart skins, and biointegrated sensors. Although their 3D molecular architecture increases flexibility and stability while intramolecular cavities sequester $Pb^{2+}$ ions, so lowering toxic leakage and enhancing safety for prolonged skin-contact applications, hyperbranched polymers (HBPs) also play a dual role in improving mechanical robustness and functionality. Moreover, HBPs improve charge mobility, reduce recombination losses, and protect perovskite films from environmental damage, so guaranteeing long-term dependability in robotic and biomedical HMIs. Particularly sensitive in perovskite solar cells is the interface separating the electron transport layer (ETL) from perovskite. With adhesive groups and dynamic hydrogen bond networks, polyamide-amine-based HBPs reinforce this interface and improve fracture energy from 1.08 to 2.13 J $m^{-2}$. These polymers absorb mechanical energy, heal microfractures, and enable considerable deformation, so strengthening the mechanical stability of flexible PSCs. Retaining 88.9% of their original efficiency after 10,000 bending cycles, the polymer-functionalized PSCs reach power conversion efficiencies of 25.05% for rigid devices and 23.86% for flexible devices. Furthermore, intramolecular cavities of HBPs bind $Pb^{2+}$, so reducing lead leakage—a major concern for flexible PSCs in wearable technologies [62]. Inspired by mussel wet adhesion via dopamine groups, a multiidentate-cross-linking strategy is proposed to directly chelate perovskite materials in multiple directions. This produces a vertical scaffold spanning most of perovskite films, binding to grains and substrates with great adhesion capacity. Under high humidity this improves mechanical durability. Low-temperature solution process produced perovskite films have internal grain boundary (GB) defects that can be resolved by dimensional engineering, cross-linking modification, and self-healing GB [63]. In another work, the authors have developed a new bio-inspired design for photodetectors based on perovskite including honeycomb polymer grids for mechanical flexibility and high-resolution performance. The honeycomb construction distributes mechanical strain, so enhancing the deformation resistance of the device. Stress on the perovskite layer released by the polymer grid increases mechanical ductility and agility of the layer. Long-term use in flexible substrates such as skin or textiles is suited for the perovskite photodetectors since they show great light sensitivity and high-resolution imaging. By using a flexible polymer grid free of photolithography, the fabrication technique produces detectivity $>10^{11}$ Jones with cross-walk-free imaging. Based on the described characteristics, the gadget is appropriate for wearable electronics, interactive displays, and advanced HMI applications in robotics, biomedical imaging, and AR/VR [64].

Recently, integration of engineered interlayers and self-healing polymers has shown great potential to improve the mechanical integrity of perovskite solar cells (PSCs). These multifunctional materials maximize device lifetime by imparting strong mechanical robustness without compromising power conversion efficiency, so optimizing photovoltaic performance. The optimal balance among mechanical strength, autonomous repair capacity, and high energy-conversion efficiency will determine whether PSCs able to practically compete with silicon-based technologies develop [65]. Thiourea-triethylene glycol (TUEG3) polymer added into the perovskite layer produced a bi-continuous interpenetrating network including perovskite crystals in a recent work. Under mild thermal conditions, this polymeric network helps the perovskite material to self-repair, so allowing the rejoining of cracked or damaged perovskite grains and hence promoting grain growth, so restoring the mechanical integrity and electronic properties of the material. Spin-coated onto a substrate, a solution of perovskite precursor and TUEG3 polymer was fabricate the self-healing perovskite solar cells. The devices were then thermally annealed at temperatures somewhat above room temperature, which let the polymer activate and fix mechanical damage. With the devices maintaining their power conversion efficiency after several bending cycles and recovering significant performance after extreme bending tests, the self-healing process proved to be quite highly effective. With flexible devices retaining 94% of their initial efficiency after 3,000 bending cycles, the resulting self-healing perovskite solar cells showed a power conversion efficiency exceeding 10%. Moreover, the cells shown both enhanced mechanical robustness and long-term stability by recovering up to 80% of their performance following severe mechanical stress. According to the study, including TUEG3 into perovskite solar cells presents a viable approach to improve their durability and hence enable their high efficiency even in mechanically demanding environments [66].

Advanced nanoengineering methods improve mechanical resilience and flexibility even more by means of perovskite quantum dot (QDs) deposition and composite perovskite–polymer matrix integration. Notwithstanding these benefits, stability problems at the interfaces between charge-carrying layers and perovskite films continue to challenge the commercialization of this exciting photovoltaic technology. Trap-aided nonradiative recombination resulting from crystal defects developed during manufacturing and operation mostly causes these difficulties. Recent developments in perovskite quantum dots show great possibility for improving human–machine interfaces by acting as interface modification layers in perovskite-based devices. Driven by the quantum size effect, their special optoelectronic characteristics enable exact tuning of electronic characteristics and efficient energy level alignment between layers, so optimizing charge extraction and transport and reducing recombining losses. Furthermore, these QDs are very important for defect passivation by reducing surface flaws in perovskite films, so lowering non-radiative recombination centers and increasing device stability under dynamic conditions including mechanical flexing and environmental stress. A major component of obtaining high spatial resolution and sensitivity in photodetector arrays, perovskite crystallization and film homogeneity are also helped to improve by them. Their multi-exciton generating characteristics also improve light absorption and charge carrier generation, so guaranteeing effective operation even in low-light conditions for uses including wearables, gesture recognition systems, and augmented reality interfaces [67]. $CH_3NH_3PbI_3$ quantum dots were synthesized and deposited between the hole transport layer (HTL) and perovskite film to passivate interface defects using a reasonably cheap ligand-assisted reprecipitation method. The QD film thickness was exactly tuned to improve hole extraction by optimizing the spin-coating speed, so raising the power conversion efficiency (PCE) of carbon-based perovskite solar cells from 14.61 to 16.11%. These devices record 26.1% PCE at an ideal spin rate of 4000 rpm. Using their low excitation energies and strong photoluminescence quantum yields to align energy levels and enable effective electron transfer, graphene quantum dots were also presented concurrently as an ultrathin electron extraction layer between the electron transport material and the perovskite. These quantum dot-based interface engineering techniques taken together provide a scalable, simple method for more stable, higher-performance perovskite photovoltaics [68].

Human–machine interface development depends critically on the design path of fabrication of wearable sensors able of real-time monitoring of physiological parameters, motion, and environmental stimuli. Using the inherent multifunctional electromechanical properties of perovskite materials—more especially, the coexistence of piezoelectricity, in which mechanical stress generates an electric charge, and piezoresistency, in which mechanical deformation results in changes in electrical resistance, Fig. 8.7—perovskite-based pressure and strain sensors are indispensable in this field. These two responses, with great accuracy, help to exactly capture subtle biomechanical signals including pulse, breathing, and joint movement. These events let perovskite materials run concurrently as energy harvesters as well as high-sensitivity mechanical sensors in flexible and wearable devices. Advanced perovskite-based flexible pressure sensors show remarkable sensitivity to detect a wide spectrum of mechanical inputs, from mild touch to high-strain deformation, by means of the synergistic interaction between their piezoelectric and piezoresistive properties. Applications in electronic skin, soft robotics, and interactive control systems where this capability has spurred development depend on high fidelity in motion and tactile sensing [69]. Highly flexible to wearable electronics and conformal biomedical interfaces, these perovskite films and composites' intrinsic mechanical flexibility lets them integrate onto challenging, curved, and dynamically moving surfaces. Leading edge of these developments are metal-free perovskite piezoelectric nanogenerators, which provide low-toxicity, environmentally friendly substitutes for conventional lead-based systems. Especially metal-free halide perovskites, such MDABCO-$NH_4I_3$, show natural piezoelectricity as confirmed by piezoresponse force microscopy (PFM) observations. Under mechanical excitation, these materials have shown strong electrical outputs, reaching voltages up to ~15.9 V and currents of ~54.5 nA enough to drive autonomous electrical stimulation devices and self-powered sensors. Their non-toxic mix and stable piezoelectric responses make them interesting prospects for future bio-integrated, environmentally friendly human–machine interfaces [53]. Beyond single-phase perovskite systems, recent developments in perovskite-based composites, especially those integrating ($N_{0.5}B_{0.5}TiO_3$–$Bi_{0.5}Na_{0.5}TiO_6$ (NBT-BT) nanocubes, have brought a paradigm change in the design of multifunctional electromechanical materials. Using nanoscale chemical heterogeneity and interfacial polarization effects, these composites finely tune piezoelectric and piezoresistive behaviors. Along with mechanical strength and fracture toughness, control over nanocube dispersion, crystallographic orientation, and matrix–nanocube strain coupling has greatly enhanced electromechanical coupling coefficients. Under cyclic load, mechanical stress-concentrated nanointerfaces, and reconfiguration of defect dipole structures under mechanical stress together provide enhanced domain wall mobility, which mechanistically increases strain sensitivity and superior fatigue resistance. Concurrent localized stress-induced conductivity paths at the nanocube–matrix interfaces generate piezoresistive signal outputs, so enabling multimodal transduction paths within a single material system [70].

Particularly for uses like electronic skins, smart contact lenses, and augmented reality displays—that which calls for subtle, unobtrusive integration—clear electrodes are absolutely essential in forming next-generation HMIs. Balancing optical transparency and electrical conductivity is a basic difficulty in this field since changes in one quality usually result in a degradation of another. While conventional materials such as indium tin oxide (ITO) have low sheet resistance and high transmittance, their brittle character, great cost, and environmental instability make them impractical for flexible devices [71]. Research has thus turned toward perovskite-based substitutes using sophisticated nanostructuring. Engineers can achieve the exact electrical performance required for high-fidelity signal processing in HMIs by embedding perovskite nanostructures into flexible conductive matrices—such as transparent conductive oxides or graphene layers—such that high optical clarity is

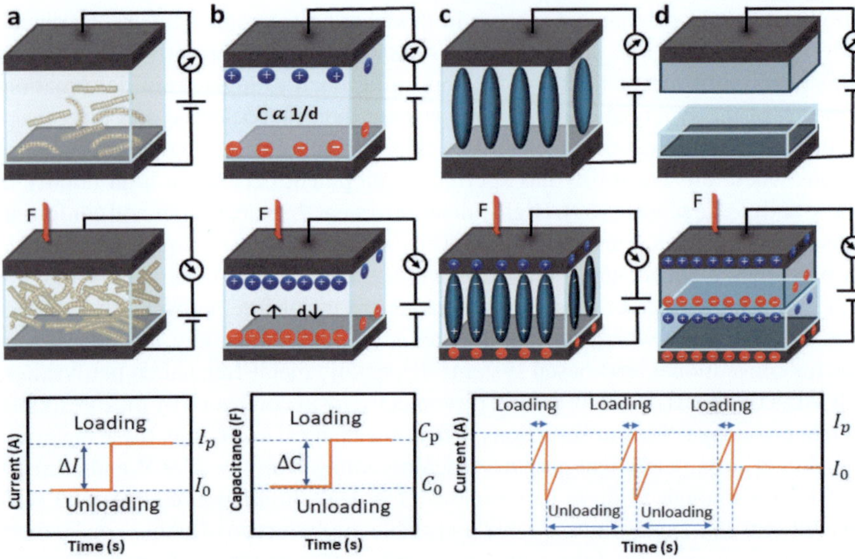


**Fig. 8.7** Schematic illustration of four common electromechanical transduction mechanisms: **a** piezoresistive, **b** capacitive, **c** piezoelectric and **d** triboelectric. Reproduced from Ref. [69] published by Materials. Licensed under Creative Commons Attribution 4.0

preserved. Thanks to their extraordinary and controllable optoelectronic characteristics, hybrid organic–inorganic halide perovskites have quickly become top materials for transparent electronics. Using compositional engineering (e.g., halide ion substitution), these materials show amazing bandgap tunability, usually ranging from 1.5 eV to over 2.3 eV, so enabling exact control over their absorption and emission properties. Especially when optimized for minimum thickness and defect density, thin films of methylammonium lead halide ($MAPbX_3$, where X = Cl, Br, I) and related compositions show high optical transparency, often exceeding 80% across the visible spectrum. Furthermore, these hybrid perovskites have very good charge carrier mobilities—usually more than 10 $cm^2V^{-1}$ $s^{-1}$—which are either equal to or better than those of many current organic semiconductors and transparent conducting oxides. Their emergence as flexible candidates for a broad spectrum of transparent optoelectronic applications—including transparent photovoltaics, photodetectors, and future-generation HMI devices—is underpinned by this special mix of high transparency, adjustable electronic properties, and excellent charge transport. Moreover, the easy low-temperature processing of these materials on flexible substrates provides a clear benefit for scalable production, so encouraging their integration into wearable transparent electronics [72].

Design of split-ring structured all-inorganic perovskite photodetector arrays finds a transforming framework in transparent optoelectronic materials in Human–Machine Interface applications. First concentrating on photodetection performance improvements, the fabrication techniques and material choice—especially all-inorganic perovskites like $CsPbBr_3$—highlight important paths for developing transparent electrode systems. Important for flexible, conformable HMIs, these materials exhibit natural synergy between high optical transparency and electrical conductivity. Through well-controlled local electromagnetic fields via micro/nanostructuring, designed perovskite layers can decouple optical transmittance from functional performance. Ultra-thin perovskite films improve the photoelectric characteristics of photodetectors by maximizing light absorption and transmission, while their lateral architecture facilitates manufacture and integration. This design enhances efficient current conduction and optimizes the use of the perovskite photoactive layer. The device produces a substantial photocurrent under bias voltage, achieving a peak responsivity of $1.44 \times 10^5$ mA $W^{-1}$ at low light intensities and a specific detectivity of $2.16 \times 10^{11}$ Jones, indicating its exceptional sensitivity to light. The photodetector has quick reaction times and stability under varying environmental circumstances, making it appropriate for applications necessitating rapid optical signal detection. Long-term durability testing of these photodetectors reveals the chemical and thermal resilience of all-inorganic perovskites, so underlining their feasibility for wearable and implantable HMIs under dynamic environmental conditions. Moreover, techniques intended to enhance optoelectronic performance in split-ring devices—such as surface passivation and interface engineering—directly solve problems in transparent electrode design by lowering defect states and enhancing carrier mobility. These techniques directly reduce scattering losses and increase crystalline alignment, so improving the transparency and efficiency required for visible-light HMIs including smart skins, transparent displays, and ocular electronics, Fig. 8.8 [73].

Apart from mechanical flexibility and optical transparency, guaranteeing long-term environmental stability remains a major design issue for perovskite-based flexible electronics and transparent human–machine interfaces. Though they have remarkable optoelectronic performance, hybrid organic–inorganic halide perovskites are intrinsically sensitive to environmental damage. While oxygen penetration under illumination can catalyze superoxide generation, leading to irreversible breakdown of the active layer, moisture causes hydrolysis of the perovskite lattice [74]. Moreover, perovskites are thermally unstable; higher temperatures can cause phase transitions, breakdown of the perovskite lattice, or volatilization of organic components such formamidinium (FA) ions, so aggravating performance losses and lowering device lifetime. Particularly problematic for flexible devices exposed to bending and mechanical heating during use, thermally accelerated degradation often shows as a reduction in crystallinity, formation of lead iodide ($PbI_2$) byproducts, and irreversible changes in optoelectronic properties [75].

Particularly important for wearable and flexible electronics, advanced hybrid encapsulation methods exhibit remarkable lifetime—maintaining structural and functional integrity over 1000 bending cycles at radii as small as 5 mm. Not only do such multilayer systems meet the rigorous mechanical, optical, and environmental criteria of next-generation human–machine interfaces [76] but also greatly extend the operational lifetime of perovskite solar cells and associated optoelectronic devices. Two main encapsulating strategies have been developed, underlining the strategic

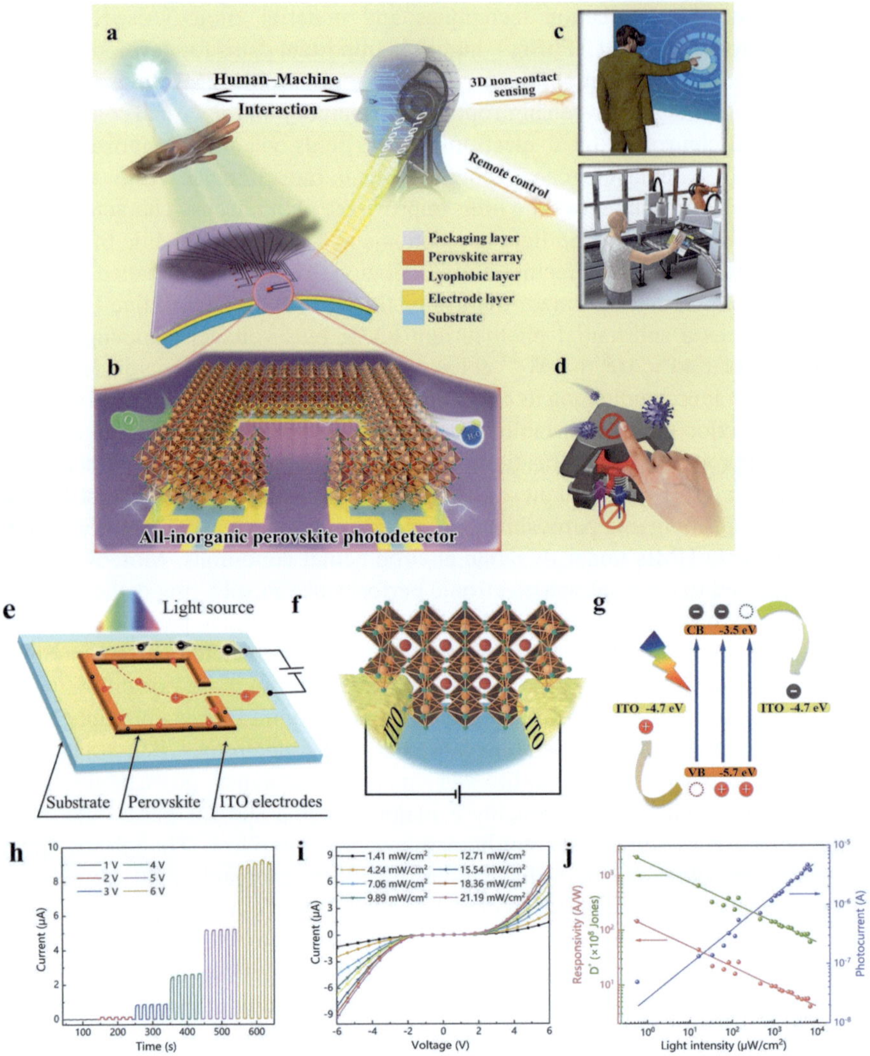


**Fig. 8.8** **a** A schematic representation of the human–machine interaction structure containing CsPbBr3 photodetector array facilitating HMI interaction. The apparatus consists of a packaging layer, a lyophilic layer, a perovskite film array, an electrode layer, and a substrate. **b** Schematic representation of the split-ring structured perovskite photodetector. **c** Applications of devices in human–machine interaction. **d** This robust and sanitary interface contrasts with the mechanical button that harbors viruses. **e** Schematic representation of the lateral configuration and operational mechanism of the photodetection capability of an individual perovskite photodetector on a glass substrate. **f** Schematic representation of the partial magnification of SRP. **g** The alignment of energy bands in structure. **h** Current–time (I–t) curves of the perovskite photodetector at varying bias voltages when exposed to a 450 nm pulsed light source. **i** I–V characteristics of the photodetector when exposed to a 450 nm light source at varying light intensities. **j** The relationship between responsivity, D*, and photocurrent with respect to light intensity at 520 nm. Reproduced from Ref. [73] published by Nano-Micro Letters. Licensed under Creative Commons Attribution 4.0

relevance of these techniques in bridging the gap between high laboratory efficiencies and commercial viability as thoroughly investigated by Ma et al. [77].

***Internal Encapsulation***: This approach targets material level stabilization of the perovskite layer. Techniques including grain boundary passivation and interface engineering are used to reduce degradation paths—including ion migration, phase transitions, and moisture-induced reactions—that arise intrinsically within the perovskite structure. These techniques greatly lower the tendency for structural breakdown under environmental stress by chemically and physically strengthening the perovskite film [78, 79].

***External (Device-Level) Encapsulation***: Complementing internal strategies, external encapsulation consists in building multilayer barrier architectures shielding the whole device from extrinsic environmental aggressors including moisture and oxygen. Usually the first line of protection is a hydrophobic polymer layer—such as fluorinated ethylene propylene, poly(dimethylsiloxane), or parylene-C. Apart from low water vapor permeability, these polymers preserve the mechanical compliance and optical transparency required for flexible and wearable uses. Many times, these polymer layers are enhanced with inorganic barrier films (such those produced by atomic layer deposition) to provide synergistic protection, so ensuring that even if one layer suffers minor breaches, additional layers maintain general device integrity. Atomic layer deposition (ALD) methods have been applied extensively to deposit ultrathin inorganic films including $Al_2O_3$, $TiO_2$, or ZnO outside polymeric barriers. These films provide pinhole-free conformality, building dense, impermeable barriers that drastically reduce moisture and oxygen ingress without adding significant mechanical stiffness. Especially, ALD layers of just 10–30 nm thickness can greatly extend device lifetime while preserving transparency above 85%, which is crucial for transparent electronics, Fig. 8.9 [80, 81].

Emerging as a strong solution for improving the dependability of perovskite-based devices are hybrid encapsulation systems combining flexible organic polymers with inorganic barrier layers. These sophisticated multilayer configurations enable water vapor transmission rates (WVTR) below $10^{-6}$ g $m^{-2}$ $day^{-1}$ by harnessing the mechanical resilience and stretchability of polymers and the extraordinary gas barrier properties of inorganic films. These structures ensure long-term durability under dynamic conditions by deliberately alternating brittle inorganic films with ductile polymer layers, so preventing the propagation of cracks during mechanical deformations—such as bending, stretching, and torsional strain. Notwithstanding these successes, major problems still exist. While the encapsulation must simultaneously provide excellent barrier performance, minimal optical loss, and resistance to mechanical fatigue over extended operation, fabrication of defect-free, large-area encapsulation over flexible substrates at low cost remains challenging. Practical deployment of perovskite devices in real-world applications depends on addressing these challenges. Future work is toward the development of self-healing encapsulation systems, the refinement of vapor-phase infiltration techniques for flexible substrates, and the construction of multifunctional encapsulation layers that not only protect but also provide additional capabilities including integrated sensing and

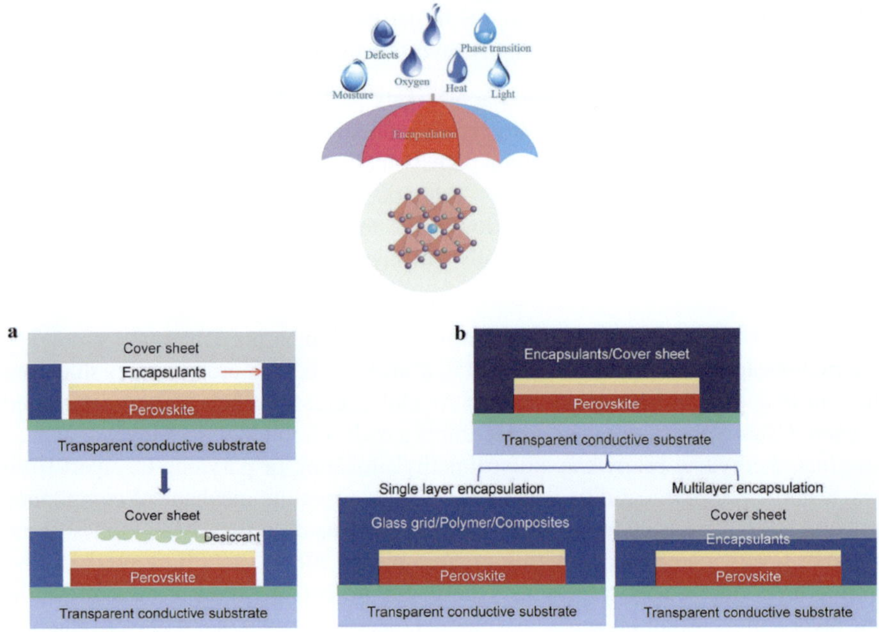


**Fig. 8.9** The external encapsulation structures of PSCs. **a** Gap encapsulation, and **b** Non-gap encapsulation. Reproduced from Ref. [80] published by Matter. Licensed under Creative Commons Attribution 4.0

dynamic transparency adjustment. Furthermore, the combined use of internal passivation and external multilayer encapsulation offers a thorough protection against both intrinsic material instabilities and extrinsic environmental stresses [77].

### *8.3.4 Challenges and Stability Issues*

The effective integration of perovskite materials into deformable and even implantable architectures opens new paths for soft electronics as the demand for intelligent, skin-conformal electronics keeps growing—especially in applications spanning health monitoring, surgical robotics, prosthesis, and interactive systems. Though their use in rigid photovoltaic and photodetection systems is well-known, their integration into mechanically dynamic, skin-compatible systems remains a major materials and engineering difficulty. Chen et al. [82] responded to this demand by introducing a novel dual-mode perovskite–graphic sensor platform that shows the viability of including halide perovskites into soft, low-power systems for HMI. Made entirely using room-temperature printing techniques, this system combines graphene–PDMS-based strain sensors in a parallel planar configuration with methylammonium lead bromide ($MAPbBr_3$) perovskite photodetector microarrays. Addressing important application needs in wearable gesture tracking, touch sensing, and remote interaction systems, the architecture lets one monitor both optical and mechanical stimuli in real-time. This work shows that while preserving high responsiveness and long-term stability, perovskites can be effectively processed into microscale optoelectronic arrays straight onto flexible substrates. Especially, template-guided crystallization helps to overcome a significant processing bottleneck by enabling controlled morphology and low defect states, so enabling deformable electronics. Processed by pneumatic printing, the graphene-based pressure sensors enhance the multifunctionality and bio-compatibility of the integrated device by further complementing the optical sensing with outstanding mechanical compliance and fast response. Halide perovskites have special optoelectronic characteristics including high carrier mobility, long diffusion lengths, and tunable bandgaps that enable smooth interfaces between biological environments and electronic systems, so enabling real-time biosignal detection, energy harvesting, and bidirectional communication. But their use on skin-conformable and implantable platforms presents important problems mostly related to mechanical durability, environmental stability, and biocompatibility [82].

#### 8.3.4.1 Environmental Degradation and Encapsulation Strategies

The sensitivity of perovskite materials to moisture, oxygen, heat, and UV radiation is one of the most important obstacles to their real-world application in wearable and implantable settings. Particularly in lead halide perovskites based on methylammonium-based or formamidinium-based exposure to these elements causes phase segregation, halide migration, and structural degradation. Natural to epidermal and subdermal interfaces, these degradation routes are sped up in humid and warm biological settings. Self-encapsulated designs have lately been presented to solve this restriction. For instance, laminating techniques combining elastomeric and thermoplastic polymer films have been developed to generate perovskite devices with mechanical flexibility and strong diffusion barrier formation against environmental agents. Long-term soak tests show that these encapsulating layers are often multi-layered to decouple mechanical stress from environmental exposure, so preserving device performance after repeated bending or even immersion in biological fluids [83].

#### 8.3.4.2 Mechanical Stability Under Dynamic Deformation

Dynamic mechanical stresses—including torsion, compression, and stretching—are routinely experienced by wearable and implantable electronics. For brittle perovskite films, which are prone to cracking or delaminating under strain, these forces bring a major failure mode. Researchers have created creative ways to improve the mechanical resilience of perovskite layers in order to meet these obstacles. These include mixing polymer precursors, engineering composite structures with nanoconfined grain boundaries, and including mechanically compliant electrode layers including graphene or silver nanowire meshes. These methods preserve the great efficiency of perovskite films and significantly increase their mechanical endurance. Furthermore used to achieve mechanical decoupling between rigid functional units and soft biological surfaces are advanced device architectures including serpentine interconnects, island-bridge configurations, and strain-relieving substrates. By allowing perovskite modules to fit perfectly to curvilinear body regions or internal tissues, these designs minimize mechanical failure and greatly increase the operational lifetime of wearable and implantable devices [84].

#### 8.3.4.3 Toxicity and Biocompatibility

The lead conundrum still a major challenge for in vivo applications is lead-based perovskite potential cytotoxicity. Although the high optoelectronic quality of perovskite films is mostly dependent on the inclusion of lead halides, any mechanical degradation or environmental exposure may cause harmful lead ions to leak into nearby tissues. This risk becomes especially important in implantable systems where extended exposure could cause bioaccumulation and consequent biological interference. Researchers have looked at multi-barrier encapsulation techniques to offset these issues. The method consists in sandwiching the perovskite active layer between several protective films, so shielding the semiconductor from environmental damage and serving as safety barriers to contain heavy metals. Excellent biocompatibility and strong barrier properties make materials including parylene-C, polyurethane, and medical-grade silicones highly sought after. Experiments simulating body fluid conditions, for example, have shown that these encapsulating layers can efficiently stop ion leakage for long running times, so reducing possible cytotoxic effects. Concurrently, research of lead-free perovskite analogues is becoming more and more popular. Studies have concentrated on replacing lead with less dangerous elements including bismuth (Bi) and tin (Sn). Although these substitutes show good optoelectronic characteristics, current versions still find it difficult to match the efficiency and long-term stability of perovskites including lead. With the aim of obtaining ideal biocompatibility without sacrificing functionality, the trade-offs between performance and safety still inspire invention in material synthesis and device engineering. These approaches—robust multi-barrier encapsulation and the creation of lead-free substitutes—are together greatly advancing the discipline of wearable and implantable electronics. Reducing the cytotoxicity and heavy metal release risks helps scientists open the path for safer, more dependable integration of perovskite-based devices into biomedical uses [85–87].

#### 8.3.4.4 Operational Stability and Interface Engineering

Long-term operational reliability is a major factor in both wearable and implantable HMI systems, especially in cases when these devices are subjected to continuous illumination or continuous electrophysiological signal transduction. In such environments, photodetectors and perovskite-based sensors are prone to degradation from intrinsic mechanisms including ion migration under applied electric fields as well as from environmental factors. Over long times, this ion movement can cause hysteresis and signal drift, so undermining device performance [88]. Improving the performance and stability of wearable and implantable human–machine interfaces based on perovskite has turned mostly on interface engineering. Important developments include the use of passivating interlayers acting as both physical and chemical barriers, so limiting undesired ion migration and physically and chemically blocking defect states. Complementing this, graded band alignments—achieved by continuously tuning energy levels across interfaces—and optimized electrode contacts that lower contact resistance have been shown to help to enable smoother charge injection and extraction. These methods taken together reduce the formation of interfacial traps, which cause most of hysteresis and signal drift. Consequently, devices show better signal fidelity and keep steady operation even under continuous, dynamic conditions [89]. Apart from these methods, the combination of ionic liquids and two-dimensional perovskite capping layers has shown great help in extending device lifetime. Ionic liquids can passivate surface states and offer moisture resistance by penetrating the perovskite structure; this is important in physiological conditions when ambient humidity might otherwise hasten degradation [90, 91]. Similar strong surface stabilization provided by 2D perovskite capping layers shields the underlying photoactive material from mechanical and chemical stress [92]. These techniques taken together help preserve performance stability in applications including photoplethysmography, neurostimulation, and bioimpedance spectroscopy, where even small signal fluctuations can greatly affect general device dependability. These interfacial engineering techniques taken together not only improve the lifetime of perovskite-based devices but also guarantee consistent signal integrity essential for accurate and dependable HMI operation in both wearable and implantable environments. These developments offer a route toward the next generation of durable, high-performance perovskite devices for biomedical uses by combining passivity with structural stabilization.

#### 8.3.4.5 System-Level Integration and Reliability Testing

Comprehensive reliability testing procedures must be embraced if perovskite-based HMIs are to move from lab-scale demonstrations to clinical or consumer-grade devices. These comprise in vitro/in vivo biocompatibility tests, mechanical cycling modeled by daily activity, and accelerated aging under physiological settings. Integration with data transmission modules, microcontroller interfaces, and wireless power management units also brings fresh failure modes connected to mechanical coupling and thermal dissipation. Emerging trends point toward system-on-skin or implantable microelectronic systems in which perovskite materials act as active components in adaptive, closed-loop bioelectronic systems rather than only passive sensors or power sources [93, 94]. Self-powered systems are employing perovskite materials not only as high-performance sensors but also as efficient energy harvesters, so defining an expanding frontier in HMIs. These HMIs can independently collect ambient energy—capturing light, mechanical movement, and thermal fluctuations—by combining perovskite-based photovoltaic cells with piezoelectric nanogenerators, so sustaining their operation without the need for external power sources. This creative solution directly meets the increasing need for wearable technologies free of maintenance that are sustainable. The difficulty of juggling mechanical integrity with material performance, user comfort, and environmental robustness forms the central focus of these advances. While new device architectures and system integration strategies guarantee that these materials remain useful even under dynamic, real-world conditions, advances in materials chemistry have produced perovskite formulations that offer enhanced stability and efficiency. Perovskite-enabled HMIs will redefine the benchmarks of intelligent human–machine communication as these multidisciplinary technologies converge, opening the path for a new generation of self-sustaining, flexible wearable devices [53].

## 8.4 Perovskite Piezoelectric-Based Flexible Energy Harvesters

Discovered by the Curie brothers in 1880, the piezoelectric effect is the conversion between mechanical and electrical variables producing charges on opposite surfaces in response to mechanical strains. This phenomenon results from the asymmetry of crystal structures or molecular chains, where electric dipole moments develop in piezoelectric crystals by arrangement of ions with different charges. Emerging to satisfy the demand for small, sustainable, and autonomous biomedical tools are self-powered systems that harness biomechanical energy from the human body. Piezoelectric materials have been extensively applied in sensing, actuation, energy conversion, and catalysis from their discovery. New biopiezoelectric materials for different biomedical uses have been made possible by the recent identification of biocompatible piezoelectric and ferroelectric materials, so transforming mechanical energy into electrical energy in biological systems. Because of their high electromechanical coupling, good biocompatibility, sensitivity to stimuli, physiological response, and mechanical flexibility, piezoelectric materials are absolutely vital for biomedical uses. Among the noteworthy developments are sensors, energy harvesters, biocompatible devices, smart textiles, tissue regeneration, and electronic skins. Four main groups define them: biomolecular piezoelectric materials, biopiezoelectric polymers, biopiezoelectric ceramics, and biopiezoelectric nanomaterials [95]. Researchers first presented piezoelectric nanowires using ZnO, capable of converting mechanical energy into electrical energy with outputs of 2 V and around 100 nA. These materials can generate electricity from various physical motions. However, challenges such as low output currents and complex fabrication methods persisted. Great promise is shown by perovskite-based flexible piezoelectric thin films including barium titanate ($BaTiO_3$), lead zirconate titanate (PZT), and lead magnesium niobate–lead titanate (PMN–PT) due of their exceptional piezoelectric properties. These materials can transform minute biomechanical motions—such as heartbeats and muscle contractions—into electrical energy, so perhaps eliminating the need for large batteries and allowing ongoing physiological monitoring. Because of their high piezoelectric coefficients and ferroelectric character, perovskites appeal especially. For instance, bulk PMN–PT can reach $d_{33}$ values of almost 2500 pC/N, much higher than lower output conventional materials like ZnO nanowires. Although $BaTiO_3$ has less electrical output, its lead-free composition improves biocompatibility, hence it is appropriate for long-term implants. By contrast, in dynamic biomedical environments PZT and PMN–PT are more effective in converting low-amplitude signals. Advancements included the development of flexible $BaTiO_3$ thin-film energy harvesters. A significant breakthrough was achieved with the creation of a flexible single-crystal PMN-PT thin-film energy harvester on a thin plastic substrate, which achieved an electric output current of up to 0.22 mA, significantly higher than previous piezoelectric NGs. Flexibility made possible by innovative fabrication methods has not compromised performance [96–98]. As the mechanism of producing flexible perovskite-based devices has been discussed in the previous section, here we focus on the biomedical applications of perovskite piezoelectric-based flexible energy harvesters.

Beyond energy harvesting, perovskite-based devices have been developed for a range of biomedical applications. Self-powered pacemakers and neurostimulators utilizing PMN–PT-based harvesters have been shown to generate enough electrical output from cardiac and respiratory motions to stimulate tissue function, potentially extending implant longevity and reducing battery replacement needs. Flexible PZT-based harvesters have also been tailored into biomimetic hearing aids that convert acoustic vibrations within the 100–1600 Hz range into electrical signals, offering a pathway for next-generation auditory implants. Additionally, ultra-thin PZT nanoribbons have been employed as nanoscale mechanosensors capable of detecting sub-nanometer deformations in individual cells, paving the way for high-resolution neural interfacing and tissue diagnostics. In following text some mentioned applications will be reviewed.

### *8.4.1 Monitoring of Physical Health and Implantable Medical Devices*

Medical diagnostics and treatment depend on creative energy collecting solutions as energy needs increase. Energy storage is still a difficulty even if implantable biomedical devices can improve patient outcomes and offer important data. Usually requiring surgical replacement upon expiration, implantable medical electronics lack structural flexibility and hence increase the risk of infection. Miniature and intelligent biopiezoelectric biosensors made from biopiezoelectric materials—which can independently generate electricity by converting mechanical energy from body motion—have lately advanced. By means of flexible, wearable health monitoring systems that can be readily attached to the skin or tissues, these self-powered devices have helped to reduce the necessity for intrusive surgeries [99]. Diagnosing chronic diseases depends on monitoring breathing, which is especially emphasized by the need for portable health monitoring solutions driven by the COVID-19 epidemic; yet, current respiratory sensors have major limits including single functions and poor smart signal analysis. Recent attempts concentrate on introducing a low-cost, integrated wearable respiratory monitoring sensor (RMS) system that uses artificial intelligence for diagnosing respiratory abnormalities by measuring multiple parameters, which can precisely track metrics such breathing frequency, apnea hypopnea index (AHI), vital capacity (VC), and peak expiratory flow (PEF) with over 95% accuracy. The data is wirelessly sent to a cloud terminal for thorough study, helping to diagnose disorders including chronic lung diseases and sleep apnea as well as supporting COVID-19 rehabilitation, Fig. 8.10a [100]. Lead-free inorganic perovskites have great energy collecting capacity and have been developed into piezoelectric nanogenerators (PENG). Using polyvinylidene difluoride (PVDF) and a polymer matrix, presynthesized $CsCuCl_3$ particles were composited. With a high sensitivity of 13.8 $kPa^{-1}$ and a good response and recovery time, the composite attained an optimal open-circuit voltage and short-circuit current of 73.2 V and 5.44 μA. With a sensing parameter reaching values as high as 98.7, the device also serves as a piezo-resistive sensor and pressure sensor. Appropriate for uses in human health care monitoring related to self-powered real-time respiratory monitoring systems, the device is efficient in the low-pressure range and can detect various bending states with a sensitivity of 0.5/degree [101].

Building on the progress in wearable sensors for healthcare, another work in this field investigates the development of a piezoelectric nanogenerator addressing the important issue of energy source lifetime. This novel work generates energy using a composite material comprising three-phase double perovskite nanoparticles—i.e., $La_2CuMnO_6$, $La_2CoMnO_6$, and $La_2NiMnO_6$—Polypyrrole, and PVDF. The PENG has possible uses in epidermal devices for skin motion and temperature sensors for medical and robotics applications as well as charges commercial capacitors and provides constant power [102, 103]. Based on a PZT thin film, Park et al. [104] built a self-powered flexible piezoelectric biosensor for real-time arterial pulse monitoring. After being laid on an ultrathin plastic substrate, the biosensor was stably fastened to a human wrist with biocompatible liquid bandage spray. The biosensor could be stably deformed in response to blood vessel movements and shown great mechanical flexibility. Characteristics of the sensor with a sensitivity ($\approx$0.018 $kPa^{-1}$), response time ($\approx$60 ms), and good mechanical stability are revealed by experimental studies. A smartphone could be wirelessly equipped with the produced electrical signals to monitor an arterial pulse in real time. After 5000 cycles the biosensor

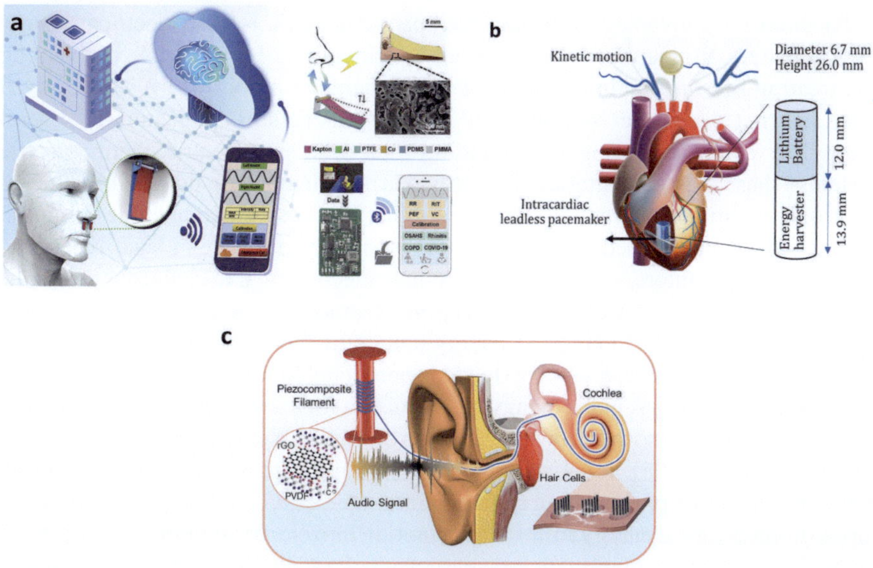


**Fig. 8.10** Schematic overview of recent advances in perovskite-based, self-powered biomedical sensors and auditory prostheses. **a** Integrated piezoelectric nanogenerator (PENG) sensors for portable respiratory and pulse monitoring enable wireless, AI-supported diagnosis of respiratory abnormalities and vital sign tracking. **b** Biocompatible, wearable pacemaker, converting cardiac motion into electric pulses for direct muscle stimulation and wireless transmission of heart health data to mobile devices, and **c** Advanced, fully implantable, miniaturized piezoelectric cochlear implants. From Refs. [100, 105, 110], respectively. Licensed under Creative Commons Attribution 4.0

showed outstanding mechanical stability and accurate piezoelectric response. This self-powered piezoelectric pulse sensor allows in vivo measurement of radial/carotid pulse signals in near-surface arteries. Applied mechanical force to the sensor—that of the faint pressure from an arterial pulse—causes the PZT film to distort. Transient electric potential results from this deformation displacing the positive and negative charge centers inside its non-centrosymmetric crystalline lattice. Under a normal pressure of around 10 kPa, finite element analysis (FEA) carried out in the study revealed that a well-distributed piezopotential of roughly 26.7 V can be produced between adjacent electrodes. This suggests that measurable voltage signals can be produced even from small pressure changes like those generated by a human pulse. The ultrathin design of the sensor guarantees appropriate mechanical coupling with the epidermis by enabling conformally attaching to the skin. This helps it to convert minute dynamic pressures from vocal cord vibrations, breathing, and arterial pulses into electrical outputs. Then, without outside power sources, these self-generated signals can be processed and wirelessly broadcast for real-time health monitoring. This mechanism not only emphasizes the potential of perovskite-based piezoelectric sensors in producing really autonomous, self-powered biomedical devices but also their efficiency in gathering energy from biomechanical motions [104].

Pacemakers, implanted electronic therapeutic devices, release electrical energy from a battery to treat heart dysfunction, so stimulating heart muscle. Preventing and treating life-threatening heart diseases has helped millions of lives to be saved since the first pacemaker was invented in 1958. Successful application of several energy harvesting technologies—including solar cells, pyroelectric, thermoelectric systems—has charged batteries and provided supplementary power for medical electronic equipment. Development of a simple, low-cost, biocompatible structural design is desperately needed to increase the output performance of lead-free piezoelectric nanogenerators, so advancing pacemaker technology. In canine models, self-powered pacemakers have shown to capture biomechanical energy from cardiac pulsations, so directly linking the energy source to the heart for efficient cardiac pacing, Fig. 8.10b. These devices' extended operational lifetime and decrease of high-risk repeated operations make them very promising for clinical use. To increase their clinical use, though, thorough investigation on their pacing efficiency at different sites is still necessary [105]. High stretchability, waterproofness, and biocompatibility define an all-in-one piezoelectric nanogenerator (A-PENG), which efficiently transforms biomechanical energy from cardiac motion into electrical output. By means of optimized distribution of piezoelectric particles in the A-PENG matrix, stress transfer and electrical generation are significantly enhanced—achieving maximum open-circuit voltage and short-circuit current values almost three times higher than those of conventional piezocomposites. Reliable pacing at several key cardiac sites (left ventricle, His bundle, right atrium) was obtained by the A-PENG-based pacemaker under canine models. For future clinical translation, this shows the viability of multifunctional, self-powered pacemakers including strong diagnostic and therapeutic capacity [106].

Conductive and sensorineural hearing loss are the two primary forms of auditory impairment, together affecting nearly 360 million people globally and representing a significant public health concern. Although traditional acoustic hearing aids are still the accepted intervention, they are sometimes constrained by disadvantages including sound distortion, feedback noise, ear canal discomfort, limited amplification, and related social stigma. Bypassing the ear canal, implantable middle ear hearing devices (IMEHDs) have been developed to address these issues and so reduce acoustic feedback and improve sound fidelity and user comfort by delivering mechanical stimulation directly to auditory structures—such as the ossicles, tympanic membrane, or round window, Fig. 8.10c. Among these, the incus-body driving type has been investigated extensively since the 1990s since it least disturbs middle ear anatomy and has limited impact on residual high-frequency hearing [107]. Turner et al. [108] reviewed material selection criteria for ultrasonic-powered implants, stressing in particular that miniaturized PZT actuators are feasible for middle-ear and intracochlear deployment and provide better energy conversion efficiencies. Implying the natural auditory transduction path, implanted piezoelectric transducers can capture ossicular vibrations or basilar membrane motion unlike external microphones. For uses like single-unit cochlear implants, these systems allow completely embedded auditory feedback and eliminate mechanical coupling loss or signal degradation resulting from ambient noise and body motion. The field is now moving toward totally implantable hearing systems (TIHS), which physiologically interact with the auditory system and remove all external hardware completely. By combining internal sound sensing, neural stimulation, and energy autonomy, these next-generation devices seek to transform auditory prosthesis. Advanced perovskite-structured piezoelectric materials—especially lead zirconate titanate (PZT), known for their exceptional electromechanical coupling—are enabling this innovation. Because they depend on outside microphones and power sources, traditional middle ear and cochlear implants are vulnerable to mechanical failure and aesthetic restrictions. Conversely, developing technologies harvest mechanical energy from tympanic membrane and ossicular vibrations using micromachined piezoelectric transducers implanted in the middle or inner ear, then translate this energy into electrical signals for auditory stimulation. Reviewed by Latif et al. [109], the main candidate for energy harvesting and sensing in such systems is PZT, with its remarkable piezoelectric coefficient (up to 374 pC/N), electromechanical coupling, and mechanical robustness. Based on microelectromechanical systems (MEMS) design concepts, their thorough study divides these into two main modules: energy harvesters and implantable microphones. PZT is a member of the $ABO_3$ perovskite family and has ferroelectric crystal structure that can be poled to produce strong electrical responses under mechanical pressure. Particularly in two configurations, these features make PZT perfect for miniaturized, MEMS-compatible acoustic transducers: One uses Cantilever Beams with Proof Masses. Designed to match human speech, these convert vibrational energy into alternating current at designated frequencies (e.g., 589 Hz), so producing voltages up to 1.5 V RMS [109]. And second Diaphragm Structure: With dual-electrode configurations to increase sensitivity, thin-film PZT diaphragms can operate as directional microphones. These devices not only detect acoustic energy with great accuracy but also act as power sources producing densities of ~1–1.5 $mW/cm^1$, which is enough to drive signal amplifiers and neural stimulators [110].

Conventional PZT has disadvantages despite its utility including lead toxicity, which calls for biocompatible encapsulation, and high processing temperatures (~1200 °C), so limiting compatibility with soft polymers or CMOS circuitry. These constraints have spurred research on lead-free substitutes or PZT modification. For instance, Xiang et al. [111] presented microcone-array flexible piezoelectric acoustic sensors based on multicomponent lead-free perovskites, so providing enhanced mechanical compliance and frequency selectivity—qualities especially important for biomimetic hearing. Designed to interface directly with cochlear tissues, Mokhtari et al. [110] developed self-powered nanostructured piezoelectric filaments made of PVDF blended with barium titanate ($BaTiO_3$) or reduced graphene oxide (rGO). These flexible filaments essentially replicate the operation of natural hair cells by converting sound-induced mechanical vibrations into electrical signals, so enabling high-fidelity acoustic capture without external power sources. Along with outstanding cytocompatibility, these materials provide a scalable and biocompatible platform for the development of next-generation, fully implantable, energy-autonomous cochlear implants. Their sensitivity of up to 6.22.2 mV $Pa^{-1}$ $cm^{-1}$ and peak voltages around 7.5 mV provide this basis. Furthermore emphasized by

Danti and Berrettini [112] are how material encapsulation and nanoscale structure might greatly increase the sensitivity and long-term stability of piezoelectric systems, so improving their fit with biological environments. Especially, nanocomposites combining polymers like PVDF-TrFE with ceramics like $BaTiO_3$ or $LiNbO_3$ solve important limits of conventional PZT-based cochlear implants, including suboptimal sensitivity and biocompatibility. By means of enhanced dipole alignment and in situ poling, electrospun nanofibers show improved piezoelectric performance—achieving voltages up to 90'mV—while maintaining mechanical flexibility required to interface with delicate cochlear structures. Their high surface area allows close contact with spiral ganglion neurons, so lowering stimulation thresholds and ensuring long-term biointegration by including nanoceramics, which has anti-inflammatory and cytoprotective effects. A significant progress in auditory neuroprosthesis, this nanotechnology-enabled approach opens the path for self-powered, totally implantable, and tissue-compliant cochlear prosthesis.

### 8.4.2 Disease Diagnosis and Treatment

Chemosensors are analytical tools used through artificial recognition elements to detect and measure chemical compounds or biological structures and macromolecules. Their great sensitivity and specificity help them to be applied in many disciplines, including environmental monitoring, medical diagnostics, and industrial process control. On the other hand, biosensors identify biological molecules by means of biological recognition elements including enzymes, antibodies, or nucleic acids. By turning biological reactions into electrical signals, these devices help to precisely measure molecules such as glucose, infections, or DNA sequences. Piezoelectric chemosensors and biosensors use the piezoelectric effect to sense changes in mass or mechanical characteristics upon target analyte interaction. Although alternative configurations can also be used, they provide the possibility for label-free assays, so enabling direct monitoring of analyte presence. In piezoelectric chemosensors, the change in resonant frequency of the piezoelectric material follows from the binding of the analyte to the chemical recognition element. Analogous effects are achieved by piezoelectric biosensors using biological recognition elements, so allowing sensitive detection of biological molecules. With the option for label-free assay design, the piezoelectric platform improves electrochemical and optical assays significantly above conventional assays requiring chemical reagents, Fig. 8.11 [113]. Emerging as transforming candidates for noninvasive, self-powered disease diagnostic platforms are perovskite materials, well-known for their adjustable crystal lattices, strong electromechanical coupling, and robust piezoelectric properties. Using the unique features of perovskite-based solar cells to produce effective, small energy harvesters, self-powered perovskite biosensors mark a novel development in wearable electronics. Because of their adjustable bandgaps, high absorption coefficients, and long carrier diffusion lengths—which let them operate well even under low-light conditions typical of indoor environments—halide perovskites, including $MAPbI_3$ and

$FAPbBr_3$, are especially effective. Recent developments in device architecture—including compositional engineering and quasi-2D layers—have produced output power densities of up to 3.3 mW for small areas and power conversion efficiencies exceeding 26%. With their photoresponse impacted by biological interactions, PSCs can serve both as active sensing materials and power sources in the framework of biosensing. This capacity lets PSCs be integrated with different sensors, including ones tracking temperature and sweat, so enabling autonomous data collecting and transmission in real-time. Practical application of these technologies, however, is hampered by material instability brought on by moisture and light exposure, toxicity from lead-based components, and scalable manufacturing challenges. Researchers are investigating ways to get beyond these challenges including scalable deposition methods, lead-free perovskite substitutes, and encapsulation with hydrophobic materials. Combining PSCs with other energy harvesting technologies—such as triboelectric nanogenerators or thermoelectric devices—may also produce hybrid systems improving operational stability and power delivery [114]. Real-time monitoring and early disease diagnosis are made possible by these sensors transforming minute biomechanical signals into electrical outputs. High sensitivity, self-powered operation, flexibility, conformability, and real-time monitoring and therapeutics are just a few of the benefits perovskite piezoelectric sensors present for clinical use. They can track dynamic physiological events acting as indirect markers of disease states and identify early-stage changes in physiological parameters linked with cardiovascular, respiratory, or neurological diseases. In neurological diseases, minute changes in tissue elasticity or blood flow dynamics can be recorded and provide important diagnostic and prognostic information. Integration with smart systems, hybrid device architectures, point-of-care and wearable applications, and advanced fabrication technologies constitute future directions for improving the impact of perovskite piezoelectric sensors in disease diagnosis. Combining sensor arrays with wireless communication modules, microcontrollers, and machine learning algorithms yields completely integrated diagnostic platforms that monitor physiological signals in real time and offer predictive analytics for early disease detection. Combining perovskite piezoelectrics with other nanomaterials could enhance mechanical robustness and signal fidelity as well as provide multi-modal sensing capabilities integrating both electrochemical and piezoelectric responses [115, 116].

Because of their special electromechanical coupling response, piezoelectric smart biomaterials have shown amazing ability in diagnosing and treating different disorders. These materials are self-powered biomedical devices since they can be used as wireless therapeutic approaches with ultrasonic stimulus. Personalized healthcare is using them for several therapeutic diagnosis including physiological signals, viruses, pathogens, and diseases like cancer and neurodegenerative ones. Offering noninvasive treatments for many diseases, piezoelectric nanoparticles combined with ultrasonic waves have shown great promise in sonodynamic therapy and piezocatalytic therapeutics. Piezoelectric energy generators included into implantable medical devices can enhance physiological performance and power cardiac pacemakers [117]. Despite these promising advances, several challenges impede the

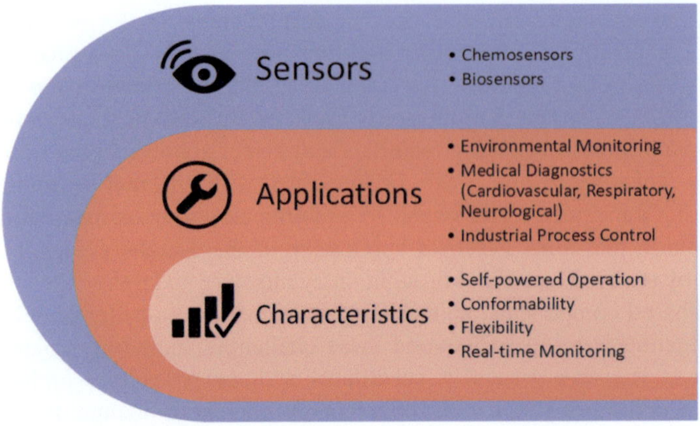


**Fig. 8.11** Sensors: types, applications, and characteristics

transition from laboratory prototypes to widespread clinical use. Lead toxicity associated with PZT and PMN–PT remains a significant concern, necessitating effective encapsulation strategies or a switch to lead-free alternatives like modified $BaTiO_3$. Moreover, the complex, multi-step fabrication processes—such as Inorganic-based Laser Lift-Off (ILLO), mechanical exfoliation, and wet etching—can limit scalability and reproducibility. Long-term mechanical fatigue, cyclic durability, and stability under the dynamic and chemically active conditions of the human body also require further investigation. Looking ahead, strategies such as additive manufacturing, hybrid organic–inorganic composites, and the integration of these materials into wireless telemetry and closed-loop biosensing systems offer promising avenues to overcome these hurdles.

## 8.5 Summary

X-ray detection and human–machine interfaces (HMIs) have been much changed by the developments in flexible perovskite technologies and ionizing radiation detection. Using perovskite materials has improved X-ray detector performance by raising sensitivity, spatial resolution, and detection limits as well as by addressing stability and biotoxicity issues. Furthermore, included are wearable devices based on highly sensitive and flexible perovskite-based technologies. Moreover, the special qualities of perovskite materials have helped to advance HMIs and enable the construction of flexible interfaces improving user interaction in many fields, including augmented reality and healthcare. Studies on gesture recognition and brain-computer interfaces show how capable these technologies are to change behavior and improve communication access. Notwithstanding good progress, questions about the environmental and health effects of perovskite materials still exist. Lead-free substitutes and efficient encapsulating techniques' dependability and safety of use will help them to be constantly developed. Artificial intelligence and flexible device engineering combined with perovskite material science puts these materials front and first in future radiation detection and human–machine interfaces. Prospects for flexible perovskite technologies in wearable electronics and human–machine interfaces are bright thanks to continuous research and invention targeted at addressing present challenges and unlocking new opportunities for intelligent, sustainable, and biocompatible electronic systems.

## References


1. L. Cai et al., Osseosurface electronics-thin, wireless, battery-free and multimodal musculoskeletal biointerfaces. Nat. Commun. **12**(1), 6707 (2021). https://doi.org/10.1038/s41467-021-27003-2
2. P.A. Thwaites et al., Review article: current status and future directions of ingestible electronic devices in gastroenterology. Aliment Pharmacol. Ther. **59**(4), 459–474 (2024). https://doi.org/10.1111/apt.17844
3. J.V. Vaghasiya, C.C. Mayorga-Martinez, M. Pumera, Wearable sensors for telehealth based on emerging materials and nanoarchitectonics. npj Flex. Electron. **7**(1), 26 (2023). https://doi.org/10.1038/s41528-023-00261-4
4. J. Min et al., An autonomous wearable biosensor powered by a perovskite solar cell. Nat. Electron. **6**(8), 630–641 (2023). https://doi.org/10.1038/s41928-023-00996-y
5. A. Lazaro, R. Villarino, M. Lazaro, N. Canellas, B. Prieto-Simon, D. Girbau, *Recent Advances in Batteryless NFC Sensors for Chemical Sensing and Biosensing* (2023). https://doi.org/10.3390/bios13080775
6. K.J. Riisnaes et al., 2D hybrid perovskite sensors for environmental and healthcare monitoring. ACS Appl. Mater. Interfaces **16**(24), 31399–31406 (2024). https://doi.org/10.1021/acsami.4c02966
7. Y. Tang et al., Enabling low-drift flexible perovskite photodetectors by electrical modulation for wearable health monitoring and weak light imaging. Nat. Commun. **14**(1), 4961 (2023). https://doi.org/10.1038/s41467-023-40711-1
8. Y. Zheng et al., Highly sensitive perovskite photoplethysmography sensor for blood glucose sensing using machine learning techniques. Adv. Sci. **11**(43), 2405681 (2024). https://doi.org/10.1002/advs.202405681
9. Z. Xu, Y. Hao, A. Luo, Y. Jiang, Technologies and applications in wireless biosensors for real-time health monitoring. Med-X **2**(1), 24 (2024). https://doi.org/10.1007/s44258-024-00041-3
10. Y. Liao et al., Ambient nano RF-Energy driven self-powered wearable multimodal real-time health monitoring. Nano Energy **128**, 109915 (2024). https://doi.org/10.1016/j.nanoen.2024.109915
11. M. Yang, Z. Ye, Y. Ren, M. Farhat, P.Y. Chen, Recent advances in nanomaterials used for wearable electronics. Micromachines (Basel) **14**(3) (2023). https://doi.org/10.3390/mi14030603
12. H. Wu, Y. Ge, G. Niu, J. Tang, Metal halide perovskites for x-ray detection and imaging. Matter **4**(1), 144–163 (2021). https://doi.org/10.1016/j.matt.2020.11.015
13. B. Wang et al., Flexible perovskite scintillators and detectors for X-ray detection. iScience **25**(12), 105593 (2022). https://doi.org/10.1016/j.isci.2022.105593

14. F. Liu et al., Recent progress in halide perovskite radiation detectors for gamma-ray spectroscopy. ACS Energy Lett. **7**(3), 1066–1085 (2022). https://doi.org/10.1021/acsenergylett.2c00031
15. A. Kyung-Chan et al., A dynamic gesture recognition algorithm using single halide perovskite photovoltaic cell for human-machine interaction, in *2024 International Conference on Electronics, Information, and Communication (ICEIC)* (2024), pp. 1–4. https://doi.org/10.1109/ICEIC61013.2024.10457182
16. G.-T. Hwang et al., Self-powered cardiac pacemaker enabled by flexible single crystalline PMN-PT piezoelectric energy harvester. Adv. Mater. **26**(28), 4880–4887 (2014). https://doi.org/10.1002/adma.201400562
17. J. Tang, Y. Li, Y. Yu, Q. Hu, W. Du, D. Lin, *Recent Progress in Flexible Piezoelectric Tactile Sensors: Materials, Structures, Fabrication, and Application.* (2025). https://doi.org/10.3390/s25030964
18. J.A. Seibert, J.M. Boone, X-ray imaging physics for nuclear medicine technologists. Part 2: x-ray interactions and image formation. J. Nucl. Med. Technol. **33**(1), 3 (2005). http://tech.snmjournals.org/content/33/1/3.abstract
19. X. Geng et al., Lead-free halide perovskites for direct x-ray detectors. Adv. Sci. **10**(23), 2300256 (2023). https://doi.org/10.1002/advs.202300256
20. A. Datta, Z. Zhong, S. Motakef, A new generation of direct X-ray detectors for medical and synchrotron imaging applications. Sci. Rep. **10**(1), 20097 (2020). https://doi.org/10.1038/s41598-020-76647-5
21. C.F. Sabottke, B.M. Spieler, The effect of image resolution on deep learning in radiography. Radiol. Artif. Intell. **2**(1), e190015 (2020). https://doi.org/10.1148/ryai.2019190015
22. E. Kraft et al., Counting and integrating readout for direct conversion x-ray imaging: concept, realization and first prototype measurements. IEEE Trans. Nucl. Sci. **54**(2), 383–390 (2007). https://doi.org/10.1109/TNS.2007.891571
23. M. Sytnyk, S. Deumel, S.F. Tedde, G.J. Matt, W. Heiss, A perspective on the bright future of metal halide perovskites for X-ray detection. Appl. Phys. Lett. **115**(19) (2019). https://doi.org/10.1063/1.5125999
24. W. Li et al., Arising 2D perovskites for ionizing radiation detection. Adv. Mater. **36**(26), 2309588 (2024). https://doi.org/10.1002/adma.202309588
25. L. Pan, S. Shrestha, N. Taylor, W. Nie, L.R. Cao, Determination of X-ray detection limit and applications in perovskite X-ray detectors. Nat. Commun. **12**(1), 5258 (2021). https://doi.org/10.1038/s41467-021-25648-7
26. P. Jin et al., Realizing nearly-zero dark current and ultrahigh signal-to-noise ratio perovskite X-ray detector and image array by dark-current-shunting strategy. Nat. Commun. **14**(1), 626 (2023). https://doi.org/10.1038/s41467-023-36313-6
27. Y. Wu, J. Feng, Z. Yang, Y. Liu, S. Liu, Halide perovskite: a promising candidate for next-generation x-ray detectors. Adv. Sci. **10**(1), 2205536 (2023). https://doi.org/10.1002/advs.202205536
28. S. Tie et al., High-performance lead-free Cs3Bi2IxBr 9–x perovskite x-ray detector based on anionic engineering. J. Phys. Chem. C (2025). https://doi.org/10.1021/acs.jpcc.4c07924
29. A. Datta, J. Fiala, S. Motakef, 2D perovskite-based high spatial resolution X-ray detectors. Sci. Rep. **11**(1), 22897 (2021). https://doi.org/10.1038/s41598-021-02378-w
30. G.-H. Dun et al., Wafer-scale photolithography-pixeled Pb-free perovskite x-ray detectors. ACS Nano **16**(7), 10199–10208 (2022). https://doi.org/10.1021/acsnano.2c01074
31. W. Zhu et al., Low-dose real-time x-ray imaging with nontoxic double perovskite scintillators. Light Sci. Appl. **9**(1), 112 (2020). https://doi.org/10.1038/s41377-020-00353-0
32. X. Liu et al., Biocompatible metal-free perovskite membranes for wearable x-ray detectors. ACS Appl. Mater. Interfaces **16**(13), 16300–16308 (2024). https://doi.org/10.1021/acsami.4c01069
33. H. Li et al., Wearable photoferroelectric perovskite x-ray detectors. Adv. Funct. Mater. **34**(46), 2407693 (2024). https://doi.org/10.1002/adfm.202407693

34. Y. Lei et al., Tailored molecular for ultra-stability and biocompatible pseudohalide metal-free perovskite towards X-ray detectors with record sensitivity. npj Flex Electron. **8**(1), 46 (2024). https://doi.org/10.1038/s41528-024-00330-2
35. Z. Li et al., PF6−pseudohalides anion based metal-free perovskite single crystal for stable x-ray detector to attain record sensitivity. Adv. Mater. **35**(25), 2300480 (2023). https://doi.org/10.1002/adma.202300480
36. H. Li et al., Flexible large-scale self-driven perovskite x-ray detector by precise heterogeneous integration. ACS Energy Lett. **9**(1), 64–74 (2024). https://doi.org/10.1021/acsenergylett.3c02152
37. I. Hirata et al., Sensing technologies for extravasation detection: a review. ACS Sens. **8**(3), 1017–1032 (2023). https://doi.org/10.1021/acssensors.2c02602
38. W. Zhang, H. Wang, Z. Chen, Y. Shao, Advancing the commercialization of perovskite-based radiation detectors for high-resolution imaging. Laser Photon. Rev. 2400470. https://doi.org/10.1002/lpor.202400470
39. Y.C. Kim et al., Printable organometallic perovskite enables large-area, low-dose X-ray imaging. Nature **550**(7674), 87–91 (2017). https://doi.org/10.1038/nature24032
40. Y. Shao, Z. Xiao, C. Bi, Y. Yuan, J. Huang, Origin and elimination of photocurrent hysteresis by fullerene passivation in CH3NH3PbI3 planar heterojunction solar cells. Nat. Commun. **5**(1), 5784 (2014). https://doi.org/10.1038/ncomms6784
41. L. Basiricò et al., Wearable perovskite films for on-line monitoring of radiotracers in nuclear medicine. Adv. Mater. Technol. 2401111 (2024). https://doi.org/10.1002/admt.202401111
42. A. Maity, S. Siraj, A.K. Raychaudhuri, A. Saha, B.D. Ghosh, Low power paper electronics based wearable radiation detector using hybrid halide perovskite (MAPbBr3): a real time monitoring of gamma ray. Flex. Print. Electron. **8** (2023)
43. A. Ullah, D.Y. Kim, S.I. Lim, H.-R. Lim, Hydrogel-based biointerfaces: recent advances, challenges, and future directions in human–machine integration. (2025). https://doi.org/10.3390/gels11040232
44. M.T. Islam, K. Sepanloo, S. Woo, S.H. Woo, Y.-J. Son, A review of the Industry 4.0 to 5.0 transition: exploring the intersection, challenges, and opportunities of technology and human–machine collaboration. (2025). https://doi.org/10.3390/machines13040267
45. P. Kulkarni, B. Pandiya, Human-machine interface in the industry 5.0 workplace, in *Expanding Operations Through Agile Principles and Sustainable Practices*, ed. by D. Thangam, H. Chittoo, I. Ghosal, R. Kandasamy, J.Y. Park (IGI Global, Hershey, PA, USA, 2025), pp. 341–366. https://doi.org/10.4018/979-8-3693-6274-7.ch018
46. V. Dakulagi, K.H. Yeap, H. Nisar, R. Dakulagi, G.N. Basavaraj, M.V. Galindo, Chapter 4—An overview of techniques and best practices to create intuitive and user-friendly human-machine interfaces, in *Artificial Intelligence and Multimodal Signal Processing in Human-Machine Interaction*, ed. by A. Subasi, S. M. Qaisar, H. Nisar (Academic Press, 2025), pp. 63–77. https://doi.org/10.1016/B978-0-443-29150-0.00002-0
47. Q. He, P. Wang, T. Peng, B. Wang, R. Tang, Human digital twin in human-centric smart manufacturing, in *Human-Centric Smart Manufacturing Towards Industry 5.0*, ed. by B. Wang, P. Zheng, L. Wang, D. Mourtzis (Springer Nature Switzerland, Cham, 2025), pp. 59–76. https://doi.org/10.1007/978-3-031-82170-7_3
48. J. Guo, Y. Du, Z. Wang, D. Wei, A paradigm shift from traditional non-contact sensors to tele-perception. J. Mater. Chem. A **13**(13), 8939–8967 (2025). https://doi.org/10.1039/D4TA09222H
49. A. Pandey, J. Joseph, Human-machine interaction: ethical and legal considerations, in *Exploring AI Implications on Law, Governance, and Industry*, ed. by S. Chakraborty (IGI Global, Hershey, PA, USA, 2025), pp. 179–202. https://doi.org/10.4018/979-8-3373-3384-7.ch007
50. S. Mahmud, X. Lin, J.H. Kim, Interface for human machine interaction for assistant devices: a review, in *2020 10th Annual Computing and Communication Workshop and Conference (CCWC)* (2020), pp. 768–773. https://doi.org/10.1109/CCWC47524.2020.9031244

51. C. Han, X. Dong, W. Zhang, X. Huang, L. Gong, C. Su, Intelligent systems for inorganic nanomaterial synthesis. (2025). https://doi.org/10.3390/nano15080631
52. Y. Su, D. Yin, X. Zhao, T. Hu, L. Liu, Exploration of advanced applications of triboelectric nanogenerator-based self-powered sensors in the era of artificial intelligence. (2025). https://doi.org/10.3390/s25082520
53. H.-S. Wu et al., Metal-free perovskite piezoelectric nanogenerators for human-machine interfaces and self-powered electrical stimulation applications. Adv. Sci. **9**(18), 2105974 (2022). https://doi.org/10.1002/advs.202105974
54. S. Aftab et al., Flexible perovskite solar cells: a revolutionary approach for wearable electronics and sensors. Mater. Today Energy **51**, 101872 (2025). https://doi.org/10.1016/j.mtener.2025.101872
55. F. Chun, F. Wang, Recent advances in perovskite-based flexible electroluminescent devices. ACS Nano **18**(38), 25939–25965 (2024). https://doi.org/10.1021/acsnano.4c06587
56. G. Li et al., Recent progress in self-healing triboelectric nanogenerators for artificial skins. (2025). https://doi.org/10.3390/bios15010037
57. S. Duan et al., Flexible mechano-optical dual-responsive perovskite molecular ferroelectric composites for advanced anticounterfeiting and encryption. Sci. Adv. **10**(48), eadr2886 (2025). https://doi.org/10.1126/sciadv.adr2886
58. H.-T. Zhang et al., Perovskite neural trees. Nat. Commun. **11**(1), 2245 (2020). https://doi.org/10.1038/s41467-020-16105-y
59. H. Zhao, J. Chen, H.-T. Zhang, Perovskite nickelate ionotronics for AI and brain-machine interfaces. Mater. Lab **1**(2), 220038 (2022). https://doi.org/10.54227/mlab.20220038
60. B. Wang, J. Zhao, F. Ni, L. Qiu, X. Wang, Intelligent gesture recognition gloves for real-time monitoring in wireless human-computer interaction. ACS Appl. Mater. Interfaces **17**(1), 790–798 (2025). https://doi.org/10.1021/acsami.4c20660
61. T. Wu et al., Wearable perovskite-based shadow recognition sensor for ambient and nonobtrusive human-computer interaction. Adv. Intell. Syst. **5**(1), 2200307 (2023). https://doi.org/10.1002/aisy.202200307
62. Z. Li et al., Hyperbranched polymer functionalized flexible perovskite solar cells with mechanical robustness and reduced lead leakage. Nat. Commun. **14**(1), 6451 (2023). https://doi.org/10.1038/s41467-023-41931-1
63. Z. Li et al., Boosting mechanical durability under high humidity by bioinspired multisite polymer for high-efficiency flexible perovskite solar cells. Nat. Commun. **16**(1), 1771 (2025). https://doi.org/10.1038/s41467-025-57102-3
64. D. Zheng et al., Ultra-flexible pixelated perovskite photodetectors enabled by honeycomb polymer grids for high-resolution imaging. Adv. Mater. 2415068 (2025). https://doi.org/10.1002/adma.202415068
65. Y. Yu, F. Zhang, H. Yu, Self-healing perovskite solar cells. Sol. Energy **209**, 408–414 (2020). https://doi.org/10.1016/j.solener.2020.09.018
66. B.P. Finkenauer et al., Mechanically robust and self-healable perovskite solar cells. Cell Rep. Phys. Sci. **2**(2), 100320 (2021). https://doi.org/10.1016/j.xcrp.2020.100320
67. Y. Zhou, X. Luo, J. Yang, Q. Qiu, T. Xie, T. Liang, Application of quantum dot interface modification layer in perovskite solar cells: progress and perspectives. (2022). https://doi.org/10.3390/nano12122102
68. P. Tipparak et al., The impact of MAPbI3 quantum dots on CsFA perovskite solar cells: interface and hole extraction improvement. ACS Appl. Energy Mater. **8**(1), 355–365 (2025). https://doi.org/10.1021/acsaem.4c02474
69. A. Naz et al., Cutting-edge perovskite-based flexible pressure sensors made possible by piezoelectric innovation. (2024). https://doi.org/10.3390/ma17174196
70. H. Yin, Y. Guan, Y. Li, Z. Zheng, Y. Guo, Modulation of piezoelectricity and mechanical strength in piezoelectric composites based on N0.5B0.51T-BNT nanocubes towards human-machine interfaces. Nano Energy **118**, 109044 (2023). https://doi.org/10.1016/j.nanoen.2023.109044

71. S. Sharma, S. Shriwastava, S. Kumar, K. Bhatt, C.C. Tripathi, Alternative transparent conducting electrode materials for flexible optoelectronic devices. Opto-Electron. Rev. **26**(3), 223–235 (2018). https://doi.org/10.1016/j.opelre.2018.06.004
72. Q. Chen et al., Under the spotlight: the organic–inorganic hybrid halide perovskite for optoelectronic applications. Nano Today **10**(3), 355–396 (2015). https://doi.org/10.1016/j.nantod.2015.04.009
73. B. Shi et al., Split-ring structured all-inorganic perovskite photodetector arrays for masterly internet of things. Nano-Micro Lett. **15**(1), 3 (2022). https://doi.org/10.1007/s40820-022-00961-y
74. Y. Zhao, K. Zhu, Organic–inorganic hybrid lead halide perovskites for optoelectronic and electronic applications. Chem. Soc. Rev. **45**(3), 655–689 (2016). https://doi.org/10.1039/C4CS00458B
75. G. Niu, X. Guo, L. Wang, Review of recent progress in chemical stability of perovskite solar cells. J. Mater. Chem. A **3**(17), 8970–8980 (2015). https://doi.org/10.1039/C4TA04994B
76. S.-M. Yi, I.-S. Choi, B.-J. Kim, Y.-C. Joo, Reliability issues and solutions in flexible electronics under mechanical fatigue. Electron. Mater. Lett. **14**(4), 387–404 (2018). https://doi.org/10.1007/s13391-018-0043-0
77. S. Ma, G. Yuan, Y. Zhang, N. Yang, Y. Li, Q. Chen, Development of encapsulation strategies towards the commercialization of perovskite solar cells. Energy Environ. Sci. **15**(1), 13–55 (2022). https://doi.org/10.1039/D1EE02882K
78. W. Li et al., Internal encapsulation enables efficient and stable perovskite solar cells. Adv. Funct. Mater. **35**(4), 2414004 (2025). https://doi.org/10.1002/adfm.202414004
79. D. Xu et al., An internal encapsulating layer for efficient, stable, repairable and low-lead-leakage perovskite solar cells. Energy Environ. Sci. **15**(9), 3891–3900 (2022). https://doi.org/10.1039/D2EE01016J
80. Q.-Q. Chu et al., Encapsulation: the path to commercialization of stable perovskite solar cells. Matter **6**(11), 3838–3863 (2023). https://doi.org/10.1016/j.matt.2023.08.016
81. H.H. Park, D.J. Fermin, Recent developments in atomic layer deposition of functional overlayers in perovskite solar cells. (2023). https://doi.org/10.3390/nano13243112
82. S. Chen et al., Printing perovskite and graphene parallel structure-based optical-mechanical sensors for human-machine interaction. Sci. China Technol. Sci. **68**(2), 1220201 (2025). https://doi.org/10.1007/s11431-024-2816-6
83. D. Wu et al., Self-encapsulated wearable perovskite photovoltaics via lamination process and its biomedical application. iScience **26**(7), 107248 (2023). https://doi.org/10.1016/j.isci.2023.107248
84. R. Zhao, Z. Gu, P. Li, Y. Zhang, Y. Song, Flexible and wearable optoelectronic devices based on perovskites. Adv. Mater. Technol. **7**(3), 2101124 (2022). https://doi.org/10.1002/admt.202101124
85. G. Schileo, G. Grancini, Lead or no lead? Availability, toxicity, sustainability and environmental impact of lead-free perovskite solar cells. J. Mater. Chem. C **9**(1), 67–76 (2021). https://doi.org/10.1039/D0TC04552G
86. F. Yang et al., Lead-free perovskites: growth, properties, and applications. Sci. China Mater. **64**(12), 2889–2914 (2021). https://doi.org/10.1007/s40843-021-1755-4
87. M. Wang et al., Lead-free perovskite materials for solar cells. Nano-Micro Lett. **13**(1), 62 (2021). https://doi.org/10.1007/s40820-020-00578-z
88. T. Lu, S. Ji, W. Jin, Q. Yang, Q. Luo, T.-L. Ren, Biocompatible and long-term monitoring strategies of wearable, ingestible and implantable biosensors: reform the next generation healthcare. (2023). https://doi.org/10.3390/s23062991
89. V. Abhinav et al., Advancements in wearable and implantable BioMEMS devices: transforming healthcare through technology. (2025). https://doi.org/10.3390/mi16050522
90. N. Sa, M. Wu, H.-Q. Wang, Review of the role of ionic liquids in two-dimensional materials. Front. Phys. **18**(4), 43601 (2023). https://doi.org/10.1007/s11467-023-1258-6
91. H. Peng et al., Ionic liquid assisted imprint for efficient and stable quasi-2D perovskite solar cells with controlled phase distribution. Nano-Micro Lett. **15**(1), 91 (2023). https://doi.org/10.1007/s40820-023-01076-8

92. J. Yuan, Y. Jiang, T. He, G. Shi, Z. Fan, M. Yuan, Two-dimensional perovskite capping layer for stable and efficient tin-lead perovskite solar cells. Sci. China Chem. **62**(5), 629–636 (2019). https://doi.org/10.1007/s11426-018-9436-1
93. J. Gong, T. Xu, Perovskite materials in biomedical applications, in *Revolution of Perovskite: Synthesis, Properties and Applications*, ed. by N.S. Arul, V.D. Nithya (Springer Singapore, Singapore, 2020), pp. 95–116. https://doi.org/10.1007/978-981-15-1267-4_4
94. S. Baumann et al., Stability and reliability of perovskite containing solar cells and modules: degradation mechanisms and mitigation strategies. Energy Environ. Sci. **17**(20), 7566–7599 (2024). https://doi.org/10.1039/D4EE01898B
95. Q. Xu et al., Construction of bio-piezoelectric platforms: from structures and synthesis to applications. Adv. Mater. **33**(27), 2008452 (2021). https://doi.org/10.1002/adma.202008452
96. Z.L. Wang, J. Song, Piezoelectric nanogenerators based on zinc oxide nanowire arrays. Science (80-) **312**(5771), 242–246 (2006). https://doi.org/10.1126/science.1124005
97. K.-I. Park et al., Piezoelectric BaTiO3 thin film nanogenerator on plastic substrates. Nano Lett. **10**(12), 4939–4943 (2010). https://doi.org/10.1021/nl102959k
98. G.-T. Hwang, M. Byun, C.K. Jeong, K.J. Lee, Flexible piezoelectric thin-film energy harvesters and nanosensors for biomedical applications. Adv. Healthc. Mater. **4**(5), 646–658 (2015). https://doi.org/10.1002/adhm.201400642
99. T.L. Shi et al., Global trends and hot topics in clinical applications of perovskite materials: a bibliometric analysis. Biomater. Transl. **4**(3), 131–141 (2023). https://doi.org/10.12336/biomatertransl.2023.03.002
100. Y. Li et al., Integrated wearable smart sensor system for real-time multi-parameter respiration health monitoring. Cell Rep. Phys. Sci. **4**(1) (2023). https://doi.org/10.1016/j.xcrp.2022.101191
101. S. Mondal, S. Poddar, S. Bhattacharjee, S. Maiti, A. Banerjee, K.K. Chattopadhyay, Lead free halide perovskite embedded PVDF based efficient mechanical energy harvester: self-driven respiratory sensor. Nano Energy **115**, 108689 (2023). https://doi.org/10.1016/j.nanoen.2023.108689
102. T. Das, M. Kumar Yadav, A. Dev, M. Kar, Double perovskite-based wearable ternary nanocomposite piezoelectric nanogenerator for self-charging, human health monitoring and temperature sensor. Chem. Eng. J. **496**, 153926 (2024). https://doi.org/10.1016/j.cej.2024.153926
103. K.K. Das, R. Pandey, A.K. Dubey, Piezo-electronics: a paradigm for self-powered bioelectronics. Biomaterials **318**, 123118 (2025). https://doi.org/10.1016/j.biomaterials.2025.123118
104. D.Y. Park et al., Self-powered real-time arterial pulse monitoring using ultrathin epidermal piezoelectric sensors. Adv. Mater. **29**(37), 1702308 (2017). https://doi.org/10.1002/adma.201702308
105. M. Khazaee, S. Riahi, A. Rezania, Conceptual piezoelectric-based energy harvester from in vivo heartbeats' cyclic kinetic motion for leadless intracardiac pacemakers. (2024). https://doi.org/10.3390/mi15091133
106. Y. Zhang et al., Self-powered pacemaker based on all-in-one flexible piezoelectric nanogenerator. Nano Energy **99**, 107420 (2022). https://doi.org/10.1016/j.nanoen.2022.107420
107. H. Liu et al., Concept and evaluation of a new piezoelectric transducer for an implantable middle ear hearing device. Sensors (Basel) **17**(11) (2017). https://doi.org/10.3390/s17112515
108. B.L. Turner et al., Ultrasound-powered implants: a critical review of piezoelectric material selection and applications. Adv. Healthc. Mater. **10**(17), 2100986 (2021). https://doi.org/10.1002/adhm.202100986
109. R. Latif, M.M. Noor, J. Yunas, A.A. Hamzah, Mechanical energy sensing and harvesting in micromachined polymer-based piezoelectric transducers for fully implanted hearing systems: a review. Polymers **13**(14) (2021). https://doi.org/10.3390/polym13142276
110. F. Mokhtari et al., Self-powered nanostructured piezoelectric filaments as advanced transducers for new cochlear implants. Energy Environ. Mater. **8**(1), e12807 (2025). https://doi.org/10.1002/eem2.12807

111. Z. Xiang et al., High-performance microcone-array flexible piezoelectric acoustic sensor based on multicomponent lead-free perovskite rods. Matter **6**(2), 554–569 (2023). https://doi.org/10.1016/j.matt.2022.11.023
112. S. Danti, S. Berrettini, New perspectives for piezoelectric material-based cochlear implants: getting to nano. Front. Mater. **11** (2024). https://www.frontiersin.org/journals/materials/articles/10.3389/fmats.2024.1362141
113. M. Pohanka, Piezoelectric chemosensors and biosensors in medical diagnostics. (2025). https://doi.org/10.3390/bios15030197
114. Q. Li et al., All-in-one self-powered wearable biosensors systems. Mater. Sci. Eng. R Rep. **163**, 100934 (2025). https://doi.org/10.1016/j.mser.2025.100934
115. L. Durai, S. Badhulika, Current challenges and developments in perovskite-based electrochemical biosensors for effective theragnostics of neurological disorders. ACS Omega **7**(44), 39491–39497 (2022). https://doi.org/10.1021/acsomega.2c05591
116. H.-S. Wu, B.T. Murti, J. Singh, P.-K. Yang, M.-L. Tsai, Prospects of metal-free perovskites for piezoelectric applications. Adv. Sci. **9**(12), 2104703 (2022). https://doi.org/10.1002/advs.202104703
117. P. Tripathi, A.K. Dubey, Role of piezoelectricity in disease diagnosis and treatment: a review. ACS Biomater. Sci. Eng. **10**(10), 6061–6077 (2024). https://doi.org/10.1021/acsbiomaterials.4c01346

# Chapter 9
# Applications of Perovskite Wearables in the Outfit Industry

Recent progress in textile engineering and materials science is facilitating the fabrication of smart clothes. By using fibers or sheets filled with photoluminescent perovskite composite, displays and environmental warnings might be able to produce light with minimal energy consumption. Physiological and mechanical data could be tracked by real-time sensor networks, combined with AI systems that analyze real-time health data and provide feedback. Moreover, modern humidity-responsive color-changing camouflage materials have been designed using nanoparticle-based multi-spectral concealment systems. Accurately mapping sports biomechanics and impact forces is possible using perovskite-doped piezoelectric composites and conductive nanowire networks. Then, using this high-resolution data, machine-learning algorithms can find ways to prevent injuries and offer personalized coaching. Multifunctional textiles have made a big difference in modern functional and health clothing. These textiles are changing from barriers that don't move to interfaces that can use energy, track biological parameters, and provide information.

## 9.1 Introduction

The ageing population and increasing healthcare expenses globally are driving demand for new wearable technology that can support the elderly and improve public health results. Traditional centralized healthcare systems may not always allow for continuous monitoring because of the need on large, immobile instruments and regular doctor visits. "Smart clothing," which transforms ordinary clothing into data-driven health interfaces controllable by the user's fingertips, has been a game-changing solution. These garments combine wearable sensors, artificial intelligence, and high-tech fabrics to transform the way we track our fitness, health, and environment. Smart textiles allow for continuous, real-time monitoring of physiological conditions by converting signals produced by the human body—biomechanical,



M. Darab et al., *Perovskites for Next-Generation Wearable Technology*, Engineering Materials, https://doi.org/10.1007/978-3-032-03931-6_9

biochemical, and thermal energy—into electrical outputs. This change in perspective opens the door for healthcare to move from a disease-centric, reactive model to a wellness- and prevention-centric, tailored one [1]. Reflective of these developments, the smart clothing market is thriving. Market projections say that rising consumer interest in health and changing demographics will drive the value from $3.5 billion in 2022 to more than $10 billion by 2026. Smart textiles are set to be a basic component of next-generation wearable healthcare solutions with the mix of artificial intelligence, IoT connectivity, and multifunctional fabric design, even if material processing, scalability, and energy efficiency remain problems [2]. The next great advance in the field's fast expansion is increasing smart textiles' multifunctionality, user acceptability, and wearability over time. Advancements in textile engineering have led to the creation of conductive, flexible fabrics that maintain sensor efficacy even after normal washing and use. By combining actuators, computers, power sources, and sensors onto flexible substrates, these systems let real-time physiological monitoring, motion detection, and temperature control [3]. AI also enhances these systems by enabling real-time individualized feedback and predictive analytics; new technologies like energy-harvesting textiles, safe data protocols, and self-healing fabrics are pushing the boundaries of technology as well [4]. Researchers are looking at several energy-harvesting technologies—including piezoelectric mechanisms, triboelectric, magnetoelectric, and electrochemical techniques—to create fully autonomous biomonitoring systems. Recent developments have concentrated on including piezoelectric fibres into garments, therefore enabling autonomous health monitoring without requiring any outside power sources. Piezoelectric polymers like PVDF and ceramic composites like barium titanate have shown to be especially well-suited because of their adaptability, biocompatibility, and ability to convert mechanical energy into electrical current. Made using techniques including weaving, knitting, braiding, printing, and electrospinning, these fibers are quite adaptable, elastic, and breathable, making them ideal for daily use, Fig. 9.1 [5].

A mentioned above, fully autonomous wearables have been significantly hampered by consistent on-body power supply. Many studies have concentrated on identifying techniques to include solar energy collecting into textile structures as a remedy for this issue. Solar E-textiles use the broad availability of solar energy and the advancement of photovoltaic (PV) technology to offer continuous power without the need for cumbersome external batteries. Innovations in the field of solar energy include flexible solar panel attachment, photovoltaic yarn embedding, organic, dye-sensitized, and perovskite solar cell direct printing or lamination onto fabric substrates. Especially in terms of power conversion efficiency, perovskite solar cells are quite promising; yet, they still have some of the flexibility and washability required for use in wearable technology [3]. Mechanical durability and comfort are improving from developments in graphene, carbon-based fibers, and printable conductive inks. Researches aim to create clothing with sensing, processing, and power producing ability for smart interfaces. Future e-textiles must nevertheless overcome challenges including secure data processing, dermal compatibility, effective power management, and launderability [6]. A new family of perovskite-based

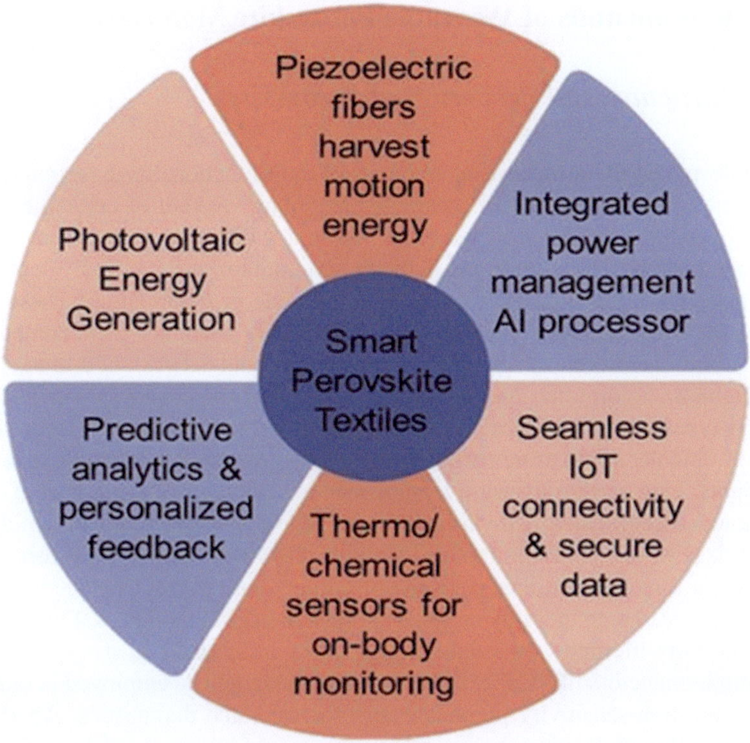


**Fig. 9.1** Schematic overview of a perovskite-based smart textile system for autonomous sensing, energy harvesting and data analytics

functional compounds has developed to satisfy the growing need for energy and health solutions in wearable electronics and the fast development of smart textiles. Their unusual optoelectronic properties are opening the door for wearable technologies of the future, which could run in solar, piezoelectric, and thermoelectric domains. Perovskites offer an attractive mix of efficiency, flexibility, and control to satisfy the industry's need for high-performance materials that could be used with porous, flexible substrates. Here we will discuss the structural characteristics of perovskites, their application in textiles, the importance of these materials in energy harvesting, the aesthetic fashion uses, the relevance of these materials in biological sensing, and the challenges related to their widespread use.

## 9.2 Fundamentals of Wearable Perovskite Materials

### 9.2.1 Structure and Properties of Perovskites

Perovskite materials' structural tunability and multifunctionality have made them very sought after in the field of wearable electronics. A kind of crystalline material, perovskites have the stoichiometric formula $ABX_3$. 'A' in this formula is a monovalent cation (like caesium or methylammonium), 'B' is a divalent metal cation (like $Pb^{2+}$, $Sn^{2+}$, or $Ti^{4+}$), and 'X' is a halide or oxide anion. These materials could show tuneable optoelectronic features by changing their composition and structure, therefore enabling their three-dimensional network of corner-sharing $BX_6$ octahedra holding the A-site cation in the interstitial areas [7]. The two most common types are perovskites composed of halides and oxides. The evolution of wearable optoelectronic systems depends on the direct bandgaps, high absorption coefficients, long carrier diffusion lengths, and excellent charge transport of halide perovskites—specifically methylammonium lead iodide ($MAPbI_3$), formamidinium lead iodide ($FAPbI_3$), and caesium lead bromide ($CsPbBr_3$). Structurally, these materials form a three-dimensional network of corner-sharing $PbX_6$ octahedra, which allows efficient charge mobility and strong light-matter interaction. $CsPbBr_3$, which shows mobility-lifetime ($\mu\tau$) values similar to or even better than those of traditional semiconductors like CdTe, is one material that could be employed in radiation sensors and high-sensitivity photodetectors for wearable diagnostics. Apart from their main solar application, halide perovskites could be improved in their piezoelectric behavior by changing their composition and using polymer matrices [8, 9]. This dual capacity, which permits the concurrent collecting of energy from mechanical and solar sources, greatly helps autonomous wearable systems. Low-temperature solution-based processing techniques such as spin-coating, inkjet printing, and vapor-phase deposition make it possible to include these materials into flexible substrates such PET and polyimide as well as deformable textile matrices suitable for conformal, on-body applications. Though their toxicity and environmental instability render them inappropriate for commercial use, lead halide perovskites have much promise. Still unresolved are ongoing problems with ion migration, photo-induced degradation, and phase separation brought on by light and moisture. Though mixed-cation and mixed-halide formulations, several other approaches that have somewhat improved stability, and encapsulation techniques have advanced, concerns about long-term use remain. Lead-free substitutes have been created to fit these limitations; among them are tin-based halide perovskites (Sn-PVKs) as $MASnI_4$, $CsSnI_4$, and $FASnI_4$. While providing a more environmentally friendly degradation route, these compounds keep desirable optoelectronic qualities—including low exciton binding energies, high absorption across the visible and near-infrared spectrum, and narrow direct bandgaps (1.1–1.4 eV). Self-doping effects and high defect densities caused by the chemical instability of $Sn^{3+}$, which readily oxidises to $Sn^{4+}$, however, degrade the performance of the device. Lattice stability and photocarrier dynamics have been enhanced by recent developments including $SnF_2$ doping and A-site cation careful tuning such

$FA^+$. Even under ambient conditions, the orthorhombic structure of $FASnI_3$, its high dielectric constant, and low effective carrier mass allow for efficient charge generation and transport. Furthermore, tin-based perovskites are perfect for use in thermoelectric wearable devices since they have a tuneable electrical conductivity and extremely low thermal conductivity. Though not as efficient as lead-based devices, Sn-PVKs are significant step toward flexible and multifunctional perovskite-based wearables. Their future success depends even more on material stability, defect passivation, and scalable textile platform integration as well as other development [10].

Oxide perovskites—multifunctional materials usually with the $ABO_3$ structure exhibiting ferroelectric, piezoelectric, and pyroelectric properties—will drive the next generation of wearable electronics. Compounds like barium titanate ($BaTiO_3$) and potassium niobate ($KNbO_3$) spontaneously polarise under the B-site cation's (e.g., $Ti^{4+}$, $Nb^{5+}$) displacement within the oxygen octahedral framework. Under various stimuli, this allows these compounds to have strong electromechanical coupling and stable sensing capabilities. Though their wide bandgaps (>3 eV) limit solar performance, oxide perovskites' remarkable thermal and chemical durability makes them ideal for smart fabrics and wearable sensing systems requiring long-term use [11]. Their remanent polarisation, chemical inertness, and high Curie temperatures greatly benefit mechanical and thermal sensing applications including pressure mapping and temperature-regulated feedback systems. Historically, oxide perovskites' inherent fragility and high temperature manufacturing needs have made flexible electronics impractical. Low-temperature deposition and composite structural design—which entails including perovskite oxide nanoparticles into elastomeric polymer matrices—have greatly improved their washability and mechanical compliance [12]. These techniques are now enabling the manufacture of stretchable films with ferroelectric and piezoelectric properties that endure multiple bending and washing cycles. Simultaneously, the production of nanostructures such quantum dots, nanowires, and nanosheets has significantly enhanced the mechanical stress adaptability and fracture endurance of perovskite oxide materials. Sustaining electromechanical responsiveness in textile-integrated systems that are always changing, this control over shape increases the likelihood of totally embedded oxide perovskite sensors in garments. When viewed as a whole, these developments show that oxide perovskites are the way to go for textile substrate integration wearable applications, long-term accurate sensing, and environmentally friendly. Wearable systems that are self-powered, sturdy, and high-performance may be built on top of their structural adaptability and multifunctionality [13].

### *9.2.2 Photovoltaic, Thermoelectric, and Piezoelectric Functionalities of Wearable Perovskite*

The electrical and structural characteristics of perovskite materials are quite varied, making them perfect for use as the basis of wearable devices able to power themselves. Their extremely high absorption coefficients and direct, controllable bandgaps make them perfect for efficient solar conversion in both indoor and outdoor environments. At the same time, certain perovskite compositions let small skin-to-air temperature gradients effectively harvest thermoelectric energy due to their tunable Seebeck coefficients and naturally low heat conductivity. Many perovskite structures have significant piezoelectric reactions in addition to heat and light because of their non-centrosymmetric crystal lattices. These structures could turn electrical outputs from mechanical deformations such bending, stretching, and impact. Low-temperature, solution-based processing compatible with flexible substrates and textile integration allows one material platform to scavenge and fuse multiple ambient energy sources. This renders all three modalities available. Combining the three main energy sources—light, heat, and motion—wearable electronics of the future will be able to operate autonomously. This multifunctional synergy enhances dependability, reduces device footprints, and greatly simplifies system topologies.

#### 9.2.2.1 Photovoltaic Functionality

The rapid expansion of wearable electronics is driving an urgent need for efficient, lightweight, mechanically and physiologically safe energy solutions. Though they have a high energy density, conventional power sources such pouch cells made of lithium-polymer or lithium-ion have significant disadvantages when used in wearable applications. These batteries have major safety and dependability problems since mechanical deformation during wear can cause fractures, electrode delamination, and toxic electrolyte leaking. Wearable systems aren't as convenient or autonomous as they may be because of their limited battery life, which demands continuous charging. These issues are driving some scientists to investigate alternative energy sources including zinc-ion batteries, which are naturally safer and less flammable, and perovskite solar cells, which have a high power-to-weight ratio and can be used with flexible substrates. Collecting energy from natural events including heat gradients, biomechanical motion, radiofrequency waves, and solar radiation could help wearables to maintain themselves, Fig. 9.2a [14]. Among them, photovoltaic energy collecting stands out for its scalability, operational independence from user mobility, and high power density. Unlike kinetic or thermoelectric harvesting methods that rely on body movement or a significant skin-to-air temperature gradient, photovoltaic systems can continuously generate electricity from ambient light, inside or outside, independent of user activity. Because they can deliver power densities over 100 mW/cm$^2$ when exposed to direct sunlight and 10–100 mW/cm$^2$ when lit by artificial light, PV cells are a dependable power source for low-power wearable sensors. The total

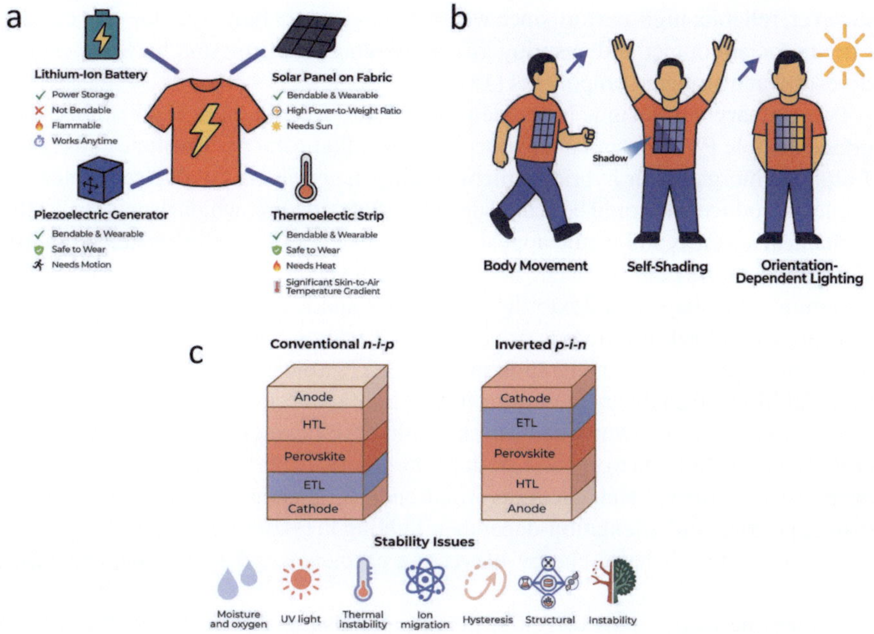


**Fig. 9.2** **a** Different energy sources for wearable technology: a comparison of rigid lithium-ion batteries with flexible solar, piezoelectric, and thermoelectric systems; **b** The significant impact of body movement and environmental factors on the energy efficiency of perovskite solar cells; **c** The architectural and stability challenges of perovskite solar cells

power output of a wearable PV system can be estimated using the Eq. (9.1):

$$\text{Energy Output (Wh)} = \text{Power (W)} \times \text{Average Hours of Light} \quad (9.1)$$

This numerical analysis highlights the relative scalability and predictability of PV systems in contrast to other approaches like piezoelectric, electromagnetic, or triboelectric generators, which frequently necessitate user participation and generally produce lower energy output, such as 45.95 mW when walking and 76.25 mW when jogging. Thermoelectric generators can also convert temperature gradients into electricity; however, they only generate roughly 10 $\mu W/cm^2$ and are very reliant on outside influences, which aren't always suitable for wearable uses. Likewise, despite their theoretical promise, biofuel-based power systems present issues with scalability, integration, and cost-effectiveness as well. Although photovoltaic technology offers some hope for powering wearable electronics, including it into textile-based platforms presents several design and technical challenges to be addressed. These challenges are fundamentally caused by the need to include into a single system parts with highly different physical, optoelectronic, and electrochemical characteristics.

However, reliable, high-performance wearable devices are built by achieving consistent interfacial contact and ensuring efficient voltage transmission between energy harvesting and storage components [15].

Two primary PV designs have been looked into for wearable applications: flat-surface flexible PV cells and fiber-based PV cells. Flat solar photovoltaic cells made of organic, inorganic, or hybrid semiconducting materials on flexible substrates are simple to produce and could be scaled up. Their flat structure, which usually lacks the mechanical softness and tactile appeal of traditional textiles, unfortunately compromises user comfort and garment integration. Conversely, fiber-based photovoltaic cells mimic the shape and flexibility of conventional textiles. Photovoltaic components are embedded into fibrous or yarn-like structures to create these cells. These structures attract the next generation of electronic textiles as they enhance mechanical flexibility and enable better absorption of light scattering. Still in its early stages of development, this method has certain limits in efficiency, scalability, and durability, however. Applied to apparel, both PV systems have the same integration issues regardless of format. Mechanical deformation, self-shading caused by body movement or posture, and orientation-dependent lighting in textile-embedded PV systems (Fig. 9.2b) could all harm energy harvesting efficiency and system dependability [16].

Among the many kinds of PSCs, flexible perovskite solar cells (F-PSCs) have risen to the front due to their remarkable adaptability and possible applications. The certified power conversion efficiency (PCE) of perovskite solar cells (PSCs) has surged from a mere 4% to an impressive 26.1% after extensive research and development. Recent developments have enabled F-PSCs to show durability—they keep 95% of their original efficiency even after 1000 bending cycles—and extraordinary efficiencies—21.73%. Haldide-based perovskites—including $MAPbI_3$, $FAPbI_3$, and $CsPbBr_3$—have a strong track record of generating great solar energy. Two distinct cell configurations—i.e., p-i-n and n-i-p—are created depending on the material and carrier transport channel employed. n-i-p designs that permit layer-by-layer deposition are built on anode, electron transport layer (ETL), light-harvesting perovskite ($FAPbI_3$, $MAPbI_3$), hole transport layer (HTL), and conducting substrates. Unique to planar cells, the inverted (p-i-n) layout is feasible for both mesoporous and regular planar perovskite solar cells (Fig. 9.2c). PSC stability problems are caused by internal and external environmental elements such humidity, oxygen content, light exposure, and thermal stability; one such problem is deterioration under moisture and oxygen. Although it has no immediate impact on perovskite, moisture does accelerate its decay. Although conventional Pb-based perovskites in oxygen have been shown to be durable, exposure to UV light causes the surface-adsorbed oxygen in the $TiO_2$ charge-transport layer to desorb, therefore generating electronic trap sites and lowering PSC efficiency. Constant irradiation-induced illumination causes phase separation, ion movement, halide segregation, and strain building, all of which significantly hinder the commercialization of PSCs [12]. PSC stability depends on not only material degradation but also major issues regarding interface and device design. By changing both the electron and hole transport layers to lower trap state density and prevent moisture penetration, complementary approaches including interface engineering have proven successful in addressing these issues. By means of effective moisture barriers created by the integration of two-dimensional (2D) perovskite layers at these interfaces, defect development is minimized and charge extraction efficiency is enhanced, so stabilizing the active layer. Phase segregation and ion migration have been significantly lowered by means of controlled annealing and compositional modification, so optimizing the perovskite crystallization process. More dependable device performance under prolonged light and thermal stress results from this control of crystallization lowering of fault concentrations and improving phase stability. Moreover, advanced encapsulation techniques based on barrier materials that offer robust protection against oxygen and moisture and allow efficient thermal control improve PSCs' long-term running lifetime [17]. One can consider the integration of noncovalent supramolecular interactions, including hydrogen bonding, π–π stacking, ionic interactions, and host–guest chemistry, together with self-healing polymers that enable intrinsic repair via reversible covalent bonds—such as disulfide exchange and Diels–Alder reactions—as a safe and consistent solution. These techniques allow the effective passivation of grain boundary defects, inhibition of ion migration, and restoration of mechanical integrity post-damage, so enhancing the optoelectronic performance and operational lifetime of PSC devices. Furthermore, the advanced polymer doping and surface passivation methods improving the crystalline quality and flexibility of the perovskite layer could finally offer higher power conversion efficiencies and extend device lifetime under useful conditions [18].

Modern methodologies such as the piezo-phototronic effect link the optical and electrical aspects of optoelectronic devices by mechanical stimulation, so improving their performance. This effect is based on structural strain-induced polarization charges generated by mechanical deformation of piezoelectric semiconductors. These charges could affect the energy band structure at important interfaces such p-n junctions and metal–semiconductor contacts, therefore increasing separation, transport, and reduction of recombination of photogenerated charge carriers. Devices using the piezo-phototronic effect—including solar cells, light-emission intensity (LEDs), and photodetectors—greatly enhance power conversion efficiency, and photoresponsivity. Materials with natural piezoelectricity and exceptional electron transport qualities, such as zinc oxide (ZnO), are especially promising in this respect. ZnO is widely used in PSCs as an ETL. As mentioned above, inherent defects and its basic surface chemistry cause the perovskite layer to deteriorate and the stability of the device to suffer. Though they have little impact on long-term stability, traditional techniques such metal-ion doping and surface passivation have been employed to get around these limitations. Building a core–shell nanostructure is a better approach. Recent studies indicate that growing ZnO nanorods on flexible substrates and then treating them with sulfide to produce a ZnS shell could greatly increase stability and PCE. The ZnS shell, as a passivation layer, increases charge transport efficiency at interfaces and lowers defect-induced recombination. Static tensile strain applied to such core–shell structures increases the energy band alignment by generating piezo-polarization charges at the ZnS/ZnO and ZnS/perovskite interfaces. In situations where wearable electronics are beneficial, this dual-modulation significantly enhances device performance [19].

#### 9.2.2.2 Thermoelectric Functionality

The idea of thermoelectric generators (TEGs), which use solid-state processes to convert heat into electricity, is an interesting new advancement in renewable energy. Ecologically friendly, they require little maintenance, run long, and create no gas emissions. There are three main types of TEG materials: ceramics with high melting points, like $BaUO_3$; semiconductors with medium melting points, like $Bi_2Te_{2.85}Se_{0.15}$; and low-melting point, or "low-temperature," substitutes, like $Bi_2Te_3$ and polymers, such as 3,4-ethylenedioxythiophene. These materials' thermoelectric efficiency is measured using the dimensionless figure of merit, ZT (Fig. 9.3a). The Eq. (9.2) shows the governing elements of the thermoelectric effect in perovskites: Seebeck coefficient (S), electrical conductivity ($\sigma$), and thermal conductivity ($\kappa$).

$$ZT = \frac{S^2 \sigma T}{\kappa} \tag{9.2}$$

Wearable heat-harvesting gadgets have to be made of low κ and high S materials. Achieving high ZT values, especially at low temperatures, is not simple. Traditionally, enhancing a material's ZT has called for altering its crystalline structure, which could lower its thermal conductivity if foreign species are introduced or if low thermal conductivity values are purposefully designed into the material. To rival conventional generators, a ZT more than three is necessary. Particularly at low temperatures, no such substance has been found despite theoretical limitations. Up to now, the largest ZT found has been 2.4 at 300 K. The development method to attain more general relevance depends on high ZT at a low cost [20]. Wearable thermoelectric generators convert body heat into electrical energy by means of a strong temperature gradient; yet, their performance is sometimes reduced by appreciable heat loss at the device-environment boundary. An impactful enhancement technique utilizes inorganic phase-change materials (IPCMs), such as calcium chloride hexahydrate ($CaCl_2 \cdot 6H_2O$), as thermal reservoirs. These IPCMs improve the temperature gradient across the thermoelectric components by using their significant latent heat around

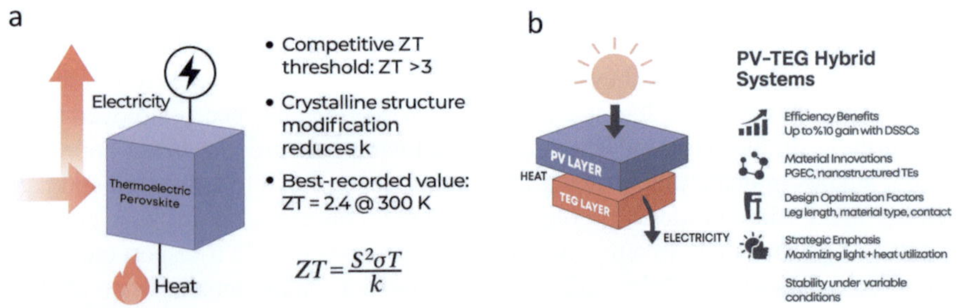


**Fig. 9.3** Illustration of **a** a perovskite thermoelectric energy converter, and **b** a hybrid PV-TEG system, showcasing innovative energy conversion methods that enhance efficiency through advanced materials and structural optimizations

ambient temperature to absorb extra thermal energy without appreciable temperature increase. The composite heat sink is built with additives—such as emulsifiers and nucleating agents—and fortified with expanded graphite powder, which increases thermal conductivity and stabilizes the material's structure. This helps to mitigate problems including supercooling, phase separation, and leakage connected with IPCMs. By precisely changing important geometric aspects, wearable thermoelectric generators could have their whole design enhanced. Appropriate dimensions for thermoelectric legs, fill factor, and IPCM particle shape (height and width) are determined using finite element simulations. This specific structural change lowers parasitic heat losses and directs more body heat towards the thermoelectric devices, so preserving a constant temperature gradient necessary for an enhanced Seebeck coefficient and electrical conductivity. Using materials like stretchy silicone or low thermal conductivity foams, or complementing flexible encapsulation methods, greatly reduce heat leakage while ensuring mechanical durability and adaptability over movement [21]. Recent research on wearable energy-harvesting devices has demonstrated that effective temperature control under sun exposure is absolutely crucial. A promising approach uses wavelength-selective metamaterial films designed to demonstrate low absorptivity in the solar spectrum (approximately 0.3–2.5 μm) while ensuring high emissivity in the atmospheric infrared window (8–13 μm), so reducing undesired heat gain on the cold side and facilitating efficient radiative cooling that maintains significant temperature gradients essential for optimal performance. Using dynamic covalent polymer networks as both the substrate and encapsulation, a new modular design approach has been developed that integrates self-healing, recyclable, and reconfigurable components while integrated conductive paths enable quick restoration of mechanical integrity and electrical connectivity and mechanical integrity after damage. By means of exact spectral management and sophisticated material design, wearable thermoelectric systems with remarkable durability, flexibility, and efficiency are generated, so enabling the development of next-generation energy-harvesting electronics maintaining performance across a range of environmental conditions [22].

Combining photovoltaic cells with thermoelectric generators allows the hybrid photovoltaic-thermoelectric generator (PV-TEG) to maximize solar energy use. By using the whole electromagnetic spectrum and concurrently absorbing visible and UV photons while converting infrared heat into electricity, the combination of perovskite solar cells with thermoelectric generators facilitates effective energy conversion (Fig. 9.3b). By means of the photoelectric effect, conventional photovoltaic solar cells transform visible and UV light into energy, so raising electrons from the valence band to the conduction band. Particularly in the infrared spectrum, a lot of solar radiation is lost in conventional photovoltaic systems and dissipated as heat, so causing efficiency losses from higher cell temperatures. Using the Seebeck effect to convert otherwise lost thermal energy into additional electrical power effectively reduces this constraint in a tandem system. The method depends on temperature differences; hence, heat collecting at the rear of the PSC generates a thermal gradient across the TEG module. Usually comprising alternating p-type and n-type semiconductor junctions, the thermoelectric material generates an electric potential as charge carriers migrate from the hot side to the cold side. This approach lowers heat accumulation in the PSC simultaneously increasing general power output, so providing more stable running conditions [23]. Apart from increasing energy output, PSC-TEG hybrid systems significantly enhance thermal management, which is crucial for lifetime and maintenance of efficiency. Higher temperatures in PSCs cause material degradation and reduced charge carrier mobility, so limiting long-term efficacy. By converting heat into practical power, the linked TEG extends the operational lifetime of the solar module and helps to stabilize its temperature. Multi-stage TEG systems improve this effect by increasing the temperature difference, so raising the power generation. Moreover, including advanced cooling techniques such water-cooled plates guarantees consistent temperature control, so stabilizing the performance of the tandem system and improving general efficiency [23, 24]. Particularly with large perovskite solar modules (PSM), it is shown that expanding PV-TEG hybrid systems is feasible outside of experimental settings. Where smallness and high power density are crucial, the change from single PSC topologies to PSM-TEG systems allows practical uses in wearable electronics and portable energy solutions. Studies show that these hybrid systems could reach up to 30% energy conversion efficiencies above those of single-sided PSCs. Furthermore, in places with high solar radiation, such as southern Europe, the thermal energy available for TEG conversion may reach 380 $W/m^2$, so indicating the great possibilities for the generation of auxiliary power [24].

Many hybrid PV-TEG designs have been developed to maximize this interaction. A 10% increase in total conversion efficiency was observed following the integration of dye-sensitized solar cells (DSSCs) with TEG modules, therefore demonstrating the potential of combining different solar technologies with thermoelectric components. These developments are dependent on optimizing TEG design parameters—including leg length, material type, and contact resistance—since they ensure good thermal-to-electric conversion without compromising PV performance. Making these hybrid systems even more efficient depends on developing more complex thermoelectric materials. Innovations such nanostructured thermoelectrics and phonon glass-electron crystal (PGEC) materials are being explored to minimize heat conductivity and boost the Seebeck coefficient, hence maximizing the thermoelectric figure of merit (ZT). Conversely, perovskite-based solar cells and multifunctional chalcogenide compounds could help next-generation PV-TEG systems as they are very compatible electrically and can be adjusted. Moreover, PV-TEG systems' long-term viability depends on structural optimization. Especially in light of temperature changes, researchers are attempting to strike a balance between power generation, thermal control, and material stability. The design approach to maintain high performance and prolong operating life comprises layers of thermal isolation, systems that effectively disperse heat, and modular arrangements. Advancements in materials science, device design, and hybrid power conversion have placed PV-TEG systems at the forefront of solar power innovation. These game-changing technologies, which capture and convert heat and light from a single source, are paving the way for more integrated, efficient, and sustainable energy harvesting solutions [25, 26].

### 9.2.2.3 Piezoelectric Functionality

Alternative energy generating methods help to integrate energy systems into flexible and conformal device topologies, therefore opening exciting possibilities for powering smart textiles and soft wearable electronics. Despite the great technological developments, though, there is no reliable internal energy generating choice in the sector that can continuously power embedded sensors or recharge integrated energy storage elements. Among the various methods under study, piezoelectric materials have attracted great attention because to their ability to convert mechanical stress into electrical energy; this property makes them especially suited to collect energy from the continuous motion of the human body. Research has focused on producing efficient, flexible, long-lasting piezoelectric devices to satisfy the growing demand for environmentally friendly, autonomous wearable systems. Piezoelectricity is a unique electromechanical event in materials with non-centrosymmetric crystal structures. Mechanical deformation shifts the centres of positive and negative charges, therefore generating surface electrical charges. Piezoelectric materials are ideal for autonomous power generation in wearable applications, as shown in Fig. 9.4, since they have natural characteristic allowing them to directly transform biomechanical energy into electrical energy. In piezoelectric materials, fundamental constitutive Eq. (9.3) controls the electric displacement D.

$$D = d \cdot \sigma + \varepsilon \cdot E \tag{9.3}$$

where d is the piezoelectric coefficient, σ is the applied stress, and ε is the permittivity [27]. Though $ABO_3$ perovskites are exceptional, both wurtzite and perovskite are popular piezoelectric material structures. A key temperature for grasping piezoelectric activity is the Curie temperature, at which piezoelectric materials change from ferroelectric to paraelectric. Below this temperature, the material retains its piezoelectric properties; above it, when the structure changes to a cubical one, they are lost. The efficiency of mechanical energy conversion to electrical energy is described by constants known as piezoelectric charge constant (d) and piezoelectric voltage constant (g), which also help to measure the performance of piezoelectric materials. Since they are anisotropic, showing the constants in a $3 \times 6$ matrix allows one to easily grasp them as it provides a full picture of the materials' reactions to stress in various directions. Of the two, $d_{33}$ denotes the electric field and charge produced in direction 3 under stress and is the most important representative parameter for piezoelectric materials [28]. While current piezoelectric materials have possible energy conversion qualities, their natural trade-offs render them unsuitable for application in wearable devices. Though lead zirconate titanate (PZT), barium titanate ($BaTiO_3$), and other ceramic-based piezoelectrics have great conversion efficiencies and high piezoelectric coefficients, their natural brittleness and lack of mechanical flexibility make them unsuitable for inclusion into stretchable or deformable platforms. Among these varieties of perovskites, oxide-based ones like $BaTiO_3$ and $KNbO_3$ have notable advantages including spontaneous polarization and strong lattice coupling,

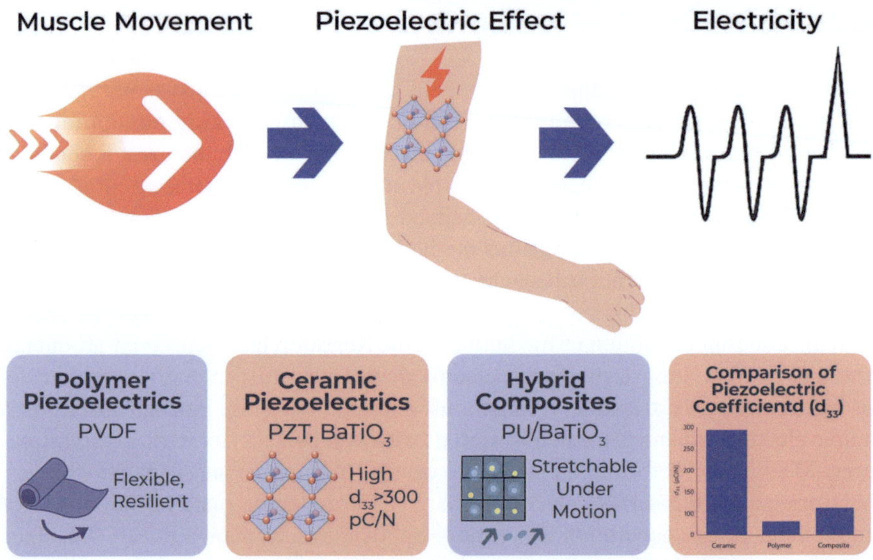


**Fig. 9.4** Transforming body movements into electrical energy: A comparative analysis of perovskite-based piezoelectric sensors highlights the trade-offs in flexibility and output among ceramic, polymer, and composite materials for optimal wearable energy harvesting

which enable piezoelectric coefficients reaching $d_{33} > 300$ pC/N. These characteristics make perovskite ceramics interesting materials for next-generation wearable energy harvesters since they can maintain their performance consistent even under mechanical distortion [29].

Wearable electronics depend on the remarkable flexibility and mechanical strength provided by piezoelectric polymers such polyvinylidene fluoride (PVDF). Regrettably, qualities like as low stretchability and naturally weaker piezoelectric efficacy limit its energy-harvesting efficiency. Recent research has focused on hybrid composite materials combining the benefits of ceramics and polymers to bridge the gap between mechanical compliance and energy generation. One such is the extraordinary stretchability and significant voltage generation under mechanical stress of a polyurethane (PU)/$BaTiO_3$ composite. These materials have great potential as biomechanical energy harvesters, paving the way for the development of wearable devices operating without reliance on outside batteries [30]. These composites not only improve flexibility, washability, and textile integration compatibility but also maintain exceptional piezoelectric performance. By changing the ceramic composition, particle shape (e.g., nanowires, platelets, flower-like particles), and matrix distribution, one may also fine-tune the piezoelectric output of the composites. Structural design innovations have increased mechanical flexibility and energy harvesting efficiency under real deformation conditions. Among these developments are aligned nanostructures, kirigami-patterned films, and 3D interconnects. Wearable piezoelectric perovskite systems offer a game-changing choice for autonomous electronic devices with these developments and well-established low-temperature manufacturing techniques (such as sol–gel, hydrothermal, and electrospinning) [29, 31]. These devices provide significant advancement in wearable energy-harvesting technology since they may both be sensors and power sources for analyzing gait and body motions. Highly sensitive and effective energy conversion systems made possible by the integration of nanotechnology in textiles are necessary for the smooth acceptance of smart fabrics into useful applications. Grain boundary manipulation and nanowire orientation are two major materials engineering developments that have enhanced fatigue resistance and raised energy conversion efficiency. Grain boundary control helps charge carrier paths to be optimized, so lowering recombination losses and raising general electrical efficiency. Concurrently, the alignment of nanowires ensures consistent stress distribution over the material, so lowering mechanical degradation and increasing the operational lifetime of textiles designed to harvest energy. Smart textiles have lately focused on including flexible energy storage devices, supercapacitors, and nanogenerators into fabric architectures. Very effective energy-harvesting and storage devices can now be developed thanks to the functionalization of textile substrates with conductive nanomaterials including graphene, carbon nanotubes, and metallic nanoparticles. Furthermore, stacked or coaxial device topologies preserve flexibility by combining complimentary energy modalities, so enabling multi-functional integration [7, 32–34]. A significant development in self-sustaining energy harvesting systems is indicated by the hybridization of piezoelectric nanogenerators (PENGs) with triboelectric nanogenerators (TENGs) and photovoltaic nanogenerators. By combining the special processes of both technologies, hybrid devices increase energy conversion efficiency and extend their application in wearable electronics, environmental sensing, and autonomous systems. PENGs work by the direct piezoelectric effect, whereby mechanical deformation produces electrical polarization leading to a potential difference. This method lets PENGs efficiently convert biomechanical energy into electrical signals, which qualifies for uses requiring precise force sensing. TENGs rely on triboelectrification and electrostatic induction; contact between materials with different electron affinities generates charge separation. TENGs can thus gather energy from many mechanical motions, including vibrations, friction, and impact forces. Using the complementing benefits of both technologies, the hybridization of PENGs and TENGs produces higher power densities than either individual PENGs or TENGs. Under continuous mechanical pressure, PENGs offer continuous, high-frequency electrical output; TENGs are effective in harnessing intermittent, low-frequency mechanical energy. This makes them particularly suitable for wearable devices, where effective small energy sources are absolutely vital. Furthermore, hybrid devices improve sensory capacities by sensing both dynamic and stationary mechanical inputs, so allowing more exact motion tracking and environmental assessment. By combining several motions with one device, hybrid systems improve energy collecting efficiency by means of both processes. Moreover, since hybrid PENG-TENG devices run effectively under various environmental conditions, they show improved endurance and flexibility [35].

## 9.3 Fashion-Oriented Perovskite

### *9.3.1 Thermochromic and Electroluminescent Perovskite Fabrics*

The inclusion of perovskite materials into wearable electronics is helping smart textiles with an eye toward style enter a new era. Figure 9.5 indicates that this new trend is a direct outcome of the growing demand for trendy apparel with modern technical qualities as well. Because to their remarkable optoelectronic properties, mechanical flexibility, and simple processability, perovskites are being thought more and more as ideal materials for next-gen wearable applications. Because they can maintain qualities like on-body energy harvesting, dynamic color change, and moisture and temperature monitoring without compromising the textile's natural softness, air permeability, or stretchability, perovskite-based materials are being considered by the fashion sector. Including perovskite structures into smart clothing changes color in response to external influences, therefore enhancing the wearer's comfort by means of built-in sensing capabilities and adding adaptability to garment design. Adding perovskite materials makes wearable electronics even more durable and long-lasting. These materials address significant issues including water exposure and mechanical deformation. Engineers are developing new structural designs like kirigami cuts, buckling topologies, and serpentine conducting channels to satisfy the mechanical needs of daily wear without compromising electrical performance or aesthetic appeal. Smart textiles made possible by perovskites offer a possible basis for the development of trendy, multifunctional clothing, which fits the fashion sector's continuous acceptance of technical innovation. Wearable technology might be transformed by the combination of creative design with state-of-the-art materials science. Rather than being only clothes, it will be able to function as interactive, smart systems that can be tailored, and even more environmentally friendly. Ultimately, perovskites could be crucial in realizing the vision of smart and responsive clothing. These fabrics have great promise for a range of applications including adaptive camouflage, flexible displays, human–machine interfaces, and real-time temperature monitoring given their dual sensing and emissive qualities [36, 37].

#### 9.3.1.1 Thermochromism in Perovskite Textiles

Thermochromic materials have a unique quality: they can alter their optical properties—such as color, transmittance, and reflectance—in response to temperature changes. Their dynamic behaviors have many possible applications, including smart windows, temperature sensors, energy-saving coatings, and flexible photovoltaics (Fig. 9.6). Though effective, conventional thermochromic solutions like hydrogels and vanadium dioxide ($VO_2$) have disadvantages such high transition temperatures, slow reaction times, and limited long-term durability under ambient conditions [38,

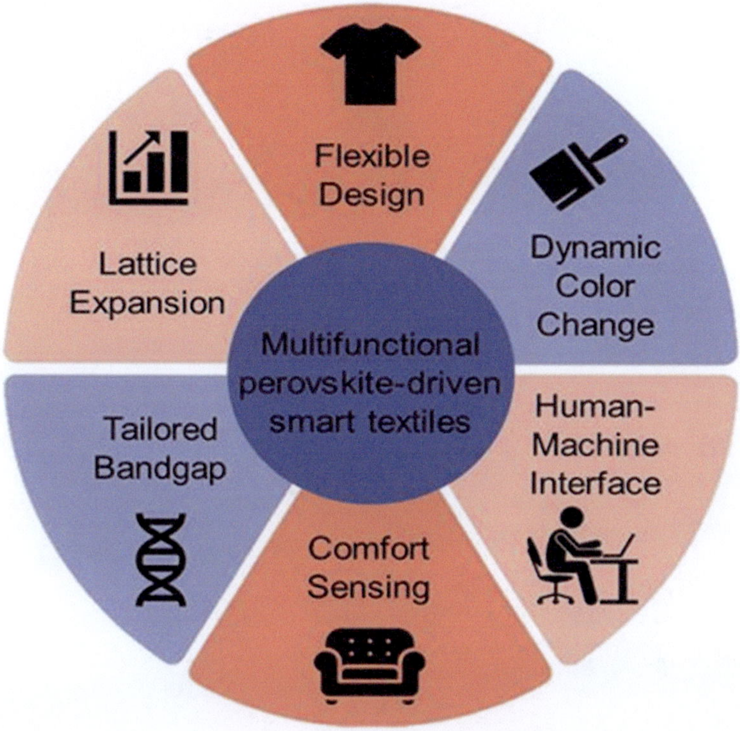


**Fig. 9.5** Multifunctional perovskite-driven smart textiles integrating adaptive sensing and energy systems, illustrating key features including lattice expansion, tailored bandgap, flexible design, dynamic color change, comfort sensing, mechanical flexibility under motion, and on-body energy harvesting

39]. Perovskite materials show promise because of their environmental responsiveness, structural tunability, optoelectronic characteristics, and thermochromism. Their crystalline structure, electrical band arrangement, and ionic mobility cause them to show adaptive optical activity in reaction to heat stimuli. Based on the flexible $ABX_3$ perovskite crystal structure, these perovskites consist of monovalent cation A, usually metal B (like $Pb^{2+}$ or $Sn^{2+}$), and halide X. Their optoelectronic characteristics are fundamentally built on the $BX_6$ octahedra, which are plentiful in the electronic band structure near the band edges [40]. Perovskites stand out when compared to more conventional semiconductors like Si and GaAs. Due to their inverted band structure and high optical absorption coefficient (~$10^5$ $cm^{-1}$), even very thin films (~hundreds of nanometers) are quite sensitive to heat stimuli. Changing the composition of the A-site cation (e.g., $MA^+ \rightarrow FA^+$), the B-site metal (e.g., $Pb^{2+} \rightarrow Sn^2$), or the halide ($I^4 \rightarrow Br^- \rightarrow Cl^-$) adjusts the bandgap from 1.1 to more than 3.0 eV, which then influences the color variations as a function of temperature [41]. Emerging

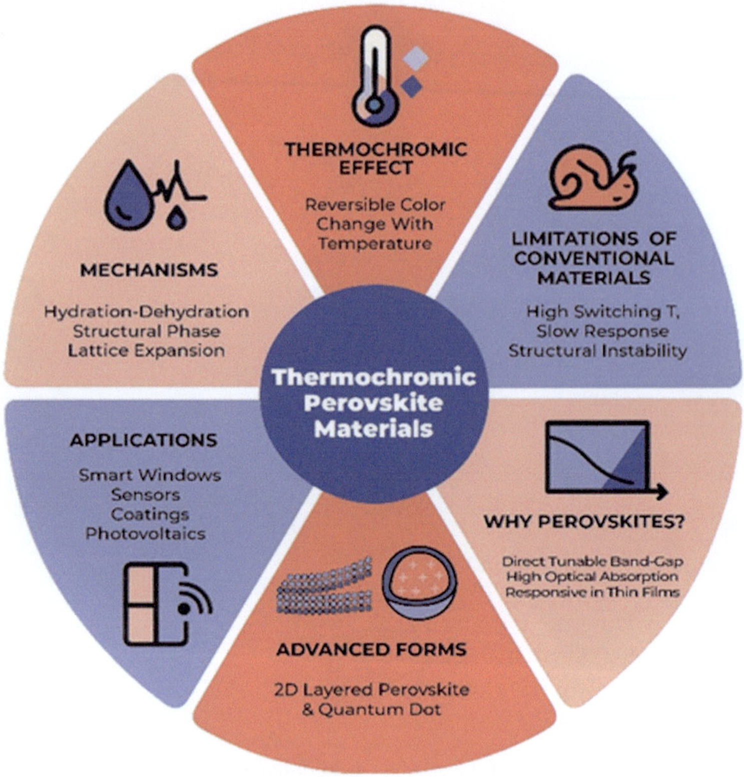


**Fig. 9.6** A comprehensive visual roadmap illustrating the key aspects of thermochromic perovskite materials

types such quantum dot perovskites and 2D layered perovskites offer stronger excitonic effects and improved stability, hence enhancing their appropriateness for long-term, reversible thermochromic applications. The compositional adaptability of these materials lets their precise engineering respond to particular thermal threshold [42].

Lattice expansion, thermally induced structural phase changes, and hydration-dehydration transitions are three primary processes governing the thermochromic behavior of halide perovskite materials. Each of these mechanisms helps to facilitate temperature-dependent alteration of the electronic band structure, which results color changes and other reversible optical responses. When these processes interact with one another, thermochromic effects are frequently amplified.

– ***Structural Phase Transitions***

Thermochromic behavior of perovskite materials is temperature-driven, a complex optical phenomenon based on structural changes in their crystalline structure. Structural phase transitions are a fundamental process key component. Thermally induced changes in a material's crystal lattice alter its absorption and scattering of light, so changing its color. Perovskites often adopt a very compact and ordered lattice structure during their low-temperature phase. The optical absorption and reflection properties are well defined here since the atomic or ionic configurations are so stable in relation to thermal motion. At some time during heating, the material hits a key threshold called the phase transition temperature ($T_c$), at which it experiences major structural rearrangement. The band structure and optical behavior of the perovskite may be changed by several processes, such as ion displacements, charge redistributions, and lattice distortions (Fig. 9.7a) [43]. Thus, the governing elements of thermochromic behavior in perovskites are usually structural transitions produced by temperature that shift the absorption edge and produce color changes. The hybrid perovskite $[2,5\text{-DCA}]_2CuCl_4$ experiences this phenomenon as the temperature rises, causing the $[2,5\text{-DCA}]^+$ cation to become disordered and the $CuCl_6$ octahedra to distort. UV–Vis spectroscopy shows the material to be yellow-green at 303 K and to absorb light below 550 nm. When the temperature rises to 413 K, the absorption edge redshifts and the bandgap narrows from 2.25 to 2.05 eV, producing a deeper hue. This change is closely related to the hydrogen bonding and mobility of organic cations, which affect lattice flexibility. Bigger and less firmly bound cations in $A_2CuX_4$ perovskites increase lattice mobility and lower transition temperatures. X-ray absorption near edge structure (XANES) studies indicate that the electrical structure is temperature invariant, implying that the thermochromic reaction is mainly caused by geometric distortions. Comparing hybrid and all-inorganic perovskites reveals more effects from the A-site cation. Unlike organic cations, which promote greater flexibility and more noticeable thermochromic effects, inorganic cations like $Cs^+$ and $Rb^+$ increase stiffness, which results in less noticeable but still controllable optical changes via lattice expansion or distortion. Reversible optical modulation in perovskites is crucial for smart devices such thermal sensors and energy-efficient windows; this can only be accomplished by means of temperature-responsive structural changes controlled by cation selection and lattice dynamics [43].

– ***Hydration–Dehydration Transitions***

Reversible hydration-dehydration cycles play a significant role in driving thermochromism in methylammonium-based halide perovskites, such as $MAPbI_3$ and its mixed-halide derivatives (e.g., $MAPbI_{3=x}Cl_x$). When the relative humidity is at 25% and the temperature is around 25 °C, $MAPbI_3$ may absorb water and change chemically into a non-perovskite phase that is hydrated:

$$MAPbI_3 + 3MAI + H_2O \rightarrow MA_4PbI_6 \cdot 2H_2O$$

This hydrated phase's drastically altered optical absorption characteristics cause a noticeable change in color; usually, it goes from being transparent or very light-colored to a deeper, reddish-brown hue (Fig. 9.7b). The material goes back to its perovskite structure when heated to roughly 80 °C, restoring its outstanding transparency and clarity of vision. This thermochromic process is usually amplified in

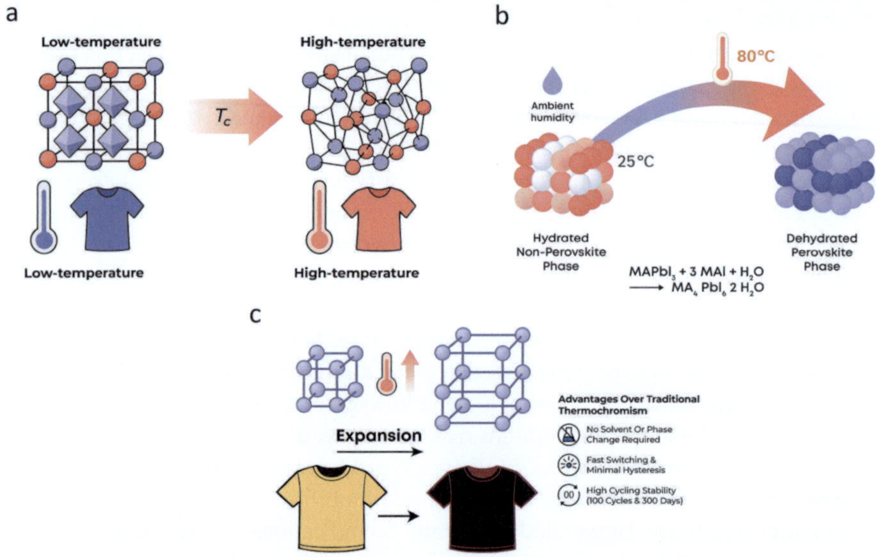


**Fig. 9.7** Exploring thermochromic mechanisms in halide perovskites: **a** Structural phase transitions driven by temperature alter lattice configurations, shifting optical properties, **b** Hydration-induced phase transformation enables reversible transitions between transparent and absorptive states, and **c** Lattice-expansion thermochromism utilizes thermal expansion for rapid, durable, and stable bandgap modulation

the presence of lattice expansion at high temperatures and it results from a cocktail of chemical phase transition, lattice distortion, and small disordering effects [44]. Adaptable fabrics reacting to environmental conditions could make use of this revolution in wearable technology. Thermonochromic perovskites smart textiles may control heat absorption and dissipation, so improving thermal comfort for users. Furthermore, wearable sensors made of hydration-sensitive materials could measure skin moisture levels, so enabling hydration tracking for medical or sporting uses. Wearables could achieve real-time responsiveness by means of the hydration-dehydration transition, so enhancing their value in health monitoring and climate-adaptive clothing. Using hydrated $MAPbI_{3-x}Cl_x$-based thermochromic perovskite smart materials as windows results in a dynamic transition between opaque and transparent states, so greatly enhancing interior thermal comfort. By means of a solar modulation efficiency ($\Delta T_{sol}$) of 23.7% and an adjustable low transition temperature, these windows effectively control heat, so reducing interior air temperatures by as much as 3.5 °C in comparison to conventional quartz glass [45]. Limited practical use of thermochromic smart materials results from manufacturing challenges in humid environments often leading to defective perovskite film morphology, increased optical haze, and reduced solar modulation. Presenting an antisolvent engineering solution to meet these difficulties has greatly improved film homogeneity and crystallinity. This method increases the integration of thermochromic and photovoltaic functions and facilitates production in conditions of variable humidity, so attaining power conversion efficiencies (PCEs) of up to 2%. These findings make hydration-responsive thermochromic perovskites practical, reasonably priced substitutes for next-generation energy-efficient materials [46]. Recent research has concentrated on $T_C$, the point at which this color change occurs, and ways to control the critical transition temperature. Researchers have developed perovskite films combining perovskites with tiny additives such chloride ions ($Cl^-$) and polymers such PEG, PVA, and PAA to achieve this. These additives interact subtly but significantly with the perovskite matrix. The additions change the perovskite crystal's surroundings, especially the reservoir of methylammonium halide, rather than the basic switching mechanism. While polymers change the material's interaction with alcohol or water molecules, chloride ions change the packing and optical absorption of crystals. The addition of these substances causes a decrease in the transition temperature, which in turn causes the material to undergo color switching within more realistic and energy-efficient ranges; for example, from over 70 °C to as low as 20–25 °C, depending on the formulation. Some additive combinations not only lower $T_C$ but also improve the visual quality of the film, reduce the switching time, and minimize optical hysteresis. Importantly, X-ray diffraction techniques have revealed these changes occur without changing the fundamental structure of the material. The capacity to precisely control the switching temperature and behavior by careful additive selection opens up novel ways to create adaptive, low-energy optical coatings that are well-suited for practical applications in climate-responsive building design [47].

– ***Lattice Expansion***

Traditionally, thermochromism—the reversible change of hue in reaction to temperature—has been explained by means of hydration/dehydration or phase transitions between crystalline forms. Though they are effective, these mechanisms are not always practical for use in devices like thermal sensors and smart windows since they are prone to structural instability, have long switching times, and degrade over repeated cycles. These circumstances might involve solvent vapours or high humidity among other things. On the other hand, lattice-expansion-induced thermochromism is becoming popular; it uses the perovskite lattice's natural flexibility instead of atmospheric stimulation or phase shift. When perovskites are heated, their atomic structures expand uniformly or anisotropically, causing some change in bond lengths and unit cell volumes. Observable changes in optical absorbance and, by extension, visible colour, may follow from the influence of these thermal deformations on the electronic band structure, particularly in systems with strong electron–phonon interaction, Fig. 9.7c [48, 49]. Lead-free double perovskites like $Cs_2NaFeCl_6$ have particularly demonstrated this behaviour; in these materials, the thermochromic properties are caused by a slow and reversible expansion rather than a collapse or rebuilding of the crystal lattice. The experimental findings reveal a striking hue shift from pale yellow to black and a notable optical bandgap redshift (up to 0.51 eV from 10 to 423 K). This change is characterized by complete reversibility with minimal hysteresis and stability over multiple heat cycles. Further studies show that when thermal expansion is the only cause, bandgap variations are minimal. The effect is greatly increased, however, when coupled with electron–phonon coupling, which includes temperature-activated lattice vibrations. The conduction and valence bands in $Cs_2NaFeCl_6$ are quite susceptible to bonding environment distortions since they are composed of localized Fe(d)–Cl(p) orbitals. The expansion along weaker Na–Cl ionic links and the preservation of stiffness along the stronger Fe–Cl covalent bonds help to maintain structural integrity and optical responsiveness. Unlike traditional lead halide perovskites (e.g., $MAPbI_4$) that show only minor bandgap changes with temperature, thermochromic materials based on lattice expansion provide faster switching, longer cycle endurance, and environmental stability. $Cs_2NaFeCl_6$ exhibits consistent optical and structural characteristics after 300 days of room temperature exposure and 100 thermal cycles [50].

Without causing significant phase changes, thermally induced lattice expansion could change the electronic structure of low-dimensional perovskites, particularly in 0D and 2D configurations. The bandgap changes slightly but reversibly with changes in bond lengths and orbital interactions. Because of their enhanced structural flexibility and variable composition, two-dimensional layered hybrid perovskites have shown especially promising for thermochromic uses. The reversible color changes between orange and red observed in single crystals like $(PEA)_2PbI_4$ and $(NEA)_2PbI_4$ (PEA = 2-phenylethylammonium, NEA = 2-(2-naphthyl)-ethanammonium) under heating and cooling cycles are caused by lattice expansion, not discrete phase transitions. Thermogravimetric-differential scanning calorimetry, X-ray diffraction, and temperature-dependent absorption spectroscopy all support this mechanism. Unlike conventional 3D perovskites that usually use structural phase changes or spin–orbit coupling effects for thermochromism, 2D systems take use of the conformational flexibility of organic spacer cations to enhance electron–phonon coupling and allow more stable bandgap tuning. Cations based on acene alkylamine have especially made possible these thermally sensitive and structurally stable optical changes [51]. Lattice-expansion-driven thermochromism is essentially a safer, more direct way to produce materials that react to temperature. It avoids the need for volatile solvents or structures susceptible to phase instability and offers variable optical qualities by means of structural and compositional engineering. Its possible applications in smart windows, temperature indicators, displays, passive heat control, and others are rather encouraging in light of this. Lead halide perovskites might be used in hybrid thermochromic-perovskite systems to heat up particular areas when exposed to light, which activates the thermochromic response since they have photothermal conversion capabilities. Such synergistic behavior offers possibilities for low-energy dynamic materials in thermal camouflage and wearable healthcare [52]. Though they are excellent at passive color changing, thermochromic devices can only change colors in reaction to ambient or body heat stimuli. Expanding the functionality of perovskite-based textiles beyond temperature alone will next be accomplished by using electroluminescence, the active conversion of electrical inputs into light emission. The following section explores electro- and mechano-luminescent perovskite films and fibers, demonstrating how this additional modality transforms perovskites into active display elements and interactive garments capable of lighting up, signaling, and interfacing with people and machines on demand.

#### 9.3.1.2 Electro- and Mechano-luminescence in Perovskite Fabrics

***Electroluminescent*** devices generate light by means of a current flowing through an emissive layer sandwiched between two electrodes. Selecting different host materials and dopants allows emission to be controlled across the visible spectrum; when charge carriers recombine inside this layer, excited molecules return to their ground state and emit photons. Typical electroluminescent displays' integration with clothing is limited by the use of stiff panels and external power sources. Wearable luminous fabrics, which include visual communication straight into the fabric we wear, are interesting substitutes. Developed to get around these limitations, new alternating-current electroluminescent (ACEL) fibers can spontaneously illuminate in response to environmental factors. In these liquid-responsive devices, the coaxial fiber electrodes are bridged and patterned lighting is activated by the simple touch of conductive fluids including water, glycerol, ethanol, or salt. Woven from these fibers, integrated fabrics are very sensitive, reversible, and able to detect fluids and display changing light patterns. Practical shows include gloves that indicate when they come into contact with liquid, athletic wear that indicates when sweat is accumulating, and umbrellas that illuminate when rain strikes them. The combination of self-powered electroluminescence with real-time environmental responsiveness on soft, flexible substrates will help to realize wearable optoelectronic systems of the future, interactive clothing and personal safety tools as well [53].

Narrow emission bands and high photoluminescence quantum yields in perovskite-based emitters—usually made up of $CsPbX_3$ (X = Cl, Br, I)—allow the very pure and vivid colors. The first efforts at creating flexible light-emitting devices used transparent conductors rather than the rigid indium-tin-oxide films and modified planar organic light-emitting diode designs to bendable substrates. Certain materials have demonstrated sufficient transparency and conductivity to resist slight bending without any performance loss. These materials consist of conductive polymers such as poly (3,4-ethylenedioxythiophene):poly (styrenesulfonate) (PEDOT:PSS), modified graphene, silver-nanowire meshes, and carbon nanotube networks. Still, the flat design restricts these devices; straining or twisting them beyond minor deformations often harms the electrodes or separates the active layers. To avoid these limitations and obtain flawless light emission integration into textiles, researchers have shifted their attention from sheets to fibers. Though this resulted in complex manufacturing and naturally rigid metal cores, the first tests achieved efficiency comparable to planar ones by vacuum depositing emissive layers of tiny molecules onto polyimide-coated wires. More scalable techniques employing coaxial wet-spinning or solution-coating processes build three concentric layers around a liquid–metal or polymer core, an emissive middle shell, and an outside transparent electrode. Woven or knitted, fabrics made from these fiber-shaped devices show voltage-dependent brightness visible with the naked eye. Still under study, though, are their long-term stability and mass manufacturing viability [37, 54]. Perfect for totally wearable electroluminescent textiles, lead-halide perovskites offer simple solution processing, narrow and tunable emission bands, and great photoluminesence quantum yields. Their natural sensitivity to heat, moisture, and UV radiation, however, has long limited their utility.

Recently, combining perovskite-polymer composites with robust fiber topologies has let scientists to overcome these challenges. Using a green wet-spinning technique to crystallize $CsPbX_3$ (where X = Cl, Br, I) inside a hydrophobic thermoplastic polyurethane (TPU) sheath will help to produce luminous filaments with a 24.5% quantum yield and stretchability near 2400%. Apart from being water-repellent, the TPU strengthens the perovskite lattice by coordinating its carbonyl groups to Pb. This lets it survive extreme pH and organic solvents, submersion in water for 30 days, continuous UV radiation, heat (50 °C annealing), and high humidity as well as heat. After five recycling cycles, it keeps 90% of its original brightness. These developments in fiber-scale device engineering, encapsulation chemistry, and filament composition all combine to produce perovskite textiles that are durable, bright, flexible, and capable of detecting strain, foil counterfeiting, and act as environmental warning systems [55]. Building on earlier advances, a new study reveals that scalable, high-performance perovskite fabrics could be produced using a straightforward, high-throughput electrospinning technique. Researchers have created composite fibers comprising multilayer polymer matrix encasing $CsPbX_3$ perovskite nanocrystals. Included in this matrix are a hydrophobic polymer core, cyclodextrin chelating layers, and a silane-enhanced outer shell. The outcome is flexible luminescent textiles with an area more than 375 $cm^2$. These electrospun textiles are really remarkable given ultra-narrow emission bandwidths below 17 nm and a record photoluminescence quantum yield of over 50%. They can survive high amounts of UV radiation, thermal annealing up to 250 °C, mechanical shocks up to 30 MPa, and continuous immersion in water for more than 3,300 h; their brightness only drops by 50% after 14,000 h of ambient storage. Importantly, the environmental safety is preserved even under strong water flow since the strong silane and cyclodextrin barriers stop the release of any detectable lead. Promising possibilities for demanding applications like marine-rescue lighting, wearable signaling, and harsh-environment smart fabrics, these electrospun perovskite textiles mix large-area manufacturability with exceptional optical purity and unmatched durability, Fig. 9.8 [56]. Skin-conformable ACEL displays with hybrid perovskite-inorganic phosphor structures have been produced by means of progress in fiber-scale and textile-integrated technologies. Uniformly coating perovskite nanocrystals onto zinc sulfide (ZnS) microparticles results in a new family of perovskite zinc sulfide (PeZS) phosphors. These phosphors show intraparticle energy transfer when driven by alternating current, producing narrowband emission with a full width at half maximum (FWHM) of less than 37 nm, adjustable across the visible spectrum from 468 to 649 nm. Using transparent electrodes, PeACEL devices are built to withstand uniaxial loads of up to 400% and achieve brightness levels above 200 cd $m^{-2}$. Easy connection with sensors and flexible circuits allows powder-based ACEL displays, fit for low-temperature, roll-to-roll manufacturing, to be connected. Their efficient real-time temperature monitoring and dynamic interactive visuals depend on their enhanced color spectrum resulting from pure red and green emissions. Potential applications in multifunctional artificial skins, wearable robotics, and biomedical displays abound from these skin-adhering panels combining mechanical flexibility with vivid electroluminescence [57]. Advancements in perovskite-based luminescent fibers and textiles have

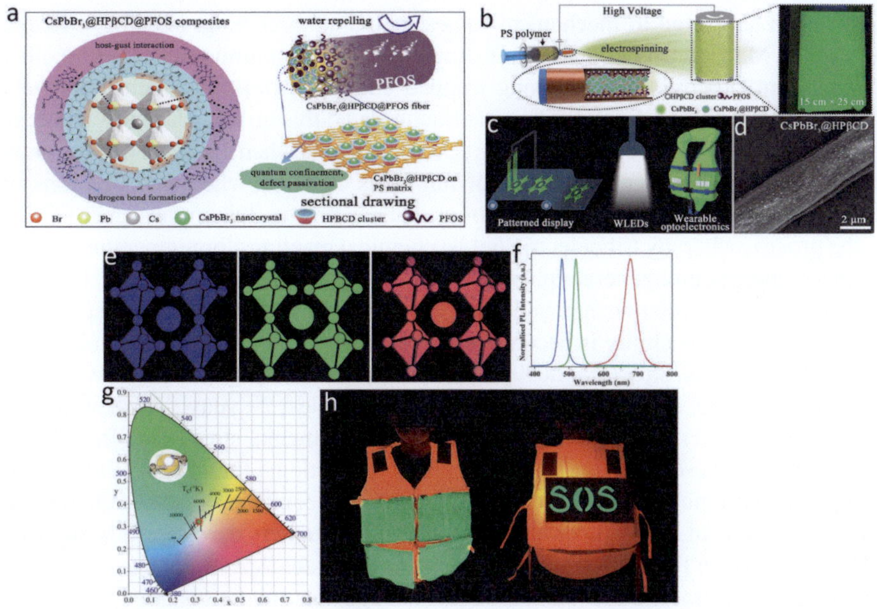


**Fig. 9.8** **a** Schematic representation of the chemical interaction between $CsPbBr_3$@HPβCD@PFOS composites and their diverse functionalities. **b** Schematic representation of the production process of PLTs and the resultant fibrous membrane under UV light exposure. **c** Utilization of $CsPbBr_3$@HPβCD@PFOS composites in patterned displays, white light-emitting diodes (WLEDs), and wearable optoelectronics. **d** SEM pictures of the $CsPbBr_3$@HPβCD fiber obtained using a YAG back-scattered electron detector. **e** Patterned fluorescent displays (10 cm × 10 cm) and **f** their respective photoluminescence spectra exhibiting various hues (blue, green, and red). The 1931 CIE chromaticity coordinates of white LEDs constructed using $CsPbBr_3$@HPbCD@PFOS composites. The operating voltage is 5 V, and the driving current is 100 mA. A demonstration of a life jacket including a "SOS" patterned PLT (10 cm × 22 cm) sewn onto it. Reproduced from Ref. [56] published by Nature Communications. 

helped them to develop from laboratory prototypes to useful applications. A multi-layer ACEL perovskite fabric has been created that can alter its color in reaction to temperature fluctuations while maintaining consistent brightness under bending. It may reversibly change between several colors when combined with thermochromic overlays, so enhancing its use for dynamic display and sensing purposes. Electro-thermochromic fibers included into perovskite-infused polymer matrices have been developed using this multimodal approach to generate light and change color in response to voltage and temperature fluctuations. These advances underline the flexibility of stimuli-reactive perovskite systems for many uses by offering fresh possibilities for wearable displays, environmental sensors, adaptive camouflage, and secure identification tags [58].

***Mechanoluminescence (ML)*** is a basically self-powered system for converting kinetic energy into optical signals; it is the process by which a solid material emits light when subjected to mechanical forces including stretching, bending, friction, or impact. Mechanical deformation distinguishes mechanoluminescence from other luminescent processes by releasing trapped charge carriers that then radiatively recombine to generate light. Sir Francis Bacon found this in 1605. Usually, there are two key kinds in ML. Deformation ML (DML) happens when elastic or plastic strain disturbs the electronic environment of luminous centers—whether doped ions or natural flaws—forcing carriers to flee traps and recombine. Tribuminescence (TML) is the process whereby triboelectric fields, chemical reactions, or temperature spikes displace charges and generate photons when materials touch, separate, or fracture. Due to the close relationship between mechanical input and optical response in both scenarios, this ability is becoming more attractive for next-generation human–machine interfaces, structural health monitoring, and safety systems. It lets one see in real time force, deformation, or motion. Over the last 20 years, much study has been done on ML materials that are robust to environmental stress and offer strong, repeatable luminescence. Inorganic phosphors' exceptional thermal stability, batch-to-batch consistency, and large-scale manufacturing capacity have pushed them to the front. By means of host lattices and dopant ions, researchers have carefully adjusted ML emission over the ultraviolet, visible, and near-infrared spectral ranges to achieve intensities measurable under ambient light. Though these materials often have lower long-term stability than inorganic ones, the work of organic and biomolecular ML systems has broadened the color palette and mechanical sensitivity profiles [59].

Interactions between lattice deformation, radiative recombination, and defect-mediated charge storage in perovskite materials lead to mechanoluminescence. The corner-sharing octahedral network ($ABX_3$ or $A'B_2O_5$) gives lead-halide and oxide perovskites structural flexibility and a high density of possible trap sites (vacancies, interstitials, or dopant-induced states), allowing them to catch carriers during prior optical or electrical charging. When the perovskite lattice is distorted by an external mechanical stimulus such elastic strain, plastic deformation, or fracture, two processes working in concert make ML possible: Strain-induced modulation of the local potential landscape lowers the activation barrier for detrapping; in non-centrosymmetric or ferroic perovskites, the deformation generates an internal piezoelectric or flexoelectric field that further aids carrier release. Released electrons or holes travel to luminous centers, where they undergo radiative recombination. Often doped with transition metals or lanthanides in oxide perovskites, these centers are in the natural excitonic states of lead-halide frameworks. This technique generates perovskite phosphors' narrowband spectra and extremely pure color output, making them easily recognizable. The near-instantaneous ML response caused by the close relationship between mechanical energy input and optical output allows quantitative mapping of force distributions in smart textiles or structural health monitors; the intensity of this response usually scales linearly with applied stress. Though there has been some advancement, our knowledge of the precise mechanism of carrier detrapping in perovskites under mechanical stress remains scant. Key issues to think about are the ways in which some defect shapes and trap depth distributions alter under strain, as well as the degree of contribution to internal field generation by ferroic domain reorientation or octahedral tilting. We must combine first-principles modeling of the strain-dependent trap energetics with in situ structural probes able to resist dynamic loading, such electron microscopy or synchrotron X-ray diffraction, to address these issues. Studies like this one will guide the creation of perovskite microstructures and compositions perfect for durable, effective mechanoluminescent devices [60].

Often using polymers like polydimethylsiloxane (PDMS), polyurethane (PU), or polyethylene terephthalate (PET), ML systems rely on the inclusion of phosphors into pliable composite matrices. Advanced designs could enhance luminosity, enable spectrum multiplexing, and incorporate hydrophobicity and self-healing among other things. Double-mode tactile displays that combine ML dyes with triboelectric nanogenerator (TENG) devices provide both electrical and light signals upon contact. These self-sustaining fibers highlight how intelligent textiles, interactive clothing, and human–machine interfaces—all of which depend on machine learning—have promise. Currently lacking is a comprehensive model clarifying the effects on ML performance of mechanical loads, crystal structure, and composite design. More general practical uses outside of first experimental demonstrations depend on closing this gap [61]. Mechanoluminescence in perovskites has been investigated in both organic–inorganic hybrid metal halides and transition-metal-doped inorganic hosts, each demonstrating unique luminous processes. Lanthanide dopants and polar lattice structures give strong luminous responses in hybrid perovskites even with minimum deformation. $Tb^{3+}$-doped perovskites are synthesized by means of chiral organic spacers integrated with corner-sharing metal-halide octahedra in a noncentrosymmetric framework. By means of internal piezoelectric forces produced by mechanical deformation, trap barriers are lowered and charge carriers are efficiently channeled to $Tb^{3+}$ centers, so enhancing luminous performance. While continuous mechanical stimulation guarantees consistent luminescence, showing the effectiveness of defect engineering and dopant selection, systematic changes to the geometry of organic cations and halide composition further affect ML intensity and consistency [62]. Similarly, high-temperature solid-state processes generate $Cr^{3+}$-activated $LaAlO_3$ that shows self-recoverable near-infrared mechanoluminescence (650–800 nm) without pre-irradiation. Unlike trap-mediated mechanisms, its luminescence mostly results from local piezoelectric forces produced by noncentrosymmetric lattice distortions. Effective elimination of residual carriers by a preheating technique strengthens this process. Combining spectroscopic characterisation with thermoluminescence, diffuse reflectance, and persistent luminescence studies supports that stress-induced internal fields help to explain carrier detrapping and radiative recombination. Notwithstanding irradiation, the cyclic stability of ML in transition-metal-doped perovskite hosts makes them interesting candidates for bio-stress sensing devices and anti-counterfeiting technologies, Fig. 9.9 [63]. By means of localized piezoelectric distortions around $Mn^{2+}$-induced defect complexes, $Mn^{2+}$-doped $MgGeO_3$ (space group Pbca) shows deep-red to near-infrared ML (600–800 nm) extending the range of mechanoluminescenceinto the red and near-infrared spectra. Mechanical stimuli such as friction, compression, or bending cause distortions that improve charge transport to $Mn^{2+}$ centers, so generating crystal-field-split d-d transitions producing extensive emission in the deep-red/NIR spectrum.

Under low stress conditions, this method effectively facilitates carrier detrapping from shallow energy levels, so increasing ML efficiency and maintaining continuous luminescence [64]. By means of the integration of perovskite quantum dots into flexible matrices, hybrid composite systems surpass conventional ML systems by including chiroptical and multifarious ML responses. Luminous cholesteric liquid crystal elastometers are produced from photopolymerized $CsPbX_3$ (X = Cl, Br, I) "nanomonomers" in situ covalent cross-linked within a cholesteric liquid crystal precursor. Under UV-irradiation, these materials show circularly polarized luminescence (CPL) with dissymmetry factors of 1.5. Transverse stretching and mechanical deformation change the helicoidal pitch and chirality so enabling reversible control of CPL signal and intensity. Particularly effective for anti-counterfeiting labeling, this mechanical operated CPL switching technique allows strain-induced polarization states to encode specific information. By including chiroptical elements into flexible machine learning systems, doped-crystal techniques are improved and new possibilities in wearable sensor technologies, strain-encoded displays, and safe data encryption are presented. These developments highlight the increasing possibility of ML materials in next-generation electrical and optical systems [65, 66].

## 9.4 Outfit Application of Perovskite

Using perovskite-based devices that transform regular textiles into platforms for ***health monitoring, human–machine interaction, and adaptive camouflage***, one may attain the comfort and aesthetics needed by current fashion. Recent discoveries have confirmed this. Made possible by perovskites' light and temperature sensitivities, ***wearable diagnostic textiles*** offer a possible answer; these textiles can generate thermal maps in real time and feature visual alerts indicating when the maps depart from the norm. Such systems could immediately react to changes in skin temperature or the presence of localized inflammation by means of color-coded perovskite emitters buried in the fabric. Apart from their use in diagnostics, perovskite coatings pave the way for interactive, wearable user interfaces. A sleeve or lapel could turn an intuitive control surface using touch or proximity triggers producing light emission or color changes from music and navigation controls to accessibility help for people with limited mobility. Moreover, high brightness yield of perovskites and simple band-gap engineering open the door for ecologically sensitive hues in ***camouflage textiles***. By always changing their emission spectra in response to environmental light or nearby colors, these fabrics could perfectly combine form and function, therefore enabling adaptive concealment or expressive fashion statements. Including ultra-thin, lightweight perovskite solar modules into textiles has addressed a significant power supply issue, which was a key obstacle to fully autonomous wearables. Smart clothes embedded with these flexible cells generated enough power to run internal systems like active cooling and health monitoring in a recent demonstration at a major international fair; they sent any surplus power to a built-in control and storage unit. Attaching the solar array to the fabric using creative wire-free sewing

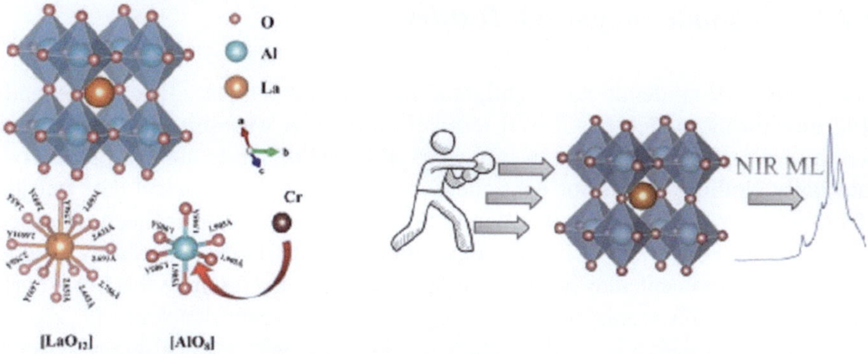


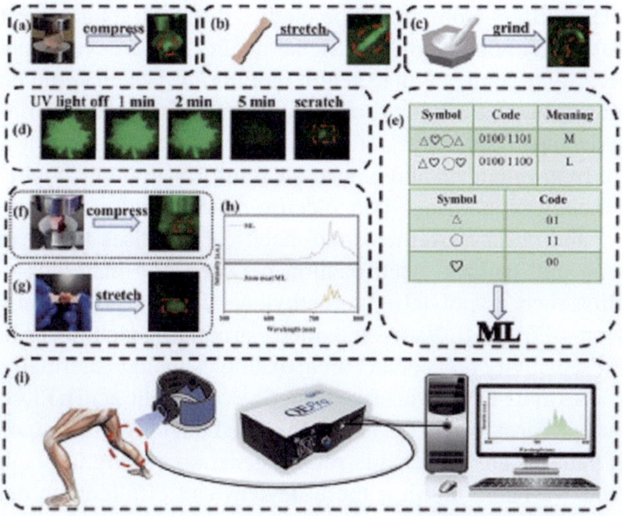


**Fig. 9.9** Possible applications in multi-modal anti-counterfeiting and biological stress detection. **a** Photographs of the pellet prior to and during loading; **b** Photographs of the PDMS elastomer prior to and during stretching; **c** Photographs of the powder prior to and during grinding; **d** PersL and ML images of the leaf film post-exposure to 254 nm UV light for 30 s; **e** Transformation of "ML" characters into ASCII codes; **f** Photographs of the pellet and pork tissue prior to and during loading; **g** Photographs of the PDMS elastomer and pork tissue prior to and during stretching; **h** ML spectra after traversing either nothing or 3 mm of meat; **i** Schematic representation of bioimaging. Reproduced from Ref. [63] published by Advanced Powder Materials. Licensed under Creative Commons Attribution 4.0

methods guarantees long-lasting and undetectable integration of energy collecting into clothes [67, 68]. Perovskite-enabled wearables integrating energy harvesting, sensing, and dynamic display into textiles are ushering in a new era in fashion technology. Clothing will no longer be viewed as passive covers but rather as smart interfaces in sync with their user and environment.

### 9.4.1 Wearable Diagnostic Textiles

The COVID-19 epidemic has highlighted the need for remote, clinically dependable monitoring, which has driven research into smart wearables for customized health services during the past twenty years. Embedding fabric-compatible sensors into daily clothing allows for continuous collection of physiological data outside of hospitals or clinics—data that would typically be limited to isolated, in-office examinations. For patients with chronic cardiopulmonary issues, this ability is quite crucial. At the moment, they must rely on sporadic visits and fast fixes that miss short-term trends in their recovering or progressing health. Newer developments record complex signals with a level of accuracy that is on par with medical-grade systems using cooperative sensor architectures—networks of densely packed, flexibly wired fabric sensors. This differs from prototype clothes that have long exhibited single-sensor functionality. Without sacrificing signal quality, these developments lay the foundation for real-time home cardiac health monitoring, which could enable early intervention and improved treatment possibilities [69]. Essential for life support is the cyclic gas exchange happening during physiological respiration; it also generates a treasure of biomechanical and biochemical information that could be applied for noninvasive health monitoring. Triboelectric nanogenerators convert tiny variations in airflow and chest movement into electrical impulses by means of electrostatic induction and contact electrification between two materials—electrodes and polymer films such as polytetrafluoroethylene (PTFE), polydimethylsiloxane (PDMS), or Nylon—Fig. 9.10. Nano- and micro-structuring techniques have significantly increased sensitivity and contact area; examples include thermally expanded hollow microspheres and PDMS with a poly (lactic-co-glycolic acid) (PLGA) pattern. Among the wearable formats that have demonstrated ongoing tracking of breathing rate, depth, and patterns are stretchable rubber generators, fully implantable core–shell devices, and washable textiles; these devices can differentiate between rest and exercise and enable breath-driven assistive controls. Exhaled breath contains hundreds of chemical species that could indicate metabolic, pulmonary, renal, and liver diseases. Two proven self-powered sensing methodologies exist for respiration monitoring: one emphasizes the observation of physical respiratory behaviors, while the other is intended for the detection of chemical reagents in exhaled breath. Driven by a TENG, the gas-sensitive interdigital electrodes in the RC circuit indicate changes in resistance when the voltage drops caused by breathing in the impedance-assisted mode. Using this technique, nitrogen dioxide has been found with a linear response up to tens of parts per million, ammonia at parts per billion, and ethanol for drunk-driving alerts. In the independent-capacitance mode, a sensor film—such as chitosan-graphene, ZnO hybrids, or polyethylenimine—is integrated straight onto the tribo-layer of the TENG. Adsorption of volatile molecules alters the dielectric permittivity, which modulates the open-circuit voltage. These systems have achieved a great deal without requiring outside power, including the linear detection of acetone as a sign of diabetes, the differentiation of inhalation from exhalation by $CO_2$ chemisorption, and the monitoring of humidity changes connected with breathing patterns [70–72].

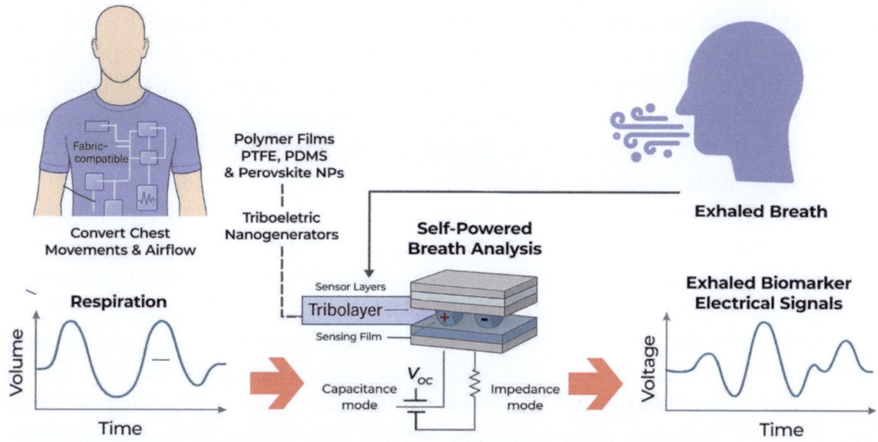


**Fig. 9.10** Self-powered wearable smart textile transforms health monitoring by incorporating soft electrodes and triboelectric nanogenerator fibers to capture chest motions and airflow, enabling continuous respiratory tracking and biochemical analysis using advanced sensing methods

The adjustable features, simple manufacture, and low price of perovskite-based humidity sensors have also aroused interest in their possible application for wearable, portable breath monitoring. Their natural material instability, especially in humid settings, has restricted their practical use. A resistive-type sensor made of bismuth-based perovskite $Cs_3Br_2Bi_9$, which has demonstrated stable operation for over 35 days at 90% relative humidity, offers one answer to this issue. The device has a large detection range, high selectivity, and very quick reaction and recovery times of 5.56 s and 6.24 s, respectively. Breathing patterns can be tracked in real-time using this strong but energy-efficient technology included into a marketable face mask. This paves the way for simple diagnostic breath analysis and the potential of wearable respiratory monitoring in the future [73]. Complementing the resistive approach, an impedance-type sensor built from ultralow-voltage (20 mV) $CsPbBr_3$ nanocrystals produced via hot injection offers an alternative route to low-power, continuous respiratory monitoring. Because to their extremely smooth surfaces and low defect density, these cubic, 100-faceted nanocrystals may undergo rapid and completely reversible impedance changes when exposed to room temperature water vapor. Electrochemical impedance spectroscopy claims that proton hopping over physisorbed water layers at low relative humidity controls the sensor response. In $CsPbBr_3$, however, natural surface flaws increase charge transport at higher humidity. This device exhibits remarkable repeatability, almost no hysteresis, and robust long-term phase stability. When integrated into a wearable face mask, its capacity to remotely and accurately track breathing rates in real time offers great promise for low-power, continuous health monitoring [74].

The COVID-19 epidemic has underlined another essential: personal protective equipment that not only filters airborne viruses but also actively inactivates them, even if these perovskite sensors are excellent in detecting humidity variations related with breath. Although they effectively block particles, conventional masks lack an inherent biocidal action and could provide a substrate for bacterial growth. By capturing the 2–5 °C temperature differentials between exhaled air and ambient air—roughly 10–20 mL every breath—a unique, self-sustaining hybrid patch reduces this problem using a pyroelectric nanorod array grown directly on a flexible paper substrate. A two-stage hydrothermal process affixes dense arrays of tens of nanometers in diameter and several micrometers in length to cellulose fibers; printed interdigitated electrodes capture the transient surface charges generated during each heat cycle. Breathing while wearing a mask alternates exposure to heated exhaled air and cooler ambient air, producing rapid polarization changes that enable surface redox reactions. Research shows that adsorbed oxygen and moisture generate hydrogen peroxide, superoxide anions, and hydroxyl radicals—which can inactivate bacterial and viral surrogates by more than 99.5% within fifteen minutes of cyclic breathing. Because the anisotropic distortion of the nanorods overlays piezoelectric fields over the pyroelectric response, additional bending of the patch increases reactive oxygen species (ROS) production by up to 40%. With ion release levels far within cytotoxic limits, the biocidal activity is totally restored upon air-drying, requires no external energy, and is safe for long dermal contact [75]. The integration of economical, reliable perovskite humidity sensors for real-time breath monitoring with a self-sustaining, ROS-generating pyroelectric patch for active disinfection indicates a progression towards advanced smart masks. Without further chemicals or outside power sources, these devices could simultaneously analyze breathing patterns and eliminate germs. Future projects will include responsive wear indicators, confirm clinical and field efficacy, and improve long-term durability under mechanical stress and humidity, so transforming personal protection equipment into flexible, multipurpose platforms.

Combining simplified manufacturing with exact crystal engineering, a new "perovskite" mask has been developed to work as a face membrane for the production of perovskite films. In this approach, thermal annealing brings a polydimethylsiloxane membrane infused with an organic passivating agent into close proximity to newly cast perovskite precursor films. This in situ "masking" directs nucleation, encourages uniform crystal formation, and controls solvent evaporation, therefore creating a limited microenvironment. Good photovoltaic performance depends on precisely controlled interfacial energy levels, vertical orientation of the (100) crystal plane, and very smooth surfaces of the resulting perovskite layers. Adding perovskite masks to combinations of methylammonium (MA) and FA-MA-Cs produced solar cells with open-circuit voltages up to 1.15 V and efficiency up to 21.4%. These cells also showed significantly improved thermal stability and humidity resistance. Apart from enabling simple control of perovskite crystallization, this multipurpose membrane-assisted technique reveals the structural and interfacial qualities of high-quality films, therefore improving and prolonging perovskite photovoltaics [76]. In labor-intensive industries such as mining and construction, unrecognized fatigue not only undermines productivity but also elevates the risk of accidents and occupational illnesses, costing U.S. employers an estimated $101 billion annually. Traditional methods, such as bulky wearable heart-rate monitors or subjective surveys, are not precise enough or not practical enough to measure physical effort in real-time on the job site. By contrast, breathing patterns could stand in for workload and noncontact humidity monitoring is proving to be a method because of its accuracy and absence of motion distortions. Most humidity sensors, however, struggle to combine basic construction with the sensitivity, stability, and rapid response/recovery required for reliable on-mask monitoring, especially in high relative humidity (RH) conditions. A lead-free perovskite sensor for breath detection and labor intensity grading was developed using $Cs_3Cu_2I_5$ nanocrystals in reaction to these challenges. Adjustable bandgap, great carrier mobility, natural electrical conductivity at normal temperature, and fault tolerance all helped $Cs_3Cu_2I_5$ to be selected. Combining the ionic liquid ethyl-methylimidazolium bis(trifluoromethylsulfonyl)imide (EMI-TFSI) helped to further passivate surface flaws and improve electrochemical stability. A reversible phase change to $CsCu2I_4$ causes humidity sensitivity; it happens when the perovskite layer comes into contact with moisture. Production and breaking of Cs–I and Cu–Cu bonds drives this transition (Fig. 9.11). With response and recovery times as short as 2.0 s and 0.5 s, respectively, electrochemical impedance spectroscopy and density functional theory calculations uncovered the mechanisms of ion migration and interfacial charge transfer that produce a notable change in resistance of 161.3 MΩ (relative ratio 0.967) when relative humidity changes from 40 to 90%. For practical evaluation, the enhanced humidity sensor readily linked to a smart-mask prototype. Real-time breathing data from 240 labeled samples was used to train a neural network model for three-level labor intensity categorization. This data was then distilled to 19 intended features comprising breathing-cycle frequency, phase-transition kinetics, and resistance dynamics. External validation verified the accuracy at 0.90. A visualization tool that goes with the app shows real-time humidity traces, predicted intensity levels, and alerts signaling when physical activity is getting too much. By combining an affordable, robust perovskite sensing layer with machine learning-driven analysis, this work offers a practical, non-contact way to gauge labor intensity. In challenging work settings, it has the capacity to increase efficiency and safety [77].

Just a few of the exciting applications for flexible multifunctional sensors combining visual input with concurrent humidity and strain measurement are electronic skin, smart clothing, and physiological monitoring. Often, in the quest of merging these several qualities, sensitivity, durability, and manufactability are sacrificed. Recently, a coaxial filament sensor was created to surpass these limitations. Constructed with dip-coating and wet-spinning techniques, this structure comprises a carbon nanotube (CNT)/sodium polyacrylate (PAAS) shell surrounding a flexible perovskite/thermoplastic polyurethane (TPU) core. Encased in polymer, the core's uniform distribution of perovskite nanocrystals generates a bright, tuneable luminescence that stays visible even after 200% mechanical deformation. Exposed to moisture, the outer PAAS/CNT layer swells radially, indicating sensitivity to humidity. At 95% RH, this results in a resistance change of 130%, a reaction time of 3.2 s, a recovery time of 4.0 s, minimum hysteresis of 3.5%, and exceptional cycle stability. Moreover, the structure has almost no thermal interference. While simultaneously supporting stresses from 95 to 200%, the elastic core gets a gauge factor of 27.0. It offers ultrafast strain detection with a reaction time of 0.2 s and a recovery time

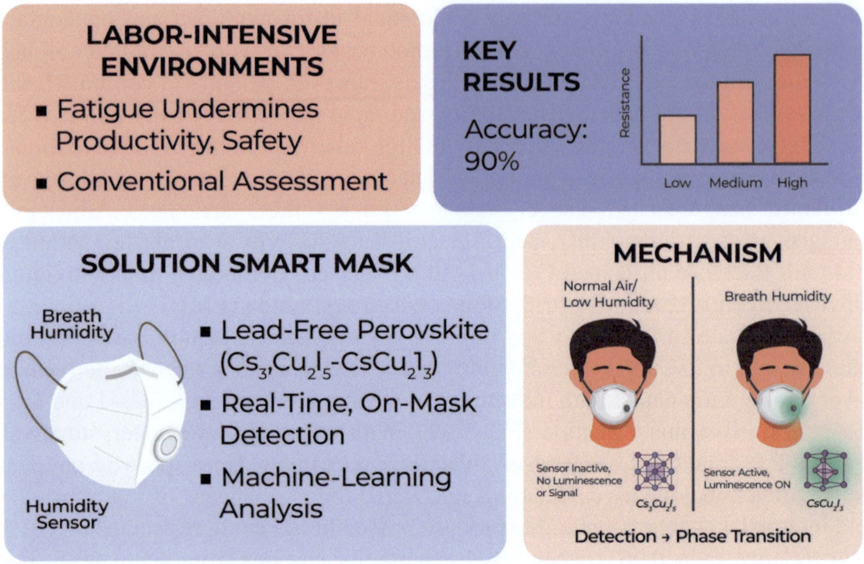


**Fig. 9.11** Smart mask integrated with $Cs_3Cu_2I_5$–$CsCu_2I_3$ perovskite humidity sensor for labor-intensity monitoring in high-exertion environments

of 0.3 s, showing again minimal temperature cross-sensitivity. Filaments, which combine luminous signaling with scalable manufacturing and strong humidity and strain sensing, could enable advanced wearable devices able to detect humidity and strain in real-time and offer early alerts of possible hazards [78]. Light-emitting hydrogels offer one more way for visual input in flexible sensors. One such is a water-resistant perovskite hydrogel. Evenly putting $CsPbBr_3$@PbBr(OH) nanocrystals into a stretchy polyacrylamide (PAAm) network helped to achieve this. This one-piece material can simultaneously detect and display the strain. At a fracture strain of 740%, it has a mechanical strength of 51.3 kPa, is very water-resistant, and glows bright green under ultraviolet exposure. Even after being stretched over 700%, this fluorescence remains. Mechanical deformation causes the electrical resistance of knitted hydrogel textiles to vary, therefore generating measurable electronic signals. At the same time, its fluorescence variations offer real-time visual input. By combining sensing and display inside a continuous matrix, this eliminates the need for distinct sensor and display modules. The system is especially reliable both on land and in water [79]. A new two-nozzle electrospinning technique was created for continuous, twisting filament manufacturing to allow scalable production of such perovskite-integrated fibers. After co-injecting perovskite and polymer precursors using opposing spinnerets, a high-voltage field is used to in situ crystallize $CsPbX_4$ quantum dots inside a polyacrylonitrile (PAN) matrix. Using a central vortex collector and a spinning mandrel, fluorescent filaments the length of a kilometer could be twisted from nanofibers. Encased in PAN, the quantum dots are shielded from moisture and oxygen. Even after being bent, stretched, or immersed in organic or pH

1–13 water for 24 h, the resulting filaments have consistent emission colours (25–27 nm FWHM). With a photoluminescence quantum yield of 53.8%, colour tunability via halide composition, and weaving-friendly mechanical flexibility, these ongoing perovskite filaments reveal a scalable approach to breathable, conformable opto-electronic textiles. They open up interesting new options for fiber-based displays, wearable luminescent clothes, and hazard-warning systems [80].

Using the inherent microfluidic structure of textiles (e.g., Janus-wetting layers, fractal channels, or cotton capillaries), flexible textile-based sweat sensors route perspiration to functionalized sensing areas, so lowering evaporative losses. These zones combine ion-selective electrodes and conductivity measurements for electrolytes ($Na^+$, $K^+$, $Cl^-$, $Ca^{2+}$), potentiometric membranes for pH, and enzymatic/electrochemical interfaces for metabolites (lactate, glucose, uric acid) and hormones including cortisol. Combining stretchable electronics, multiplexed sensor networks, and wireless communication systems creates coherent platforms that allow real-time, multianalyte monitoring of hydration, metabolic activity, and stress free of invasive blood draws (Fig. 9.12) [81, 82]. By including Joule heating components into textiles, recent innovations overcome the limits of conventional sweat collecting methods including activity-dependent iontophoresis or heating systems. A suitable skin patch with a microheater increases localised skin temperature by 5–7 °C, so enabling controlled, on-demand perspiration independent of physical activity. From inactive activities to cold environments, the generated sweat is delivered via a microfluidic system to sensor arrays including temperature-compensated biochemical detectors, guaranteeing homogeneous fluid volume and analytical precision across many conditions. This system improves user comfort by autonomous, noninvasive biofluid collecting while facilitating continuous metabolic assessment (e.g., hydration, electrolyte equilibrium, uric acid). Eliminating reliance on labor-intensive stimulation techniques helps the system link laboratory-scale discoveries to useful, wearable health-monitoring devices, so enabling custom diagnostics, early disease detection, and remote patient management [83].

Achieving power autonomy is a critical consideration for multi-analyte wearable systems. Addressing this, a flexible quasi-2D perovskite solar cell can be seamlessly integrated onto the fabric, enabling the self-powered biosensor to convert over 31% of incident ambient light into usable energy. Capillary-driven microfluidic networks within the textile ensure continuous sweat delivery to various sensing elements, including glucose oxidase amperometric electrodes, ion-selective membranes for pH and sodium, capacitive flow-rate sensors, and thermistors for temperature monitoring. A low-power microcontroller paired with a Bluetooth Low Energy (BLE) radio handles on-board data processing and wireless transmission. In trials encompassing both sedentary indoor activities and outdoor exercise, the device successfully tracked postprandial glucose spikes, pH dynamics, electrolyte depletion, sweat rate, and skin temperature changes over 12 continuous hours. The integrated solar cell demonstrated resilience under motion, retaining over 80% of its initial output after repeated bending. This fusion of high-efficiency perovskite photovoltaics and multiplexed textile sensors yields wearable diagnostic platforms that are entirely self-sufficient in both power generation and fluid management, opening avenues for

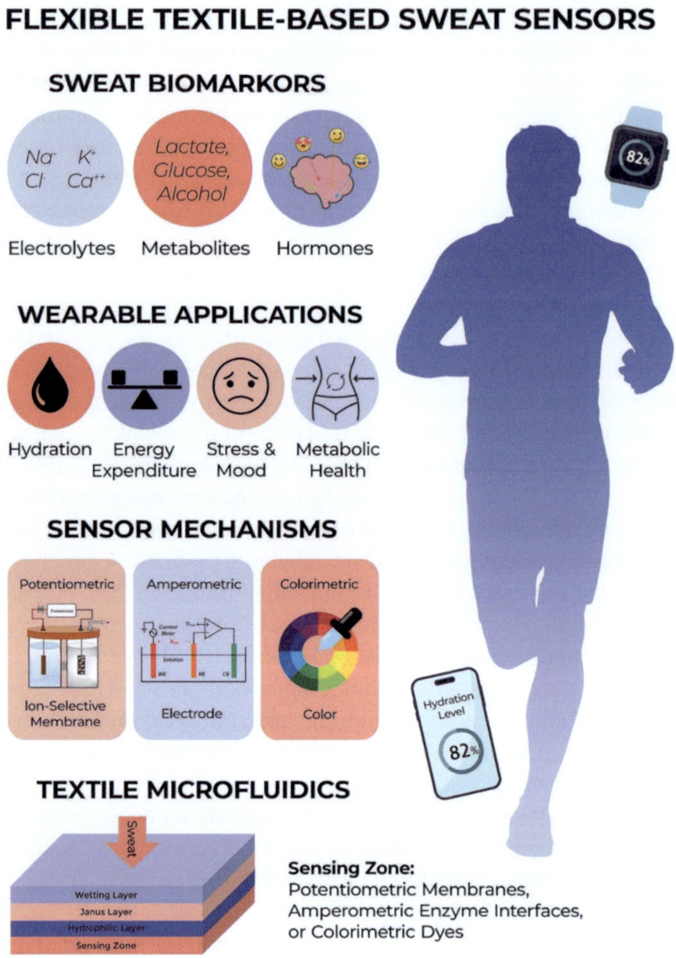


**Fig. 9.12** Flexible textile-based sweat sensors for real-time, noninvasive physiological monitoring

long-term metabolic monitoring, early disease detection, and personalized fitness tracking [84].

Though fabric-based sensors are widely used for environmental monitoring, biometric data collecting, and energy harvesting, no ionizing radiation detector has yet been included into porous, soft substrates. Neither traditional dosimeters' rigid silicon diodes nor their laborious flat-panel arrays are comfortable or conformable for long-term use, nor are they suitable for assembly into large, pixelated matrices using low-cost textile methods. Flexible plastics or membranes have shown promise for solution-processed organic semiconductors and lead halide perovskites under X-ray and gamma-ray detection (Fig. 9.13). Lead halide perovskites have several desirable qualities, including low-temperature deposition, long carrier lifetimes, and great charge-carrier mobility. Still, textile substrates are incompatible with mechanically rigid single crystals or thick-film detectors requiring vacuum or high-temperature processes. Recent developments in perovskite-infiltrated membranes and ultraflexible photodiodes have attained single-crystal performance under zero-bias operation using complex multilayer structures and inert-atmosphere manufacturing. Everyday textiles can now be used to make an entirely textile X-ray detector. Screen-printed onto cotton and conductive polymer patterns form electrodes. Unlike prior methods that merely coat textiles with wires or planar films, this technique grows perovskite crystals directly within silk-satin fibers, thereby transforming the fabric itself into the active X-ray–absorbing layer. While reaching low detection limits, consistent operation at low voltages, and high sensitivity, this woven-material assembly preserves the scalability, comfort, and breathability of the original material. Industry-compatible techniques that produce a continuous, conformal detector are screen printing, drop casting, and conductive-fiber stitching. Placing the perovskite layer between two fabric electrodes solves health and environmental concerns and lets natural, recyclable materials be used. This rules out direct skin contact with lead. A large-area, pixelated matrix of these textile sensors reveals a practical solution to wearable dosimeters that can track radiation exposure over wide areas. These conformal clothes or belts let real-time tracking of ionising radiation exposure when medical personnel or patients are undergoing radiotherapy or diagnostic imaging [85]. Further progress in addressing lead toxicity and charge transfer limitations in perovskite films, as well as the rigidity of bulk crystal detectors, has led to the development of a new class of lead-free photoferroelectric hybrid metal halide perovskites. These materials combine robust mechanical properties with spontaneous polarization to facilitate efficient charge separation and transport within the membrane structure. Sensitivities reach up to $7872 \pm 517$ μC $Gy^{-1}$ $cm^{-2}$ under a 50 V bias and $394 \pm 67$ μC $Gy^{-1}$ $cm^{-2}$ in self-driven mode, with a detection limit below 77 nGy $s^{-1}$. Spontaneous polarization inside the membrane structure enables effective charge separation and transport; imaging performance is on line with or beyond industry standards. All have been demonstrated in air and water stability, mechanical wear resistance, and long-term radiation exposure. The angular and energy responses match the criteria established by the International Electrotechnical Commission for personal dose equivalent meters that read directly. This mix of high sensitivity, safety, and durability has opened up a promising path for next-generation flexible, wearable radiation detectors able to monitor in difficult conditions in real-time [86]. Flat-panel imagers made from quasi-2D Dion-Jacobson perovskites with flexible dosimeters have recently demonstrated record-level performance under drastically reduced X-ray dosages. By adding diammonium interlayers into the perovskite lattice, these devices combine very efficient charge transfer with enhanced structural stiffness (which prevents ion migration). Producing high-quality pictures at just one-fifth the dosage of commercial systems, the panels that are created are able to attain sensitivities around $1.8 \times 10^4$ μC $Gy^{-1}$ $cm^{-2}$, detection limits below 6 nGy $s^{-1}$, and spatial resolutions below megahertz. These advances show the effectiveness of tailored perovskite structures in X-ray detection, where they offer both stability and extremely high sensitivity,

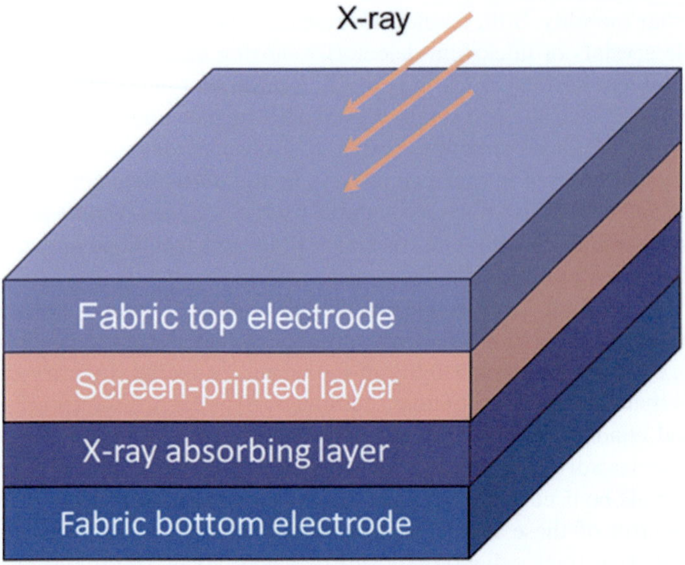


**Fig. 9.13** Wearable textile radiation dosimeter with conformal pixelated detection

and set out a clear route for using similar material and device ideas to the area of conformal dosimetry in wearables [87].

### *9.4.2 Camouflage Textiles*

Many animals in nature, including chameleons, cephalopods, frogs, and butterflies, have perfected the art of dynamic camouflage by changing their surface optics to fit in with their surroundings. These techniques have inspired the creation of adaptive materials actively controlling the interactions between light and substance, including mechanochromic elastomers, thermochromic coatings, and electrochromic films. However, most existing platforms either depend on external power supplies or are limited to a single spectral band, thereby compromising their reliability under fluctuating illumination, humidity, and temperature conditions. Recent advances in passive, dual-mode textiles harness microencapsulated cholesteric liquid crystals and nanostructured hygroscopic films to achieve reversible, self-adaptive camouflage across both visible and infrared spectra. In these systems, modest temperature changes ($\Delta T = 5$–$15$ °C) modulate the helical pitch of polymer-embedded TLCs—augmented by thermochromic leuco dyes—to effect vivid, tunable color shifts from red through blue, while an underlying Fabry–Pérot interferometric cavity autonomously adjusts mid-IR emissivity in response to ambient moisture and temperature. Spectroscopic reflectance/emissivity measurements and Bragg-law pitch analyses confirm fine, loss-free control of optical response, and durability tests demonstrate stable performance over hundreds of thermal and humidity cycles. By integrating these two passive sensing modalities without external power, the fabric "learns" environmental cues to deliver energy-efficient camouflage, with significant implications for military uniforms, protective apparel and adaptive architectural surfaces, Fig. 9.14 [88, 89]. Scalable microfabrication techniques are critical to translating dual-mode camouflage textiles from the laboratory to practical use while preserving wearability and durability. Embedding thermochromic and emissivity-tuning components directly into polymer melts during fiber spinning yields inherently adaptive yarns without post-treatment. For high-throughput batch production, roll-to-roll processes—such as screen printing or slot-die coating of microencapsulated cholesteric liquid crystals and hygroscopic Fabry–Pérot stacks—can uniformly deposit functional layers with sub-micrometer thickness control, ensuring rapid optical response and minimal stiffness. Applications requiring fine spatial control of visible reflectance and mid-IR emissivity—for example, urban or desert digital camouflage—can leverage inkjet or 3D printing of metamaterial-inspired microstructures at defined textile loci. By patterning these regions, one can selectively tailor spectral signatures across the fabric surface. Further integration of low-power environmental sensors and microcontroller units would enable semi-active systems that dynamically adjust color shifts and emissivity limits in response to real-time stimuli. Rigorous validation through standardized tests of color fidelity, infrared signature attenuation, wash durability, and mechanical endurance under realistic wear conditions is essential to establish long-term performance. Ultimately, the convergence of material innovations, scalable textile processing, and digital fabrication is poised to deliver fully autonomous, self-learning camouflage fabrics for advanced military and security applications, adaptive architectural skins, and responsive fashion systems [90]. Complete, power-free camouflage is possible with two more textile techniques based on the thermoregulation and spectrum hiding mechanisms present in plant leaves. Using multilayered electrospun fibres with a gradient-porous structure that conducts fluids and allows passive evaporative cooling, the first one imitates the absorption and reflection characteristics of foliage. Embedded hues finish the image. This design not only changes its surface temperature to ambient conditions but also deceives hyperspectral detection by means of reversible color changes reflecting natural aging and by mimicking the red-edge feature of healthy leaves. Simply printing organic dyes onto fabric and then scattering particles produces plant-like variations in total spectral response in the second approach. This produces light- and dark-green types. Lightweight, breathable, and mechanically robust, these printed textiles are perfect for use as clothing. Combining these bioinspired techniques—moisture-driven cooling for dynamic thermal management and pigment formulation for exact spectral matching—we can produce scalable, totally passive textiles that blend into plant-dominated environments across several detection modalities, all without external power or sophisticated controls [91, 92]. In conclusion, the integration of passive thermal-control layers with responsive chromic systems establishes a transformative paradigm in textile camouflage, enabling adaptive concealment across both visible and infrared spectra. Without requiring external power or laborious controls, this new system operates across many spectral bands. By employing scalable manufacturing and nature's methods, these smart materials offer practical manufacturability, comfortable wear, and efficient concealment. These developments open the door for radical applications in personal safety, defense, and other sectors.

### 9.4.3 Sport Equipment

Fiber-based wearable strain sensors transduce athletes' biomechanical movements into quantitative data streams, underpinning performance optimization, injury risk mitigation, and personalized coaching in next-generation sports equipment. By combining the comfort, breathability, and flexibility of textiles with the high sensitivity and fast reaction needed to detect everything from tiny muscle contractions to explosive joint extensions, these textile-integrated devices exceed stiff platforms. Staple textile fibers are rendered electrically conductive by either surface deposition or bulk incorporation of functional nanomaterials—such as carbon nanotubes, graphene flakes, perovskite nanocrystals, metallic nanowires, and conjugated polymers—thereby enabling robust charge transport within flexible fiber-based architectures. Regular yarns become sensor threads with capacitance or resistance that changes as they are bent, compressed, or stretched. By leveraging scalable, cost-effective fabrication methods—such as electrospinning, dry and wet fiber spinning, and surface coating—these strain sensors achieve conformal integration into sportswear, ensuring a snug fit with minimal perceptibility. They also withstand repeated deformation. For example, in combat sports, a sensor array built into a flexible glove or training suit could track punch speed, contact force, and hit position to the millisecond, therefore enabling players and trainers to use this real-time data to fine-tune their technique. Embedded in running clothes or compression sleeves, sensors track endurance sports metrics including muscle exhaustion, joint angles, and gait symmetry, and they send alerts when overuse or bad form causes damage. Apart from performance analysis, fiber sensors enable constant health monitoring that preserves athletic flow. This covers even sweat composition, heart rhythm, and breathing rate. Recent developments in materials science and textile engineering are bringing fully autonomous, intelligent sportswear that learns from and amplifies human movement in real time within reach. The next great thing is to embed wireless data transmission, multi-modal sensing (pressure, temperature, chemical biomarkers), and self-powered energy harvesters directly into textiles [93, 94]. The digital training systems of today's smart sports require the same level of accuracy and responsiveness that were once reserved for laboratory settings. Traditional vision-based and motion-capture systems struggle to accurately resolve the fast-paced, high-frequency interactions seen in combat sports like boxing and martial arts, as well as to identify precise sites of contact or immediate pressures during live exercises. Designed to

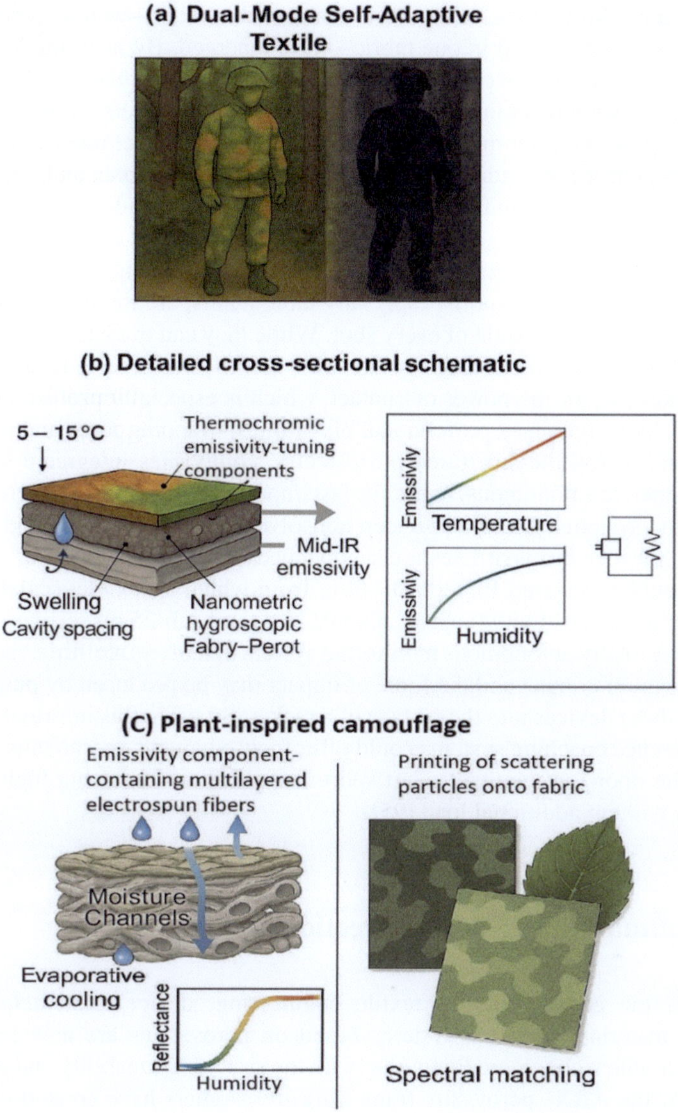


**Fig. 9.14** Bioinspired dual-mode adaptive camouflage textiles: **a** Visible-light and thermal-infrared images reveal dynamic color adaptation and heat signature blending through microencapsulated cholesteric liquid crystals and hygroscopic layers. **b** Self-adaptive garments harmonize with forest environments, utilizing temperature-responsive materials for seamless integration. **c** Innovative plant-inspired designs incorporate gradient-porous fibers and organic dyes, offering scalable, energy-efficient camouflage solutions for woodland wear

overcome these limitations, fully textile-integrated sensors—which combine proximity and pressure sensing in one fabric—have emerged. By inserting a structured dielectric spacer layer between conductive woven electrodes, this bimodal all-textile sensor detects an approaching limb without touch and records the force of each strike at the same time. This approach transforms ordinary training clothes into a wireless monitoring platform providing real-time data on punching force and speed, therefore opening new opportunities for data-driven optimization, adaptive feedback, and scientific coaching in a range of high-intensity sports in a cost-effective manner that preserves breathability, flexibility, and durability [93]. Among the most important measurements in racquet sports, especially table tennis, are the precise location of contact and the power profile of every shot. While they can track the ball and player paths, traditional vision-based systems cannot identify the precise location of a hit on the racket face or the power of contact, which is especially problematic during fast-paced, high-frequency professional play rallies. Mapping force and impact site in real-time, perovskite-structured piezoelectric composites integrated into sports clothing generate autonomous, lightning-fast impact sensors that outperform conventional vision systems. Using a two-step nonsolvent phase-separation and hot-press technique, piezoelectric particles are sandwiched between thin lamellar crystals to form a "baklava"-layered PZT/PVDF film, from which a flexible, scalable sensor with exceptional sensitivity (approx. 6.4 mV $N^{-1}$) and sub-25 ms response could be produced. A totally autonomous monitoring system that offers real-time feedback on both the point of contact and the force of impact may be produced by putting a 6 $\times$ 6 array of these devices into the surface of a racket. By using this information, artificial intelligence coaching systems could offer focused, precise comments, therefore opening the door for creative "smart sports" applications requiring high temporal resolution without additional load [95].

## 9.5 Summary and Future Direction

Thanks to the convergence of textile engineering, device manufacturing, and advanced materials research, systems based on perovskites are now leading the way in wearable technology. Beginning with the structural tunability and multifunctionality of the $ABX_3$ perovskite framework, researchers have created conformal, on-body platforms capable of collecting mechanical, thermal, and light energy using halide and oxide variations. Halide perovskites (e.g., $MAPbI_3$ and $CsPbBr_3$) combine high power-conversion efficiencies, tunable bandgaps and intrinsic piezoelectricity, making them well suited for strain sensors, electroluminescent fibers and flexible solar cells; in contrast, oxide perovskites (e.g., $BaTiO_3$ and $KNbO_3$) exhibit pronounced ferroelectric and pyroelectric responses, rendering them ideal for long-term mechanical and thermal sensing. Improvements in device designs have made these material qualities now accessible in fully integrated systems. Fiber-based perovskite solar yarns and fabric-laminated modules cause on-garment power generation. Layered photonic and hygroscopic stacks provide dual-mode visible/

infrared camouflage. Composite filaments combine humidity sensing, mechanoluminescence, and luminescence into one stretchable fiber. Hybrid energy harvesters such PV-TEG tandems and piezo-phototronic interfaces increase both power densities and operational autonomy. Smart masks with pyroelectric disinfection, conformal X-ray dosimeters, and neural-network-assisted labor intensity monitoring are just a few examples of the real-world demands that perovskite-enabled sensing systems have been able to satisfy. Important challenges remain to be surmounted, however. The halide perovskites' need for better encapsulation and lead-free formulations to handle moisture and heat, the oxide perovskite films' need for continuous innovation in composite and nanostructured designs to handle mechanical resilience, and the need for printing, roll-to-roll, and fiber-spinning processes to handle large-scale manufacturability while maintaining performance and comfort. Connecting lab-scale prototypes to commercial products will depend on establishing uniform criteria for washability, biocompatibility, and long-term durability. Combining perovskite technology with artificial intelligence, safe data protocols, and sustainable manufacturing will help future interactive clothing to be possible. Through seamless incorporation of sensing, energy harvesting, signal processing, and display functionalities into textile architectures, perovskite-infused smart fabrics are poised to transform diverse sectors—including healthcare diagnostics, human–machine interfaces, and adaptive camouflage systems for military/civilian applications. Leveraging scalable textile platforms and the unique optoelectronic/electromechanical properties of perovskites, these garments could enable responsive, multifunctional capabilities—such as autonomous strain monitoring, environmental adaptation, and integrated energy management—within the next decade.

## References


1. A. Yadav, K. Yadav, Transforming healthcare and fitness with AI powered next-generation smart clothing. Discov. Electrochem. **2**(1), 2 (2025). https://doi.org/10.1007/s44373-025-00015-z
2. R. Huang, Y. Yan, Review of research status and development expectation of health smart clothing, in *Digital Human Modeling and Applications in Health, Safety, Ergonomics and Risk Management*, ed. by V.G. Duffy (Springer Nature Switzerland, Cham, 2024), pp. 70–86
3. A. Satharasinghe, T. Hughes-Riley, T. Dias, A review of solar energy harvesting electronic textiles. (2020). https://doi.org/10.3390/s20205938
4. J. Yin et al., Smart textiles for self-powered biomonitoring. Med-X **1**(1), 3 (2023). https://doi.org/10.1007/s44258-023-00001-3
5. F. Mokhtari, Z. Cheng, R. Raad, J. Xi, J. Foroughi, Piezofibers to smart textiles: a review on recent advances and future outlook for wearable technology. J. Mater. Chem. A **8**(19), 9496–9522 (2020). https://doi.org/10.1039/D0TA00227E
6. C.-W. Kan, Y.-L. Lam, Future trend in wearable electronics in the textile industry. (2021). https://doi.org/10.3390/app11093914
7. X. Yin, H. Fu, B. Xu, Advanced design of light-assisted nanogenerators with multifunctional perovskites. Adv. Energy Mater. **14**(13), 2304355 (2024). https://doi.org/10.1002/aenm.202304355

8. K. Sakhatskyi, A. Bhardwaj, G.J. Matt, S. Yakunin, M.V. Kovalenko, A decade of lead halide perovskites for direct-conversion X-ray and gamma detection: technology readiness level and challenges. Adv. Mater. 2418465 (2025). https://doi.org/10.1002/adma.202418465
9. L. Chen, Y. Chuang, W.-D. Yang, K.-C. Tsai, C.-W. Chen, C.-D. Dong, All-inorganic perovskite CsPbX3 electrospun nanofibers with color-tunable photoluminescence and high performance optoelectronic applications. J. Alloys Compd. **856**, 157426 (2021). https://doi.org/10.1016/j.jallcom.2020.157426
10. A.K. Baranwal, S. Hayase, Recent advancements in tin halide perovskite-based solar cells and thermoelectric devices. (2022). https://doi.org/10.3390/nano12224055
11. W. Gao, Y. Zhu, Y. Wang, G. Yuan, J.-M. Liu, A review of flexible perovskite oxide ferroelectric films and their application. J. Mater. **6**(1), 1–16 (2020). https://doi.org/10.1016/j.jmat.2019.11.001
12. S. Aftab et al., Flexible perovskite solar cells: a revolutionary approach for wearable electronics and sensors. Mater. Today Energy **51**, 101872 (2025). https://doi.org/10.1016/j.mtener.2025.101872
13. R. Zhao, Z. Gu, P. Li, Y. Zhang, Y. Song, Flexible and wearable optoelectronic devices based on perovskites. Adv. Mater. Technol. **7**(3), 2101124 (2022). https://doi.org/10.1002/admt.202101124
14. S. Saifi et al., An ultraflexible energy harvesting-storage system for wearable applications. Nat. Commun. **15**(1), 6546 (2024). https://doi.org/10.1038/s41467-024-50894-w
15. D.B. Niranjan, J. Jacob, B.R. Vaidehi, M. Peter, J. Medikonda, P.K. Namboothiri, Current status and applications of photovoltaic technology in wearable sensors: a review. Front. Nanotechnol. **5** (2023). https://doi.org/10.3389/fnano.2023.1268931
16. K.A. Kim, F.S. Bagci, K.L. Dorsey, Design considerations for photovoltaic energy harvesting in wearable devices. Sci. Rep. **12**(1), 18143 (2022). https://doi.org/10.1038/s41598-022-22232-x
17. G. Nazir et al., Stabilization of perovskite solar cells: recent developments and future perspectives. Adv. Mater. **34**(50), 2204380 (2022). https://doi.org/10.1002/adma.202204380
18. F.-C. Liang et al., Self-healing polymers in rigid and flexible perovskite photovoltaics. InfoMat **7**(1), e12628 (2025). https://doi.org/10.1002/inf2.12628
19. M. Fahim, I. Firdous, W. Zhang, W.A. Daoud, Bifunctional interfacial engineering for piezo-phototronic enhanced photovoltaic performance of wearable perovskite solar cells. Nano Energy **86**, 106127 (2021). https://doi.org/10.1016/j.nanoen.2021.106127
20. G. Li, S. Shittu, T.M.O. Diallo, M. Yu, X. Zhao, J. Ji, A review of solar photovoltaic-thermoelectric hybrid system for electricity generation. Energy **158**, 41–58 (2018). https://doi.org/10.1016/j.energy.2018.06.021
21. S. Zhu et al., Persistently self-powered wearable thermoelectric generator enabled by phase-change inorganics as the heat sink. Mater. Today Phys. **32**, 101011 (2023). https://doi.org/10.1016/j.mtphys.2023.101011
22. W. Ren et al., High-performance wearable thermoelectric generator with self-healing, recycling, and Lego-like reconfiguring capabilities. Sci. Adv. **7**(7), eabe0586 (2025). https://doi.org/10.1126/sciadv.abe0586
23. L. Mo, Y. Hu, T. Wu, G. Liu, C. Fan, M. Song, Perovskite solar module-thermoelectric generator tandem system with high power output. Sol. Energy Mater. Sol. Cells **290**, 113737 (2025). https://doi.org/10.1016/j.solmat.2025.113737
24. J. Zhang, Y. Xuan, L. Yang, A novel choice for the photovoltaic–thermoelectric hybrid system: the perovskite solar cell. Int. J. Energy Res. **40**(10), 1400–1409 (2016). https://doi.org/10.1002/er.3532
25. J. Tang, H. Ni, R.-L. Peng, N. Wang, L. Zuo, A review on energy conversion using hybrid photovoltaic and thermoelectric systems. J. Power Sources **562**, 232785 (2023). https://doi.org/10.1016/j.jpowsour.2023.232785
26. Y. Zhou et al., A highly-efficient concentrated perovskite solar cell-thermoelectric generator tandem system. J. Energy Chem. **59**, 730–735 (2021). https://doi.org/10.1016/j.jechem.2020.12.020

27. D. Fang, J. Liu, Basic equations of piezoelectric materials, in *Fracture Mechanics of Piezoelectric and Ferroelectric Solids*, ed. by D. Fang, J. Liu (Springer Berlin Heidelberg, Berlin, Heidelberg, 2013), pp. 77–95. https://doi.org/10.1007/978-3-642-30087-5_4
28. B. Jaffe, W.R. Cook, H. Jaffe, Chapter 4—The perovskite structure, in *Piezoelectric Ceramics*, ed. by B. Jaffe, W.R. Cook, H. Jaffe (Academic Press, 1971), pp. 49–51. https://doi.org/10.1016/B978-0-12-379550-2.50008-9
29. X. Shi, Y. Sun, D. Li, H. Liu, W. Xie, X. Luo, Advances in wearable flexible piezoelectric energy harvesters: materials, structures, and fabrication. J. Mater. Sci. Mater. Electron. **34**(3), 220 (2023). https://doi.org/10.1007/s10854-022-09536-4
30. P. Wojciak, J.G. Smith, T. Hallam, S. Seyedin, Piezoelectric polyurethane/barium titanate composites with high stretchability and voltage output for energy harvesting in soft wearable electronics. Adv. Mater. Technol. 2500248 (2025). https://doi.org/10.1002/admt.202500248
31. G. Rong, Y. Zheng, M. Sawan, Energy solutions for wearable sensors: a review. Sensors (Basel) **21**(11) (2021). https://doi.org/10.3390/s21113806
32. M.A. Shah, B.M. Pirzada, G. Price, A.L. Shibiru, A. Qurashi, Applications of nanotechnology in smart textile industry: a critical review. J. Adv. Res. **38**, 55–75 (2022). https://doi.org/10.1016/j.jare.2022.01.008
33. L. Fan et al., Engineering of grain boundary in Pb(Zr0.52Ti0.48)O3 epitaxial films for tunable piezoelectric properties. J. Phys. Chem. C **129**(9), 4664–4672 (2025). https://doi.org/10.1021/acs.jpcc.4c07866
34. J. Han et al., Multifunctional coaxial energy fiber toward energy harvesting, storage, and utilization. ACS Nano **15**(1), 1597–1607 (2021). https://doi.org/10.1021/acsnano.0c09146
35. J. Park, J. Shin, S. Hur, C.-Y. Kang, K.-H. Cho, H.-C. Song, Cutting-edge piezo/triboelectric-based wearable physical sensor platforms (The Korean Sensors Society, 2022). https://doi.org/10.46670/JSST.2022.31.5.301
36. X. Yu et al., Structural design of light-emitting fibers and fabrics for wearable and smart devices. Sci. Bull. **69**(15), 2439–2455 (2024). https://doi.org/10.1016/j.scib.2024.05.042
37. W. Weng, P. Chen, S. He, X. Sun, H. Peng, Smart electronic textiles. Angew. Chem. Int. Ed. **55**(21), 6140–6169 (2016). https://doi.org/10.1002/anie.201507333
38. Z. Zhang et al., Thermochromic energy efficient windows: fundamentals, recent advances, and perspectives. Chem. Rev. **123**(11), 7025–7080 (2023). https://doi.org/10.1021/acs.chemrev.2c00762
39. B. AlShehhi, O. Albadwawi, A.A.B. Baloch, V. Alberts, A brief review on thermochromic halide perovskites for solar cells, in *2022 IEEE 49th Photovoltaics Specialists Conference (PVSC)* (2022), pp. 1441–1445. https://doi.org/10.1109/PVSC48317.2022.10638246
40. L. Fu, B. Li, S. Li, L. Yin, Magnetic, electronic, and optical properties of perovskite materials, in *Revolution of Perovskite: Synthesis, Properties and Applications*, ed. by N.S. Arul, V.D. Nithya (Springer Singapore, Singapore, 2020), pp. 43–59. https://doi.org/10.1007/978-981-15-1267-4_2
41. F. Hao, C.C. Stoumpos, R.P.H. Chang, M.G. Kanatzidis, Anomalous band gap behavior in mixed Sn and Pb perovskites enables broadening of absorption spectrum in solar cells. J. Am. Chem. Soc. **136**(22), 8094–8099 (2014). https://doi.org/10.1021/ja5033259
42. Y. Zhang, Z. Wang, S. Hu, P. Yan, H. Li, C. Sheng, Robust and swiftly reversible thermochromic behavior of a 2D perovskite of (C6H4(CH2NH3)2)(CH3NH3)[Pb2I7] for smart window and photovoltaic smart window applications. ACS Appl. Mater. Interfaces **13**(10), 12042–12048 (2021). https://doi.org/10.1021/acsami.1c00163
43. J. Li et al., Lead-free thermochromic perovskites with tunable transition temperatures for smart window applications. Sci. China Chem. **62**(9), 1257–1262 (2019). https://doi.org/10.1007/s11426-019-9487-0
44. Y. Liu et al., Emerging thermochromic perovskite materials: insights into fundamentals, recent advances and applications. Adv. Funct. Mater. **34**(37), 2402234 (2024). https://doi.org/10.1002/adfm.202402234
45. S. Liu et al., Organic hybrid perovskite (MAPbI3−Cl) for thermochromic smart window with strong optical regulation ability, low transition temperature, and narrow hysteresis width. Adv. Funct. Mater. **31**(26), 2010426 (2021). https://doi.org/10.1002/adfm.202010426

46. R. Zhang et al., Ultra low-haze and high transparency thermochromic perovskite smart windows with high solar modulation ability. Nano Energy **139**, 110978 (2025). https://doi.org/10.1016/j.nanoen.2025.110978
47. B.A. Rosales et al., Thermochromic halide perovskite windows with ideal transition temperatures. Adv. Energy Mater. **13**(12), 2203331 (2023). https://doi.org/10.1002/aenm.202203331
48. B. Wei et al., Thermochromic structural color based on pressure storage-release toward emerging applications. Chem. Eng. J. **497**, 154637 (2024). https://doi.org/10.1016/j.cej.2024.154637
49. J. Lin et al., Thermochromic halide perovskite solar cells. Nat. Mater. **17**(3), 261–267 (2018). https://doi.org/10.1038/s41563-017-0006-0
50. F. Ji et al., Remarkable thermochromism in the double perovskite Cs2NaFeCl6. Adv. Opt. Mater. **12**(8), 2301102 (2024). https://doi.org/10.1002/adom.202301102
51. Y. Cao, Y. Wang, Y. Zhang, X. Liu, J. Lin, Stable and rapid thermochromic reversibility in acene alkylamine intercalated layered hybrid perovskites. Mater. Today Commun. **37**, 107544 (2023). https://doi.org/10.1016/j.mtcomm.2023.107544
52. X. Sun, J. Zhang, X. Li, X. Gong, D. Kong, L. Zhi, Thermal response color-tunable electroluminescent device for real-time visual temperature monitoring. Giant **20**, 100339 (2024). https://doi.org/10.1016/j.giant.2024.100339
53. G. Li et al., Autonomous electroluminescent textile for visual interaction and environmental warning. Nano Lett. **23**(18), 8436–8444 (2023). https://doi.org/10.1021/acs.nanolett.3c01653
54. J. Shi et al., Smart textile-integrated microelectronic systems for wearable applications. Adv. Mater. **32**(5), 1901958 (2020). https://doi.org/10.1002/adma.201901958
55. L. Yang et al., Large-scale, stretchable, self-protective, and multifunctional perovskite luminescent filament with ultra-high stability. Adv. Mater. **36**(25), 2400919 (2024). https://doi.org/10.1002/adma.202400919
56. T. Tian et al., Large-area waterproof and durable perovskite luminescent textiles. Nat. Commun. **14**(1), 234 (2023). https://doi.org/10.1038/s41467-023-35830-8
57. F. Chun et al., Multicolour stretchable perovskite electroluminescent devices for user-interactive displays. Nat. Photon. **18**(8), 856–863 (2024). https://doi.org/10.1038/s41566-024-01455-6
58. M. Duan, X. Wang, W. Xu, Y. Ma, J. Yu, Electro-thermochromic luminescent fibers controlled by self-crystallinity phase change for advanced smart textiles. ACS Appl. Mater. Interfaces **13**(48), 57943–57951 (2021). https://doi.org/10.1021/acsami.1c17232
59. H. Liu, Y. Shi, Y. Pan, Z. Wang, B. Wang, Sensory interactive fibers and textiles. npj Flex. Electron. **9**(1), 23 (2025). https://doi.org/10.1038/s41528-025-00398-4
60. A. Feng, A.P.F. Smet, A review of mechanoluminescence in inorganic solids: compounds, mechanisms, models and applications. Materials **11**(4) (2018). https://doi.org/10.3390/ma11040484
61. S. Chang, K. Zhang, D. Peng, Y. Deng, C.-X. Shan, L. Dong, Mechanoluminescent functional devices: developments, applications and prospects. Nano Energy **122**, 109325 (2024). https://doi.org/10.1016/j.nanoen.2024.109325
62. H. Zhao et al., Mechanoluminescence from organic-inorganic metal halide perovskite derivative. Adv. Sci. **12**(9), 2414588 (2025). https://doi.org/10.1002/advs.202414588
63. P. Shao, P. Xiong, Y. Xiao, Z. Chen, D. Chen, Z. Yang, Self-recoverable NIR mechanoluminescence from Cr3+ doped perovskite type aluminate. Adv. Powder Mater. **3**(2), 100165 (2024). https://doi.org/10.1016/j.apmate.2023.100165
64. Y. Xiao et al., Deep-red to NIR mechanoluminescence in centrosymmetric perovskite MgGeO3: Mn2+ for potential dynamic signature anti-counterfeiting. Chem. Eng. J. **453**, 139671 (2023). https://doi.org/10.1016/j.cej.2022.139671
65. X. Zhang et al., Mechanically tunable circularly polarized luminescence of liquid crystal-templated chiral perovskite quantum dots. Angew. Chem. Int. Ed. **63**(22), e202404202 (2024). https://doi.org/10.1002/anie.202404202
66. G. Chen, L. Meng, S. Liu, L. Peng, Circularly polarized luminescence in composite films: a combination of perovskites and chiral nematic liquid crystals. Molecules **29**(22) (2024). https://doi.org/10.3390/molecules29225347

67. Y. Tian, R. Ding, S.S. Yoon, S. Zhang, J. Yu, B. Ding, Recent advances in next-generation textiles. Adv. Mater. **37**(8), 2417022 (2025). https://doi.org/10.1002/adma.202417022
68. A. Pandiyan et al., A comprehensive review on perovskite and its functional composites in smart textiles: progress, challenges, opportunities, and future directions. Prog. Mater. Sci. **140**, 101206 (2023). https://doi.org/10.1016/j.pmatsci.2023.101206
69. R. Paradiso, L. Caldani, Chapter 4—Textiles and smart materials for wearable monitoring systems, in *Wearable Sensing and Intelligent Data Analysis for Respiratory Management*, ed. by R.P. Paiva, P. de Carvalho, V. Kilintzis (Academic Press, 2022), pp. 95–120. https://doi.org/10.1016/B978-0-12-823447-1.00005-1
70. Y. Su et al., Self-powered respiration monitoring enabled by a triboelectric nanogenerator. Adv. Mater. **33**(35), 2170277 (2021). https://doi.org/10.1002/adma.202170277
71. G. Cay, D. Solanki, M.A. Al Rumon, V. Ravichandran, K. O. Fapohunda, K. Mankodiya, SolunumWear: a smart textile system for dynamic respiration monitoring across various postures. iScience **27**(7), 110223 (2024). https://doi.org/10.1016/j.isci.2024.110223
72. B. Mondal, R. Sarkar, D. Saini, V. Gupta, T.K. Kundu, D. Mandal, All-electrospun, water-resistant, breathable, wearable, and stable metal halide perovskite engineered electroactive polymer textiles for flexible piezoelectric nanogenerator. Adv. Mater. Technol. **8**(18), 2300614 (2023). https://doi.org/10.1002/admt.202300614
73. C. Pi et al., Highly stable humidity sensor based on lead-free Cs3Bi2Br9 perovskite for breath monitoring. J. Mater. Chem. C **9**(34), 11299–11305 (2021). https://doi.org/10.1039/D1TC02339J
74. Z. Wu et al., An excellent impedance-type humidity sensor based on halide perovskite CsPbBr3 nanoparticles for human respiration monitoring. Sens. Actuators B Chem. **337**, 129772 (2021). https://doi.org/10.1016/j.snb.2021.129772
75. M.-J. Kim et al., Breathing-driven self-powered pyroelectric ZnO integrated face mask for bioprotection. Small **19**(2), 2200712 (2023). https://doi.org/10.1002/smll.202200712
76. S. Tang et al., Incubating high-quality perovskite film via facial mask technique for efficient and stable solar cells. Adv. Funct. Mater. **33**(6), 2210417 (2023). https://doi.org/10.1002/adfm.202210417
77. W. Xuan, Y. Chen, D. Hu, X. Gao, S. Huang, Smart mask based on lead-free perovskite humidity sensor for labor intensity grading by breath monitoring. Sens. Actuators B Chem. **397**, 134622 (2023). https://doi.org/10.1016/j.snb.2023.134622
78. X. Li et al., Flexible, visual, and multifunctional humidity-strain sensors based on ultra-stable perovskite luminescent filaments. Adv. Fiber Mater. (2025). https://doi.org/10.1007/s42765-025-00518-9
79. X. Li et al., Stretchable luminescent perovskite-polymer hydrogels for visual-digital wearable strain sensor textiles. Adv. Fiber Mater. **5**(5), 1671–1684 (2023). https://doi.org/10.1007/s42765-023-00302-7
80. F. Gao et al., Large-scale and continuous production of perovskite filament for wearable fluorescent textiles. Chem. Eng. J. **502**, 158191 (2024). https://doi.org/10.1016/j.cej.2024.158191
81. J. Yin, J. Li, V.S. Reddy, D. Ji, S. Ramakrishna, L. Xu, Flexible textile-based sweat sensors for wearable applications. (2023). https://doi.org/10.3390/bios13010127
82. X. Ma, X. Wu, W. Luo, Z. Liu, F. Wang, H. Yu, Large-scale wearable textile-based sweat sensor with high sensitivity, rapid response, and stable electrochemical performance. ACS Appl. Mater. Interfaces **16**(14), 18202–18212 (2024). https://doi.org/10.1021/acsami.4c01521
83. X. Li, X. He, X. Yang, G. Tian, C. Liu, T. Xu, A wearable sensor patch for joule-heating sweating and comfortable biofluid monitoring. Sens. Actuators B Chem. **419**, 136399 (2024). https://doi.org/10.1016/j.snb.2024.136399
84. J. Min et al., An autonomous wearable biosensor powered by a perovskite solar cell. Nat. Electron. **6**(8), 630–641 (2023). https://doi.org/10.1038/s41928-023-00996-y
85. L. Possanzini, L. Basiricò, A. Ciavatti, M. Tessarolo, B. Fraboni, Fully textile X-Ray detectors based on fabric-embedded perovskite crystals. Adv. Mater. Interfaces **9**(15), 2101417 (2022). https://doi.org/10.1002/admi.202101417

86. H. Li et al., Wearable photoferroelectric perovskite X-Ray detectors. Adv. Funct. Mater. **34**(46), 2407693 (2024). https://doi.org/10.1002/adfm.202407693
87. D. Xin et al., Low-dose and stable X-Ray imaging enabled by low-dimensional Dion-Jacobson perovskites. Adv. Funct. Mater. **34**(38), 2402480 (2024). https://doi.org/10.1002/adfm.202402480
88. M.A. Hossain, Adaptive camouflage textiles with thermochromic colorant and liquid crystal for multidimensional combat background, a technical approach for advancement in defence protection. Am. J. Mater. Eng. Technol. **9**(1), 31–47 (2021). https://doi.org/10.12691/materials-9-1-3
89. H. Liu et al., Moisture assisted photo-engineered textiles for visible and self-adaptive infrared dual camouflage. Nano Energy **93**, 106855 (2022). https://doi.org/10.1016/j.nanoen.2021.106855
90. L.M. Degenstein, D. Sameoto, J.D. Hogan, A. Asad, P.I. Dolez, Smart textiles for visible and IR camouflage application: state-of-the-art and microfabrication path forward. (2021). https://doi.org/10.3390/mi12070773
91. Q. Lu, M. Li, A. Tian, S. Fu, Green plant leaf-inspired smart camouflage fabrics for visible light and near-infrared stealth. J. Bionic Eng. **19**(3), 788–798 (2022). https://doi.org/10.1007/s42235-022-00156-6
92. C. Jiang, P. Wei, L. Yuan, X. Qing, X. Ma, X. Weng, An intelligent multi-band camouflage textile inspired by natural leaves. Adv. Funct. Mater. 2424325 (2025). https://doi.org/10.1002/adfm.202424325
93. X. Ye et al., All-textile sensors for boxing punch force and velocity detection. Nano Energy **97**, 107114 (2022). https://doi.org/10.1016/j.nanoen.2022.107114
94. Z. Liu et al., Functionalized fiber-based strain sensors: pathway to next-generation wearable electronics. Nano-Micro Lett. **14**(1), 61 (2022). https://doi.org/10.1007/s40820-022-00806-8
95. G. Tian et al., Rich lamellar crystal baklava-structured PZT/PVDF piezoelectric sensor toward individual table tennis training. Nano Energy **59**, 574–581 (2019). https://doi.org/10.1016/j.nanoen.2019.03.013

MIX
Papier aus verantwortungsvollen Quellen
Paper from responsible sources
FSC® C105338

FSC
www.fsc.org

If you have any concerns about our products,
you can contact us on
**ProductSafety@springernature.com**

In case Publisher is established outside the EU,
the EU authorized representative is:
**Springer Nature Customer Service Center GmbH**
**Europaplatz 3, 69115 Heidelberg, Germany**

Printed by Libri Plureos GmbH
in Hamburg, Germany